Introduction to Experimental Methods

Introduction to Experimental Methods succinctly explains fundamental engineering concepts in mechanics, dynamics, heat transfer, and fluid dynamics. From conceptualizing an engineering experiment to conducting a comprehensive lab, this book enables students to work through the entire experimental design process.

Offering a complete overview of instruction for engineering lab methodology, the book includes practical lab manuals for student use, directly complementing the instruction. Numerous worked examples and problems are presented along with several hands-on experiments in individual lab manuals. This book discusses how to write lab reports, how to configure a variety of instruments and equipment, and how to work through failures in experimentation.

Introduction to Experimental Methods is intended for senior undergraduate engineering students taking courses in Experimental Methods.

Instructors will be able to utilize a Solutions Manual for their course.

Features:

- Provides an overview of experimental methods in mechanics, dynamics, heat transfer, and fluid dynamics
- Covers design of experiments, instruments, and statistics
- Discusses SolidWorks and PASCO Capstone software
- Includes numerous end-of-chapter problems and worked problems
- Features a Solutions Manual for instructor use

Introduction to Experimental Methods

Terry W. Armstrong

CRC Press is an imprint of the
Taylor & Francis Group, an **informa** business

Designed cover image: Terry W. Armstrong

First edition published 2024
by CRC Press
6000 Broken Sound Parkway NW, Suite 300, Boca Raton, FL 33487-2742

and by CRC Press
4 Park Square, Milton Park, Abingdon, Oxon, OX14 4RN

CRC Press is an imprint of Taylor & Francis Group, LLC

Library of Congress Cataloging-in-Publication Data
Names: Armstrong, Terry W., author.
Title: Introduction to experimental methods / Terry W. Armstrong.
Description: First edition. | Boca Raton, FL : CRC Press, 2023. |
Includes bibliographical references and index.
Identifiers: LCCN 2023000958 | ISBN 9781032358932 (hbk) |
ISBN 9781032358956 (pbk) | ISBN 9781003329237 (ebk)
Subjects: LCSH: Engineering–Laboratory manuals. | Engineering–Study and teaching. |
Education–Experimental methods.
Classification: LCC TA152.A76 2023 | DDC 620.0078–dc23/eng/20230124
LC record available at https://lccn.loc.gov/2023000958

ISBN: 978-1-032-35893-2 (hbk)
ISBN: 978-1-032-35895-6 (pbk)
ISBN: 978-1-003-32923-7 (ebk)

DOI: 10.1201/9781003329237

Typeset in Times
by codeMantra

Access the Instructor Resources: routledge.com/9781032358932

Contents

Preface

Experimentation is essential to engineering and scientific studies. Theories and mathematical models provide broad predictive power, but it is experimental data that connects conceptual theory to the real world.

As important as experimental methods should be to engineering studies, the educational approach is often inconsistent. Of the three primary goals for this textbook, the first goal is to provide a consistent "*arc of instruction*" across several useful topics for all experimental methods curricula. Some topics, such as *design of experiments* and *statistics*, warrant volumes of study in their own right. The goal is to provide enough concise content in these fields of study while breaking through some of the difficult concepts. Gaining a baseline understanding is singularly important for follow-up studies.

The second primary goal of this textbook is to provide a lecture- and study-friendly text. This textbook is as much an outline in bulleted form as it is a narrative. The format is intended to focus on the essentials, making content succinct and easier to scan for both students and instructors. Narratives expand on topics when appropriate. The lab manuals are fully narrative where traditional writing styles are demonstrated for report writing.

The third primary goal of this textbook is to provide comprehensive, well-refined lab experiments with sufficient detail to avoid frustration due to missing technical information. These labs cover the typical engineering disciplines of mechanics, dynamics, heat transfer and fluid mechanics. The emphasis is on consistency, lab plan development and expected results. Planning is always the most important part of any successful endeavor, and a well-written lab plan with all formulas worked out to the maximum extent possible is heavily emphasized.

The lab manuals included in this textbook are integral to the overall text. Instructional content is balanced both between the chapters of the textbook and the individual lab manuals. There is some overlap of material, but most of the content is complementary and provides additional information. The lab manuals can individually serve as instructional resources while also serving as stand-alone documents.

As an "arc of instruction" textbook, topics range from conceptualizing an experiment, designing an experiment, modeling a system, performing statistical analysis and reporting results. Most of the system modeling instruction is mechanics-oriented since all mechanical engineering curricula require these topics. The heat transfer and fluid dynamics topics are generally relegated to the lab manual narratives.

The chapters on report writing and presentations cannot be emphasized enough. While technical competence is required for all engineers to succeed, the ability to both write and speak opens the doors for true success within an organization. Technical writing is usually passive, and this textbook follows suit. However, knowing how to write actively can always clean up passive styles. Conciseness, clarity and simplicity should be the goals for writing, not page counts. For spoken presentations, practice and preparation are always key. In the age of online presentations, preparing the online environment is every bit as important as a professional in-person presentation.

Most of the chapter exercises are limited to a dozen or half-dozen questions to assist in study and exam composition. The true homework associated with lab courses is preparing lab plans with expected results and the report writing.

Educationally, technical and mathematical problems can usually be divided into two categories: application and process. Application problems begin with a governing equation and generally require only proper application of the equation. Process problems require understanding a sequence of steps to arrive at a solution. Computing a temperature decay time constant is an example of the former while creating a shear-moment diagram is an example of the latter. My preferred implementation style for this textbook is to provide the bottom-line upfront by first identifying the problem category. Introduce a topic with relevant and interesting information, then provide the governing equation of interest. Show how to use the governing equation in an example. From there, follow up with any necessary derivations. Alternative approaches beginning with derivations can be difficult to follow without first knowing the end goal and utility. For a process-type problem, knowing both the end goal and that *the process* is important greatly facilitates student educational retention.

My desire with this textbook is to bring together the right mix of theory and hands-on engineering experimentation in the most concise manner while hopefully contributing to the field of experimental methods.

Terry W. Armstrong, PhD
Lt Col, USAF (Ret)

Author

Dr. Terry W. Armstrong served as an Assistant Professor of Mechanical and Aerospace Engineering at New Mexico State University. He developed and taught courses in Experimental Methods I (mechanics), Experimental Methods II (heat transfer and fluid mechanics), Aerospace Systems Engineering, Engineering Analysis (advanced engineering mathematics), Thermodynamics, Dynamics, and he created a new lab-based engineering calculus course. Dr. Armstrong holds a PhD in Aerospace Engineering from New Mexico State University, a master's degree in Aeronautical Science from Embry-Riddle Aeronautical University, and a bachelor's degree in Engineering Science from the United States Air Force Academy. His teaching awards include the Mechanical and Aerospace Academy Outstanding Teaching Professor, Donald C. Roush Award for Teaching Excellence and Frank Bromilow Engineering Teaching Excellence.

Prior to university instruction, Lt. Colonel Armstrong served 21 years in the US Air Force, first as a T-38 instructor and then as an F-15C combat pilot and instructor pilot. Lt. Colonel Armstrong led combat missions over Iraq during Operation SOUTHERN WATCH with 150 combat flight hours. His duties have included command of Detachment 1, 85th Test and Evaluation Squadron, Tyndall AFB, FL, Chief of Flight Safety, USAFE, Ramstein AB, Germany, and Director of Special Test programs at Kirtland AFB, NM, before retiring in 2006. Lt. Colonel Armstrong holds several aviation top gun awards, distinguished graduate awards, flight instructor awards, and his service medals include the Air Medal and Aerial Achievement Medals. "*How much better to get wisdom than gold; to get insight rather than silver!*" (Proverbs 16:16)

About the Cover

The left image is the photoelastic stress field from an edge-cut stress concentration in a polymer. The right image is the theoretical edge-cut stress concentration distribution. The left image was configured and photographed by the author. The right image was generated from the author's code presented in Lab #5 with increased resolution, increased bit depth and filtering to enhance appearance.

1 Approach to Experimentation

Engineers typically prefer the technical details of setting up an experiment. However, mapping out a larger approach to an experiment and considering the processes and procedures first will ensure the greatest chance of success when getting to the technical details.

People conduct small experiments all the time. A cook might add new spices to a dish, and if it turns out well, the experiment is documented on the recipe. A homeowner might mix up a few different colors or types of paint in small batches to see results before committing to a final color choice. Some experiments are much more elaborate and are designed to improve our understanding of the world around us. While not every experiment requires a particularly systematic approach, experiments with scientific insight bound for publication should follow well-established approaches and guidelines. Systematic approaches and documentation help ensure experiments can be replicated in the future with consistent results. While Grandma's meatloaf *might be the best*, if the ingredient proportions and sequence of actions are not properly documented in a recipe, no one else has the benefit of consistently recreating the *loaf of meat* (still seems like an oxymoron) and verifying it *is* the best.

- The systematic approach to experimentation offered in this text may be of benefit to
  - Academic students working on a master's or doctoral degree
  - Industry personnel seeking to improve the quality of a factory process
  - Government employees responsible for planning the test of a new system

EXPERIMENT OR TEST

While generally synonymous and interchangeable, the terms "*experiment*" and "*test*" can be more technically defined in the context of experimental methods and design of experiments (DOEs). An experiment differs from observation because an experiment is an *active process to deliberately determine causal relationships*. This concept immediately implies that an experiment compares an altered condition to an unaltered condition. The unaltered condition is called a *control* or reference, and the altered condition is called the *treatment* or test parameter.

A *test* differs from an experiment by comparing a condition to a *standard* rather than a control condition, without concern for causal relationships. In the opening example, a cook adds new spices to a dish and compares the result to the previous recipe version. The previous recipe version is the control, and the new spice is the treatment for an improved taste effect. A specific causal relationship is investigated,

DOI: 10.1201/9781003329237-1

so the activity meets the definition of an experiment. However, the homeowner mixes different colors of paint and compares the result to his or her personal color preference or a color specification. If the color fails to meet the specification, a new batch is attempted without investigating *why* the color failed. Since no causal effect is investigated and only the *standard* matters, this activity is a test. If the homeowner was trying to match a color by adding different proportions of specific colors to determine the effects, the activity becomes an experiment where the match color is the control.

A test can also be part of an experiment, and the preceding definitions still hold up. Prescribing several conditions, parameters and constraints within a large experiment means individual test runs must meet the parameter standards, whether it is the treatment or control, to ensure an overall successful experiment.

- An experiment is an active process to investigate cause-and-effect relationships
  - An experiment has a control reference
  - An experiment has a treatment or test parameter
  - An experiment requires at *least two activities*
    - One activity establishes the control condition
    - One activity checks the treatment effect
  - An experiment provides predictability with cause-and-effect relationships
- A test is similar to an experiment but compares a result relative to a standard
  - A test does not seek to establish a cause-and-effect relationship
  - A test requires only *one activity* to check results against a standard
  - A test can verify expectations
  - A test can support a hypothesis by testing a condition for expected results
  - A test can be part of an experiment
  - An experiment may be composed of several tests

Based on the definition of a test compared to an experiment, many pre-configured academic lab activities are better classified as tests rather than experiments. However, the systematic approach for both an experiment and a test is essentially the same, so the terms are still relatively interchangeable.

PURPOSE

Observation of nature or activities often reveals a *correlation* between phenomena. However, just because events or parameters are correlated, there may or may not be a *causal relationship*. It is the task of an experiment to methodically investigate whether there is a mere coincidence between phenomena or whether a causal relationship exists. There is a fundamental difference between correlation and causality. Causality is predictive of the future and nature of things, while simple correlation may be happenstance. Predictive capability means we can expect certain physical phenomena to behave in a consistent manner based on the results of an experiment. If experimentation demonstrates water consistently boils at 100°C on a standard day at sea level, we have a measure of predictability to design and build kitchen appliances that deliver sufficient heat to boil water. Substitute aerospace-grade aluminum yield stress for boiling water, and the systematic results of good experimentation with reliable predictive ability about aircraft strength become much less trivial.

If it is the job of an experiment to investigate causality, it is the job of the *hypothesis* to clearly state the contended causal relationship. An experiment delivers data to either support or refute a hypothesis. Experiments may also support an *objective* rather than a hypothesis. An objective seeks to demonstrate an accepted principle, an existing relationship or verify results. In this sense, experiments supporting an objective are more closely defined as tests.

- Experiments may support either
  - A hypothesis (supporting contended cause-and-effect relationships)
  - An objective (an experimental test to validate expectations)

HYPOTHESIS

A hypothesis statement strips away all confusion and narrowly defines the relationship of experimental investigation. All planning, experimental activities and reporting are meant to support the hypothesis. The hypothesis is fundamentally a statistical statement. Experiments not only provide data to build the case for accepting a hypothesis, the data *themselves are* the hypothesis. Refer to Chapter 17: Evaluating the Hypothesis, to clarify how a data set mean is (and should be) equivalent to a hypothesis statement. Having this data-oriented hypothesis mindset helps when developing a hypothesis.

Terminology such as *demonstrated* or *accepted* is more correct than *proven*. A hypothesis may be accepted for a period of time until new data requires modifying or rejecting the hypothesis. The very nature of experimentation is to provide evidence in the form of data. Data sets are inherently finite and cannot test every conceivable instance of a hypothesized relationship. *Proven* is an inappropriate word since it implies an incontrovertible truth that experimentation cannot supply. Any hypothesis might be struck down by the next untested data point just beyond the existing data set. However, the word "proven" is also initially more straightforward than *alternative hypothesis* or *null hypothesis* and is loosely used in the following to help explain the correct terminology.

- The hypothesis
  - Technically and more correctly known as the *alternative hypothesis*
  - *Your* version of reality
  - A statement to support or demonstrate (loosely, "*prove*" until otherwise proven wrong)
  - There *is* a relationship between phenomena
  - A *statistically supported statement* (data should support the hypothesis)
    - May require discerning a data set mean from another data set mean
  - A hypothesis can be formed based on observations, questions or ideas
    - *What do we think will happen?*
    - *What is the causal element of cause and effect?*
    - *What might be the cause of an observable effect?*
  - EXAMPLE: Atmospheric pressure change causes water to boil at different temperatures
    - Observable effect: water boils at different temperatures
    - Cause: changes in atmospheric pressure

    - Hypothesis: *water boils at temperatures other than 100°C based on pressure change*
- The null hypothesis
  - Contrary to the hypothesis
  - A statement we're left with if the hypothesis is not "proven"
  - A statistically supported statement
    - A data set already exists with a mean supporting the null hypothesis
  - Data from the *alternative hypothesis* must be significant enough from the *null hypothesis* before people will accept the proposed alternative
  - The null assumes there is *no* relationship between phenomena
    - It is the burden of the alternative to demonstrate there *is* a proposed relationship
    - Null always proposes no difference in experiments by changing the test parameter
  - EXAMPLE: Water boils at the same temperature regardless of atmospheric pressure
    - Also, atmospheric pressure change causes *no* change to boiling water temperature
    - A null hypothesis can often be constructed just by adding *NO* or *NOT* to alternative
- The characteristics of a hypothesis (the alternative hypothesis)[1]
  - It is a statement, not a question
  - It is clear
  - It is testable
  - It can be falsified
  - It has a narrow scope
    - A hypothesis should examine one well-defined item

Sometimes it can be difficult to discern the hypothesis in a written report. A hypothesis should always *contend* a condition or relationship rather than state a mere fact. Terminology such as "*we demonstrate*" and "*data supports*" usually indicate the hypothesis statement. Statements such as "*the algorithm uses the formula for fifth order temperature effects*" are usually informative statements but do not indicate a hypothesis. This statement indicates a particular approach or method that appears well-established and does not contend a unique relationship. Statements such as "*thermal diffusion is more difficult to measure*" are informative but unlikely to be a hypothesis. Broad, unquantifiable terms like "*difficult*" are not suitable for a hypothesis.

OBJECTIVE

Experiments supporting an objective are usually meant to verify conditions or parameters and more accurately meet the definition of a *test*, rather than an *experiment*. Fiberglass and composite structures have published yield and ultimate stresses, which can be used directly for design purposes. However, the actual performance of composites is very dependent upon fiber-to-resin ratios, humidity and lay-up quality. Prior to committing to a composite design parameter, good engineering practice requires setting up an experimental *test* with an *objective* to ensure the planned composite configuration has the expected properties.

- The objective
  - The purpose, goal or statement of experimental intent
  - An experiment may only have an *objective* and be considered as a *test* (experimental test)
  - A verification of performance, values or results
  - Experiments may also have several objectives to break down a complex effort

DISCOVERY VERSUS STRUCTURE

It is a misnomer that all scientific progress is the result of neatly planned and reported experimental processes. However, it is also a misnomer that true scientific advances can be made without, at some point, systematically processing and packaging creative processes into a *verifiable product. Product* has the broadest definition, where the result of a well-planned, well-executed and properly reported experiment has some measure of predictive capability for the benefit of all. *Verifiable* means other researchers or experimenters have full opportunity to replicate the experiment and results from the report.

The *scientific method* is the traditional standard for a disciplined approach to establishing causal relationships. There is generally no distinct *experimental* method different from the scientific method. The scientific method and the experimental method are essentially the same organized process for supporting scientific discoveries.

There are contrarian opinions about adherence to the scientific method. In particular, the scientific method sometimes implies *all scientific discoveries come about through an organized six-step process.* Stuart Firestein's book on failure (*Failure: Why science is so successful*)[2] indicates the scientific method bears little reality to actual science and discovery. Instead, Firestein contends scientists (engineers and experimenters) are curious about ideas and approach understanding and explanation through a variety of methods, *not* a convenient *six-step process.*

Accordingly, the creative and disjointed approach to scientific discovery and experimentation is better categorized as a *pre-scientific method* step. As a process, the scientific method has greatest value when rigor is required as a follow-up to discovery. Even if the initial discovery process and hypothesis testing are far less organized and systematic than the traditional scientific method implies, at some point, a rigorous approach is useful and necessary.

Consider the example of Alexander Fleming's accidental discovery of penicillin, where a bacterial culture petri dish was inadvertently contaminated with mold. This discovery was a medical breakthrough, but it did not follow the scientific method whatsoever; it was a haphazard discovery. However, to *verify* the action of penicillin against bacteria and test his observation, systematic experimentation was ultimately necessary.[3]

This "pre-scientific method" phase of experimentation is an unstructured *investigate phase.* This phase simply acknowledges any and all activities that might culminate in the need for a rigorous experimental approach. The phase consists of thought experiments, mini-experiments, tests, research, discussions and brainstorming. The investigate phase is distinctly different from a planning phase based on the focus of activities. The broad range of unstructured activities must be narrowed to a particular hypothesis or objective for the planning phase to begin.

METHODOLOGY

An experimentation model offered here, INVESTIGATE, PLAN, CONDUCT and REPORT (IPCR), is a simplification of the traditional six-step scientific method. The investigate phase acknowledges the unpredictable path of discovering new ideas bound for experimentation. Figure 1.1 connects the IPCR model to the traditional six-step scientific model process. Figure 1.1 also correlates with the DOE methodology discussed in Chapter 4: Design of Experiments. Note that in the IPCR model, failures in the experimental process often require returning to the planning phase to refine or improve the experimental design before re-conducting the experiment. The traditional six-step model does not clearly depict the failure steps and repeated efforts inherent to the scientific process.

- Prior to a disciplined, methodical experimental design, ideas are explored in the *first phase*
  - INVESTIGATE
- For a rigorous experimental process, experimentation occurs in three main phases
  - PLAN
  - CONDUCT
  - REPORT

To be clear, not all experiments require the same level of effort and detail. A one-person effort to determine how much vacuum a Shrader valve can hold (refer to Lab #12) might have a plan written as a few notes for measurements and force calculations. This small experiment contrasts with the experimental test program of the F-35 Lightning II jet fighter, which required years of planning, millions of dollars, multiple agencies and a host of personnel.

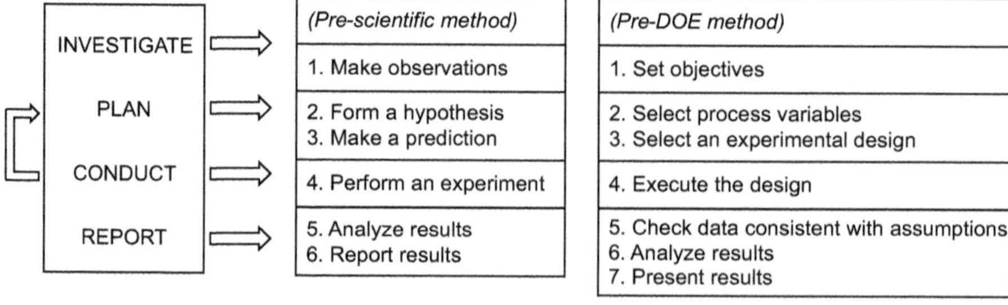

FIGURE 1.1 Approach to experimentation model compared to both the traditional scientific method and DOE method.

Investigate

The *investigate, discover* or *explore* phase is purposely without structure. This is where curiosity, observation, investigation and impromptu experimentation reign. Mini-experiments during this phase are part of the investigation process and help determine whether a more deliberate and methodical experiment is warranted. At some point, a decision is made to drop an idea or investigate it further. If the investigate phase points to an interesting phenomenon, the more formal subsequent steps should be considered. The *product* of the investigate phase is a hypothesis or an objective. It is the hypothesis or objective statement that signals the transition into a rigorous experimental process.

- This is an *unstructured* period of investigation, discovery or brainstorming of ideas
- Not every experimental investigation progresses into the following formal phases
- An objective is devised or a hypothesis is formulated prior to the next phase

Plan

Entering into the planning phase of an experiment assumes a formal approach is useful. A transition into this phase might imply an interesting correlation has been discovered and a causal relationship is theorized. This phase also recognizes that the experiment may be published, and care should be taken to document the process. Organization and planning consume most of the effort in many endeavors, while actual hands-on activity can go quickly in comparison. A well-organized project has a much greater opportunity for success compared to one with insufficient planning and an ill-considered structure.

- Proper preparation of an experiment is always the most important phase
- Never request the use of facilities or resources without also offering a basic written plan
- Clearly identify the hypothesis or objective to begin detailed experiment planning
  - Hypothesis: a theory or causal relationship to be demonstrated
  - Objective: a goal to achieve or results to verify (may be considered a *test*)
- Determine whether the experiment is necessary in full or in part
  - Perform background research, literature review and check existing data
  - Consider building upon previous or similar experiments
- Organize thoughts by considering each *phase* of experimental activities
  - Refer to Figure 1.1 for IPCR, scientific method or DOE process
  - Phases require different resources; the experiment is not complete until all phases are complete
- Outline the conceptual details of the experiment to better assess necessary resources
  - Identify a relevant mathematical model, parameter relationship or published data

  - Identify primary variable of investigation (test parameter) for hypothesis or objective
  - Determine units and range of values for the primary variable
- Work out all formulas and expected results to the maximum extent to establish measurements
- Identify the primary instrumentation, sensitivity and frequency response to measure the test variable
  - Data acquisition systems, oscilloscopes, sensors and voltage ranges
- Identify support equipment requirements
  - Power supplies, multimeters, computers, cables, soldering, and fixtures
- Identify test specimen materials, dimensions, fabrication requirements and quantities
- Assess the availability, costs and time required for each phase to plan, conduct and report
  - Do not underestimate time requirements (Rule of Thumb: multiply estimates by 3–5)
- Develop an initial checklist, a sequence of activities and a schedule of tasks
- Design the engineering experiment
  - Establish the control case
  - Choose a DOE technique (refer to Chapter 4)
    - One Factor at a Time (OFAT)
    - Full factorial
    - Fractional factorial
  - Identify potential sources of error and develop a mitigation plan
- Write the plan (refer to Chapter 3: Deliverables for details and Appendix A for an example)
  - Follow an established guide, template, or administrative or journal guidelines
  - Write in future tense since some experiments can span weeks, months or years
  - Plan needs to be sufficiently comprehensive for all involved
- Determine waste disposal requirements and cleanup
- Brief the experimental plan, as required
  - Large experiments or test programs often require presenting plan to stakeholders

Conduct

The actual hands-on activities of an experiment are often quick compared to all the planning activities. In-lab experimental activities generally follow three basic steps, which are helpful both for planning purposes and organizing the flow of lab activities:

- <u>Validate</u> the experimental setup and equipment using known test specimens
- Obtain a <u>control</u> data set for comparison as the baseline or truth data
- Perform the test parameter experiment by applying the *<u>treatment</u>* parameter of interest

Aside from these three basic steps, a successful lab experience involves several considerations, such as a final review of planned events, following proper lab procedures, maintaining an organized workspace and safeguarding captured data.

- Conducting an experiment must be efficient since lab time and resources are limited
- Safety procedures, standard protocols and etiquette must be followed
- Be familiar with typical equipment and instrumentation in the lab environment
- Follow the plan, adjust as necessary and record notes and all data
- Prepare the workspace
    - Don required personal protection equipment (PPE)
    - Ensure unique safety procedures are understood
    - Review the test plan
    - Ensure the work area is clean and organized
    - Ensure lab test equipment is available and functional
    - Ensure consumables are available
    - Ensure personnel are ready and duties are assigned and understood
    - Arrange the experimental setup and equipment for use
- Calibrate equipment
    - Turn on and warm up necessary equipment
    - Follow any calibration procedures to ensure accuracy
        - Modern digital equipment often requires little to no calibration for basic efforts
        - Check the fidelity of equipment compared to the required fidelity of experimental data
    - Perform test runs as necessary
        - Run a small-scale or a single event to ensure the experiment works as planned
        - Ensure the experimental setup is correct and the results make sense
        - Ensure data recording devices are operating and recording
- Execute the plan
- Record data
    - Save data on multiple resources
        - Portable drives can be lost or overwritten
        - Local host computer facilities can be locked and inaccessible
    - Ensure manual readings are recorded at proper intervals
        - Hand-written recordings or photos can make up for lost files
- Monitor progress and adjust the plan as necessary
    - Things do not always go as planned
    - Change procedures or techniques if necessary to adapt to the nature of the experiment
    - Document changes so they are not forgotten
- Troubleshoot problems in real-time
    - Unexpected results might be due to the experimental setup <u>or</u> subject of the experiment
    - An experimental setup might be incorrectly configured or poorly designed

  - Software settings might be wrong
- Conclude and secure the workspace
  - Always reserve at least 10 minutes of lab time for cleanup
  - Return equipment to original locations and clean up all activities
- If you borrow, *RETURN*!
  - Be a responsible borrower or no one will lend again
  - It is *your* responsibility to *actively* return borrowed equipment at the first opportunity
  - Do not wait for the lender to ask for a return
  - *Always* establish a return not-later-than (NLT) deadline to alleviate questions

Report

All formalized experiments culminate in a report of some form, either written or spoken. A well-planned experiment should have a properly written plan, which becomes the basis for a well-written report. Pre-planned spreadsheets, labeled variables with identified units and good documentation during the experimental lab activities will prevent wasting time trying to recall details during the report-writing phase.

- A report is the *product* of an experiment
  - A "*report*" may be a formal published document or merely journal log notes for the future
- The reporting phase begins with data analysis
  - Data is the product of an experiment
  - Data provides the basis for supporting a hypothesis
  - Data analysis might uncover unexpected influences and correlations
  - Use accepted statistical analysis techniques on data sets
    - Explore any outliers in the data
  - Consider the influence of errors
    - Errors are either unknown or uncontrollable influences
    - Errors would be corrected and minimized if they were known and it was possible to do so
    - There are three *types* of error: blunders (human), systematic (fixed) and random
- Write objectively for body of report
  - Assessments should be a logical consequence of results
  - Objective writing does not *necessarily* imply third person or passive styles
  - Present data and data discussions factually
  - Do not *taint* the body of the report with unnecessary opinions, conjectures or bias
    - A reader may dismiss material as a whole based on a single unsupported statement
- Subjective inferences should be a logical result of presented data
  - Save subjective interpretations for the discussion or conclusion section, after factual results

    - For reports written in the third person, consider shifting to the first person for ownership of conclusions
    - The reader is welcome to disagree with the conclusions if the bulk of the document is factual
- Write the report (refer to Chapters 3, 20 and 21 and Appendix A)
    - Use a template, appropriate guidelines or follow the structure of the lab manuals in this textbook
    - Use the IMRD standard – Introduction, Methods, Results, and Discussion[4]
- Provide a briefing or presentation of the results, as required
    - Conferences, peers, supervisor or instructor
    - Know your audience and their expectations
    - *Always* remain within the allotted time

GOOD LABORATORY PRACTICES

Adhering to proper practices within the lab is both a planning and a conduct consideration and is integral to obtaining reliable results from experiments. In the 1970s, the US Food and Drug Administration codified (21 CFR 58) *good laboratory practices* (GLPs) due to substandard products as a result of poor planning, incompetence, insufficient documentation and fraud.[5] The result was to provide a system to *promote the quality and validity of test data to help scientists obtain results that are reliable, repeatable, auditable and recognized worldwide.* While not regulatory for engineers outside the scope of FDA-regulated foods and medicines, the following five guiding principles, according to the Organisation for Economic Co-operation and Development (OECD) are suitable for all lab environments.[6]

- *Resources (organization, personnel, facilities and equipment)*
    - Establish an organizational chart to understand responsible personnel and authority
    - Ensure sufficient personnel with clearly defined responsibilities
    - Ensure proper training for all personnel
    - Ensure sufficient facilities and equipment with proper maintenance
- *Characterization (test items and test systems)*
    - Know as much as possible about the materials and systems in use
- *Rules (protocols and standard operating procedures, SOPs)*
    - Ensure proper planning (test plan), procedures and proper approvals
- *Results (documentation, raw data, final report and archives)*
    - Maintain and archive raw data
    - Results must faithfully reflect raw data
    - Accurately report results
- *Quality assurance (independent monitoring of research processes)*
    - Independent oversight functions as a witness of proper procedures

STANDARD OPERATING PROCEDURES

SOPs differ from GLPs as an organizational-specific set of documents. GLPs provide guidelines to ensure the integrity of laboratory and experimental results in general, while SOPs ensure personnel within an organization follow documented processes and procedures. It is the responsibility of the experiment team to be aware of and follow all applicable SOPs when using facilities and equipment.

- SOPs[7] are as follows
  - Written procedures for a specific laboratory program
  - Define how to carry out protocol-specific activities
  - Often written in a chronological sequence of action steps
- SOPs may be available and required for
  - Qualifications of personnel and named individuals authorized access to equipment
  - Health and safety protocols and actions in case of equipment failure or emergencies
  - Inspection, cleaning, maintenance, testing, calibration and standardization of instruments
  - Definition of raw data, analytical methods, data handling, storage and data retrieval
  - Receipt, identification and method sampling of test and control articles

COMMON PROCEDURAL FAILURES

Several resources and guidelines are available to maximize opportunities for success in the laboratory environment, some of which have been presented in this chapter. Since experiments consume time and resources, mistakes can be inconvenient or outright costly. Mistakes happen due to lack of preparation, straying from established routines, forgetfulness or taking small shortcuts. The following list, outlined by experienced lab researcher Jason Erk at Oregon Health & Science University,[8] covers ten of the most commonly BROKEN laboratory practices that should be considered precautionary to avoid mistakes.

- *Wear PPE sometimes, not all the time* (it only takes one accident to cause injury)
- *Decide not to run a control sample* (instead, check the setup with a known parameter)
- *Only record abbreviated details* (reports come later, when details are easily forgotten)
- *Failure to write anything down* (have a backup since data and files get lost or corrupted)
- *Failure to calibrate equipment* (do not get complacent with modern equipment)
- *Use the wrong tool for the job* (work within the sensitivities of equipment)

- *Work through math units only once* (instead, always do a sanity check of all values)
- *Failure to quality check* (make sure test specimens conform to expected standards)
- *Failure to train personnel* (may result in forgotten procedures or denied lab access)
- *Communicate with lab mates sparingly* (communication is vital for collaborative efforts)

All of the detail in GLPs and established procedures can be summarized in a good quote by Ludwig Huber from Agilent Technologies,[9]

> *If experimental work is conducted in compliance with GLP, it should be possible for an inspector, maybe four or five years hence, to look at the records of the work and determine easily why, how and by whom the work was done, who was in control, what equipment was used, the results obtained, any problems that were encountered and how they were overcome.*

EXERCISES

1. List any five considerations you have when planning an experiment.
2. Should all experiments follow the six-step scientific method or the IPCR model? Explain.
3. What is the difference between an experimental objective and a hypothesis?
4. What is an alternative hypothesis?
5. What is a null hypothesis?
6. Identify three characteristics of a hypothesis.
7. Identify some of the possible problems when recording data and some safeguard measures.
8. When should borrowed equipment be returned?
9. Who is responsible for ensuring borrowed equipment is returned?
10. What does IMRD stand for?
11. What are the three basic activities to accomplish when conducting an experiment?
12. Why should important data be written down, regardless of whether it was saved electronically?

NOTES

1 Kabir, M. (2016). *Basic guidelines for research: An introductory approach for all disciplines*. Book Zone Publication. p.57.
2 Firestein, S. (2016). *Failure: Why science is so successful*. Oxford. p.119.
3 Fleming, A. (1945, December 11). *Penicillin: Nobel lecture*.
4 P. K. R. Nair & V. D. Nair (2014). *Scientific Writing and Communication in Agriculture and Natural Resources*. Springer. p.13.

5 World Health Organization. *Handbook: good laboratory practice (GLP).* (2009). World Health Organization Special Programme for Research & Training in Tropical Diseases (TDR).
6 World Health Organization. *Training manual: good laboratory practice (GLP).* (2008). World Health Organization Special Programme for Research & Training in Tropical Diseases (TDR).
7 Huber, Ludwig. (2000–2002). *A primer; Good laboratory practice and current good manufacturing practice*, Agilent Technologies. p.16.
8 Erk, Jason. (2012). *10 Commonly Broken Good Laboratory Practices*, BiteSizeBio. https://bitesizebio.com/8862/10-commonly-broken-good-laboratory-practices/
9 Huber, Ludwig. (2000–2002). *A primer; Good laboratory practice and current good manufacturing practice*, Agilent Technologies. p.iii.

2 The Team

Experimental activities usually require a team of people working together. Choosing the correct teammates, setting clear expectations, employing good interpersonal skills and maintaining ethical behavior are critical to a well-functioning team.

TEAM DYNAMICS

Although a single individual may conduct small experimental research projects, most professional and corporate experimental projects involve a team of personnel. A team must function well together to be effective, and above all, the team must communicate well. Teams should consist only of essential personnel. It is far better to have every team member fully engaged in activities than extra personnel who are only partially engaged and lack commitment to team goals. Team members for any project should be carefully considered.

- Definition of a team[1]
  - *Teams* differ from *groups* as action-oriented since groups can exist and do nothing
  - People working together toward a specific objective or goal
  - People with complementary skillsets that are well-suited for a particular goal
  - People with certain levels of responsibility for managing themselves
- Teams have clear objectives or goals
  - Goals must be identified and understood by all team members
  - End products and intermediate products must be clearly identified and understood
  - Teams must set clear *internal* deadlines and goals to meet final product deadlines
  - Lab plans should clearly identify team member deadlines and products required
- Teams possess a shared responsibility and commitment to goals
  - All team internal deadlines and products must be agreed upon by team members
  - Team members must review and check each other's work for accuracy and consistency
  - Each team member must ensure their own work is aligned with team objectives and standards
  - Team members must place team success above personal success
  - Each team member must be willing to invest time and effort toward team goals

DOI: 10.1201/9781003329237-2

- Key attributes of an effective team[2]
  - An agreed upon team lead
  - Effective leadership from the team lead
  - Effective communication
  - Clear goals and objectives, such as internal deadlines rather than just the final deadline

LEADERSHIP

Every team requires a designated leader, and lab teams are no exception. Leadership *is a role* and the focal point for management and activities. Effective leadership is essential for any organization, company or team. In many ways, leadership should be considered as performance on a stage. Not that leadership is an act; rather, it is the best version of oneself for the benefit of others and the overall team. Team members are always watching how leaders perform and what a leader does, especially compared to established guidelines. Personal issues and poor moods should not be on display for the team. Professionalism should rule supreme. People are not necessarily born leaders. Leadership is a skill, and it can be learned, cultivated and grown with practice and introspection.

- Effective team leadership
  - A team leader should be selected based on their suitability to achieve goals
    - Leadership should not assigned merely based on a person's formal title
    - Leadership is a different team role with different responsibilities
  - The best leaders *endear* others to team goals
  - Leaders take responsibility for the outcome of the team
  - Leadership is not about winning friendships
    - No favoritism
    - Prevent cliques from forming
  - Always make expectations clear from the beginning
    - Responsibility then falls upon subordinates
  - Communicates for the team to instructor or supervisor
  - Organizes and manages lab activities
  - Resolves issues within the team
  - Submits the final cohesive document for all team deliverables
  - Establishes standard of performance for the team
    - Completes own tasks on time
    - Maintains personal integrity
    - Does not expect more from others than he/she is personally willing to do
    - Understands that words have value and communicates deliberately
      - Knows others will invest time based on the leader's words
      - Does not assign haphazard or last-minute tasks or meetings
      - Meetings <u>require at least one day prior notice</u> as respect for time and schedules

    - Ensures rules and deadlines are clear with specific dates and times
    - Sets the climate and mood of the team
        - Keeps a positive demeanor despite personal issues
        - Aggressive, depressed and worried traits will negatively impact team
    - Delegates duties and does not do everything alone
    - Solicits outside help as required
    - Adjusts approach to fit the circumstances
        - Switches to autocratic or directive mode when specific results are required
        - Switches to the facilitator when ideas are required
    - Provides feedback to team members
    - Available for team member interaction
    - Makes difficult decisions
    - Addresses issues up front
    - Best to be tougher and more disciplined at the beginning of team activities
        - If you are undisciplined in the beginning, it is very difficult to be tougher later
- Attributes of poor leadership
    - Controlling others
    - Acting from a *position of power* rather than a *position of responsibility*
    - Failing to adhere to standards set for everyone else
    - Lack of respect for others
    - Assuming everyone thinks and works the same way
    - Assuming *your* personal approach is always the only correct approach
    - Failure to enforce standards
    - Failure to recognize success
    - Failure to *step out front* for the team
    - Failure to maintain a positive attitude
    - Expecting everyone to *get it right* the first time
    - Unwillingness to admit errors, accept alternatives or adjust approach
    - Failure to maintain and enforce a clear chain of command and authority
    - Allowing multiple bosses with no clear authority
- How to work with challenging team members[3]
    - Reassign duties or roles since some people are more motivated with different tasks
    - Reconsider communication methods and opt for meetings, emails or one-on-one visits
    - Reconsider meeting schedules in the morning or afternoon based on attentiveness
    - Reset or clarify punctuality requirements for deadlines
        - Set deadlines everyone can agree with and abide by
            - Deadlines include necessary rework time
        - A good technique is to *ask* what deadline works for an individual, within reason
            - People almost always provide a reasonable response
            - Helps ensure personal buy-in and commitment to deadline

    - Have a backup plan for reworking
    - Silence or collect phones prior to meetings to ensure attention
- Resolve conflicts constructively
    - Address issues early and do not let issues get to a volatile point
    - Address the issue, not the person
    - Do not waste time on drama
    - Be matter-of-fact about issues and resolution
    - Do not be overly polite
        - This can cause resentment by failing to properly address issues
    - Establish rules for debate, if necessary
    - Leaders must know when to cut things off and move on
    - People have limited patience; get to the points quickly
    - Keep decisions open and transparent
- Provide feedback
    - Be clear and direct
    - Negative feedback can often be supplemented with positive feedback
        - However, do not cloud the issue at hand

COMMUNICATION

Effective communication is essential for good teamwork in an academic or professional work environment. Fortunately, modern society has a multitude of communication tools available, with email at the top of the list. However, exactly as stated, email is a *tool* for communication and not communication by itself. Messages, whether delivered verbally, electronically or by other means, must reach the recipient and be understood as intended. Unfortunately, "*I sent an email...*" has become an all-too-common phrase as an excuse for action. Complete the sentence as desired, but *sending an email* (or multiple emails) does not somehow absolve the sender of responsibility to follow through. Nothing could be further from the truth, and nothing is likely to irritate a supervisor more. An email has no value in itself since it is the content and action associated with the email that truly matter.

A classic academic example of "*I sent an email*" is when students have been required to form lab teams of four. When this assignment is late, common responses have included:

- *"I have sent out several messages... but I have not received any responses..."*
- *"I'm having difficulty finding a third person for my team. I emailed everyone..."*

Email is a tool. If it doesn't work, a new tool or approach is required. In this case, the requirement was to form a team. Sending emails is not an acceptable reason for failing the assignment. Find time before or after class to meet with others, or meet with the instructor ahead of time if the situation remains unresolved.

- Effective communication
    - Method is agreed, established and understood
    - Follow up as necessary to ensure messages are truly understood as intended

- Ineffective communication
  - Assuming an email, text or phone message is singularly sufficient
  - An emotionally reactive message or reply, especially with emails
    - Follow the rule: compose, wait, review and revise before sending an emotional reply
  - Email exchanges have a limit of usefulness
    - Two or maybe three attempts to clarify or communicate a single issue is the limit
    - Switch to verbal, phone or in-person communication to avoid an email spiral

ETHICS

Ethical issues are both an individual and a team concern. Clearly, an individual is responsible for their own ethical decisions, but team members are also affected by the actions of a single individual within the team. All team members' names on a document bear responsibility for the document's contents. At the very least, the unethical choices made by a single member of the team cast suspicion on the rest of the team. Even worse, the persuasions of a single team member can precipitate compromises in others who may not have been otherwise inclined.

Unfortunately, experience indicates academic discussions about ethical requirements often have little impact. Ethical training begins at an early age at home and is a continuous development process. Individuals who are inclined to plagiarize or cheat will do so when the situation is ripe. Students with strong ethical standards will hold much tighter to ethical ideals, even under pressure. Yet, even with this pessimistic but realistic view of ethical behavior, discussion is still required. Demanding high ethical standards gives respect to the ethical students, ensuring an even playing field where cheats will not receive an unfair advantage. Those who do plagiarize or cheat will fully appreciate their situation when the consequences are appropriately made very clear ahead of time.

Ethical issues in the academic environment range from outright cheating to plagiarism to "fudging" numbers. The pressure to compromise can be higher than expected, and no one should consider themselves or others to be completely immune to the pressure. Consider that a PhD student might invest two or more years in research only to get inconclusive results. There is a very real temptation to tweak the data and rationalize a possible relationship. Even if a person's moral compass may be strained, it is vitally important to remember that submitted and published work lives on, even after a compromise in integrity. Published documents and academic records can and have been reviewed many years later to discover ethical failures. No one wants to live with this infamy. The consequences usually include embarrassment, failing grades, loss of funding and loss of position or status. Ironically, pressure to compromise is usually self-created and can often be remedied by approaching the instructor or supervisor. Simply state the dilemma, and a new course of action can often be established.

- Ethical principles common to many codes of ethics as transcribed from a National Institute of Environmental Health Sciences article by David Resnik, PhD.[4]

- *Honesty: Strive for honesty in all scientific communications. Honestly report data, results, methods and procedures, and publication status. Do not fabricate, falsify, or misrepresent data. Do not deceive colleagues, research sponsors, or the public.*
- *Objectivity: Strive to avoid bias in experimental design, data analysis, data interpretation, peer review, personnel decisions, grant writing, expert testimony, and other aspects of research where objectivity is expected or required. Avoid or minimize bias or self-deception. Disclose personal or financial interests that may affect research.*
- *Integrity: Keep your promises and agreements; act with sincerity; strive for consistency of thought and action.*
- *Carefulness: Avoid careless errors and negligence; carefully and critically examine your own work and the work of your peers. Keep good records of research activities, such as data collection, research design, and correspondence with agencies or journals.*
- *Openness: Share data, results, ideas, tools and resources. Be open to criticism and new ideas.*
- *Respect for Intellectual Property: Honor patents, copyrights, and other forms of intellectual property. Do not use unpublished data, methods, or results without permission. Give proper acknowledgment or credit for all contributions to research. Never plagiarize.*
- *Confidentiality: Protect confidential communications, such as papers or grants submitted for publication, personnel records, trade or military secrets, and patient records.*
- *Responsible Publication: Publish in order to advance research and scholarship, not to advance just your own career. Avoid wasteful and duplicative publications.*
- *Responsible Mentoring: Help to educate, mentor, and advise students. Promote their welfare and allow them to make their own decisions.*
- *Respect for colleagues: Respect your colleagues and treat them fairly.*
- *Social Responsibility: Strive to promote social good and prevent or mitigate social harms through research, public education, and advocacy.*
- *Non-Discrimination: Avoid discrimination against colleagues or students on the basis of sex, race, ethnicity, or other factors not related to scientific competence and integrity.*
- *Competence: Maintain and improve your own professional competence and expertise through lifelong education and learning; take steps to promote competence in science as a whole.*
- *Legality: Know and obey relevant laws and institutional and governmental policies.*

DATA MANIPULATION

Not every experiment goes as planned. Even failed experiments have value and must be reported accurately. Proper reporting ensures integrity, and it can also serve as a guide for future researchers to be aware of pitfalls if re-attempting a similar experiment. For academic purposes, if an experiment fails, check with the instructor to determine if an alternate data set should be used. Only the instructor will make this determination and provide a data set for the report, if warranted. If an alternate data set is used, it must be documented as such in the report. Also, include a brief discussion about the nature of the failure and the unusable data. Transparency and completeness are keys to strong reports, no matter what happened.

- It is plagiaristic to claim another lab team's data, past or present, as your own
- It is an ethical violation and academic misconduct to manipulate or "fudge" data
- For failed or unsuccessful experiments, report data exactly as it is
  - Explain failed experiments and data as best as possible in the lab report

PLAGIARISM AND GHOSTWRITING

Plagiarism, as an ethical issue, requires unique consideration in the academic environment due to the multitude of written submissions on common assignments. Despite warnings against plagiarism, some students continue to commit the offense. Sometimes individuals do not expect to get caught. However, just as it is easy to copy and paste from an online source, it is just as easy to search for suspect text. Online searches and tools for checking plagiarism are readily available. More often than not, students are simply lured into the act of plagiarism through justification or by not stopping long enough to think their actions fully through under the stress of the moment. The written record stands as a constant reference to check. If plagiarism is not discovered initially, it is forever available to check in the future. The consequences are simply not worth it.

Ghostwriting is the close cousin of plagiarism. Rather than copying another source and calling it your own, ghostwriting is soliciting someone else to write the material and then submitting it as your own. Unfortunately, online "homework mill" services abound where individuals can pay a fee and have an assignment completed and ready for turn-in. Just as plagiarized sources are easy to search for, writing styles are easy to detect. Only the actual student, writing their own work, uses content commensurate with course material. Outsourced writers only care about finishing the document, whether it is a friend, relative or an online fee service. Style and content are easily discernible in these cases.

- Definition of plagiarism: *To steal and use the ideas or writings of another as one's own.*[5]
- New Mexico State University academic guidelines provide clarity into plagiaristic issues[6]
  - *An idea, an opinion, even when put into one's own words*
  - *A few well-said words, if these are unique insights*
  - *Many words, even if one changes most of them*
  - *Materials assembled by others, for instance quotes or bibliographies*
  - *An argument*
  - *A pattern or idea*

    - *Graphs, pictures or other illustrations*
    - *Facts*
    - *All or part of an existing paper or other resource*
- Plagiarism includes
    - Using an uncited source to construct part of a document with no original development
    - Copying and rearranging words does not ameliorate the offense
    - "Reusing" uncited research data
    - "Recycling" or using a classmate's or friend's report to "work from" as a "template"
- Plagiarism is
    - Academic dishonesty
    - A breach of journalistic ethics
    - Subject to sanctions such as penalties, suspension, and expulsion
- Plagiarism often results from
    - Panicked students stressed from time constraints
    - Students looking for a short cut on an assignment
    - Being lured into the act of plagiarism through justification
    - Not stopping to think actions fully through under the stress of the moment
- Solutions to avoid plagiarism
    - Ask for a deadline extension
    - Do not work from an existing document, unless cited
    - Do not submit the assignment; it is better to lose one grade than to lose much more

ARTIFICIAL INTELLIGENCE

With continued rapid advances in artificial intelligence (AI), writing student reports with the aid of AI will be tempting and more likely. The ethics of computer-generated text as a suitable student aid or as outright cheating will be debated in the upcoming years. Instructors, institutions and organizations must clearly establish policy and limits of AI use from the outset of any course or undertaking.

In the academic environment, the emphasis should always be on original thought and content. The single-most difficult part of writing a cohesive document is beginning with a clean page, organizing thoughts and building up a logically constructed narrative. Any tool or activity which shortcuts this process also shortcuts this very critical educational process and diminishes personal intellectual growth. At the expense of losing critical thinking skills, it is far easier to write by plagiarizing an existing document, just as it is far easier to feed key lines of text into an AI algorithm and generate a document to work from.

Style, consistency of writing competency and appropriate use of specific terminology are all key elements to help distinguish among truly original student content, outsourced documents or AI-generated reports. In-class writing exercises may also be required as style references for students' future written submissions.

CAUTIONARY EXAMPLES

Consider the following cautionary open letter from one student caught for plagiarism. The student was asked to provide this letter from a personal perspective and agreed to re-publication.

To my fellow classmates and future engineers:
I'm sure that as you have taken your seat and hold that still warm, recently printed [course] syllabus, you dive right into what seem to be the most important sections "grading scale", "schedule", "important dates", and so on, as did I. Sadly, when I got to the section titled "Plagiarism" I thought: Another unnecessary block of paragraph that I will never use. Little did I know that, that one paragraph was about to cost me my graduation and the most embarrassing moment of my existence.

At the time, I justified myself thinking that I would just base my work on that one belonging to my friend. "I will just follow his outline, rephrase a couple of important points, and fill in the blanks for the information that is missing" I thought. Of course, if I had paid any attention to the syllabus that you are now holding, or taken that first quiz seriously, I would've known that I had just defined plagiarism. I was drowning in work and homework, I desperately needed a shortcut and it was just so easy, my friend's work was right there whereas I had nothing.

Two weeks later I was emailed by my professor requesting a meeting. To my utter surprise, the professor didn't arrange the meeting to discuss my performance on the course, he wanted to discuss plagiarism. I was caught. Shock, panic, embarrassment, and tears all alternatively came flooding my face. Before he even said a word, flashbacks of that first class came to my mind "Zero plagiarism tolerance" he had said. As specified on the syllabus that you probably have already put away, as specified on mine, a whole letter grade can be deducted and an academic misconduct can be filed. For me, a letter grade deduction meant failing and so I had doomed my last semester. Retrospectively, it would've been better not to turn in anything at all than losing the job offer lined up upon graduation. In the midst of my chaos my professor said something that offered some insight to the true meaning of life: "Your good name and ability to trust your actions are among life's greatest values."

I am writing to you today to urge you not to be me. Do not skip that section, do not take it lightly, and do not let this cautionary tale become yours.

Sincerely,

Somebody working towards rebuilding [their] good name

Consider another open letter from a student caught for ghostwriting by his brother who agreed to the re-publication of their letter.

To whom it may concern,
I am writing this letter regarding a mistake I have done that I should not have done in the first place. When someone feels stressed, he does things that he is not aware of and what the consequences are afterwards. At a specific period of time in [semester], I had a really tough week to go through, there were exams, reports and quizzes in that week. I had a report due in that week. It is expected for students to feel worried at this period of time. Hence, I felt that I was in pressure and was going through hard times. I did a mistake, that is telling my brother to do the report instead of me writing it, which is the right thing to do as a responsible student. When I received the report from my brother, I did a one-time review on the assignment and submitted online in [LMS]. After the

grades showed up, I got an email from the professor asking me to come and meet him in the office. When I stepped in his office, I realized that I got caught that I plagiarized in that report. At that moment when I felt I got caught, I felt guilty of what I did I felt like an irresponsible student. We had a small conversation to each other and I told the truth to the professor that I felt stressed and told my brother to do the assignment for me. The only thing that I did not see it coming was how I will be punished after getting caught. The consequences were tough to take but absolutely deserved. They had a negative impact on me on my other classes too. This stupid mistake that I did ruined the whole course for me and destructed my mind which got me overthinking about other stuff and not focusing on my study. This was a stupid mistake and I feel guilty about it. I am 100% sure that this whole situation will not happen again in my future study. Even I feel stressed at some point, I will ask the professor for an extension so I can chill out and relax. This is a promise to myself, I am 100% sure that this whole situation will not happen again in my future study. Lastly, I just don't apologize to the professor in this case; I apologize to myself as being an irresponsible student in this situation. I encourage other students to do their own work. I finally realized that doing my own work is a really good thing that benefits me as a student. If you are not doing your work as a student and not learning anything, then what is the point of being a student?

In conclusion, I wrote this letter not just to myself but also to other students to encourage them to do their work and to let them know that doing so can lead them to a successful life after they finish their study.

In this open letter, the student was caught for ghostwriting from an online service and agreed to re-publication for the benefit of others. This is an increasingly common problem where students simply pay someone to do their work.

In mid-October specifically the fall of [year] I was under a lot of stress and anxiety by the amount of schoolwork I had ahead of me. It was towards the end of the semester and assignments and exams were piling up on me. I don't have many friends to ask for help and those who I did ask all had their own problems to worry about. Then someone came up to me and suggested this website that would help you out by doing your assignment for a fee, and I chose the cowards way out. I went in that website and hired someone to do the assignment. Little did I know that it would haunt me and be the worst thing that I have ever done in my whole undergraduate studies. My assignment was done, and I was relieved for a bit because I had more time to do my other assignments and study for my midterms. But I received an email from [professor] to schedule a meeting and I did. It was the most shameful meeting I have ever been in. I failed my mentor, he confronted me about my plagiarism, and I wanted at that time for the floor to open up and swallow me. I never felt this little and weak in my whole life. My relationship with the professor was amazing and I ruined it and for what just an assignment. This mistake hurt my name, got me an F in [course], delayed my graduation and I recently got married so it didn't just affect me but indeed affected my whole family. It was totally not worth it, as the professor explained to me that I should've talked with him about my stress and all of these assignments, he could've helped postponing some of them. But I chose not and got what I deserve because of that. My dad always says, "it's not a defect to mistake, it's a defect to repeat that mistake." My modifications for the mere future are to work harder and prove to the professor that I am a good student and hope that he will look past what shameful thing I did. My advice to my fellow friends and students is: Never choose the easy way out, because you are only fooling yourself and you will get caught. Work hard and if you feel the pressure of the semester and can't finish an assignment in time. Just talk with your professor, I promise they will help. All they want is for us students to succeed.

Several case studies associated with academic research and experimentation also highlight infamous ethics violations. These are not just examples of ethical violations; they are people who have a singular event defining their names in history.

- William Summerlin, 1973[7]
  - Immunology researcher who transplanted skin from black mice to white mice
    - If the transplant remained distinctly black rather than fading into white, it would have been an advance in tissue transplant procedures
  - He used a black marker pen to ensure the mouse had a distinctly black patch
  - Excuse: exhaustion, heavy workload, pressure to publish
- Vijay Soman, 1978[8]
  - Researcher Helena Wachslicht-Rodbard submitted a paper for publication
  - Soman was a peer reviewer and rejected the submitted paper
  - Soman then submitted a paper based on Helena's rejected paper
  - Helena was a peer reviewer of that paper and noticed it was her work
  - Investigations showed all of Soman's 14 papers were fraudulent
- Andrew Wakefield, 1998[9]
  - Published study linking the MMR vaccine to autism
  - Sample size was very small (12), uncontrolled and speculative
  - Picked-and-chose data, then falsified facts
  - Caused world-wide fear of MMR vaccinations
  - Subsequent measles outbreaks in 2008, 2009, USA, UK and Canada
- Vipul Bhrigu[10]
  - Sabotaged a fellow student's lab cultures with ethanol
  - Caught on secret security video
  - He was trying to slow down other students' success for his own benefit
  - Court-ordered to pay over $30,000 in fines and restitution

EXERCISES

1. TRUE or FALSE: Rewriting a source and changing a few words <u>without</u> reference is plagiarism.
2. TRUE or FALSE: Cutting and pasting an online source for submission <u>with</u> a reference but little to no original effort results in zero assignment credit.
3. TRUE or FALSE: Using another student's lab report "*as a template*" and changing words and details is still plagiarism.
4. TRUE or FALSE: Using data from another lab report without specific approval and reference is plagiarism.
5. TRUE or FALSE: Once a report has been submitted and the due date has passed, it cannot be "re-submitted" to resolve ethical issues.
6. TRUE or FALSE: All members of a team must review each other's work prior to submission.

7. TRUE or FALSE: All members of a team are responsible for an entire submitted document.
8. TRUE or FALSE: Plagiarism results in an academic misconduct report.
9. TRUE or FALSE: Getting help to improve the technical quality of a report is plagiarism.

Consider the following case of admitted plagiarism in Abstract #1 and Abstract #2 below which was agreed for publication. This is a classic example of rewording material from another source. There is a substantial difference in the level of effort required by the original thought of the first student versus the simple rewording of material by the second student. This disparity in effort constitutes the *cheating* associated with plagiarism. Read both abstracts and answer the following questions.

10. Highlight the sameness in structure and content between Abstract #1 and Abstract #2.
11. How much time do you estimate for Student #1 to write Abstract #1?
12. How much time do you estimate for Student #2 to write Abstract #2 using Abstract #1?
13. Do you consider the actions of Student #2 fair to other students in the course? YES or NO
14. Does Abstract #2 plagiarize Abstract #1? YES or NO, Explain.
15. What grade would you assign Student #2 for this assignment?

ABSTRACT #1 (Written First by Student #1)

A spring-mass-damper system can be described by a second-order ODE. It is an object attached to a spring and a damper that oscillates about its equilibrium position after a force has been applied. The purpose of this experiment was to calculate the theoretical and experimental values for the natural frequency of this system. Our experimental approach was to place a cart with mass on an inclined ramp. We then attached a spring to the back of the cart and the top of the ramp. We placed an ultrasonic sensor at the bottom of the ramp to measure the position of the cart. A magnet was placed on the cart to act as a damper for the system. The cart was pulled approximately 15 cm away from its equilibrium position before it was released. The oscillations of the cart were recorded by the ultrasonic sensor, yielding data that could be plotted in Excel and used to find the natural frequency. After comparing our experimental frequency with our theoretical frequency, we found an error of approximately 24.31%. This was caused by an incorrect calculation of the spring constant and friction losses between the cart and the ramp.

ABSTRACT #2 (Written Later by Student #2)

Generally, a spring-mass-damper system can be described as a second-order ordinary differential equation. The object is attached to a spring and a damper that allow it to oscillate about its equilibrium point once a force has been applied. Our goal for the experiment was to calculate a theoretical and experimental value for the natural frequency of the system. The experimental setup was to place a cart with mass attached to it and then place the cart on an inclined ramp. From there, a spring is hitched onto the back of the cart all the way to the top of the ramp; we also placed an ultrasonic sensor at the bottom of the ramp so that we may measure the position of the cart as it moves. A magnet is also attached to the cart to act as the damper for the system. The cart is pulled roughly 15 cm from its equilibrium point before being released. The ultrasonic sensor recorded the oscillations of the cart, which thereby allowed us to plot the data in Excel in order to help us find the natural frequency. After obtaining our theoretical and experimental frequencies, we calculated an error of approximately 76.63%. The result of such a large error is due to an incorrect calculation of the spring constant as well as any frictional losses from the cart and ramp.

NOTES

1 RMIT University Melbourne. (n.d.). *Achieving team goals.* https://www.dlsweb.rmit.edu.au/toolbox/legal/OFFICE/T07/T07_A/T7_LIB/T5_ATG.HTML.
2 Nazir, M. (2014, October 14). *Build an effective team to achieve your goals and objectives.* https://www.linkedin.com/pulse/20141014050236-138725871-built-a-team-to-achieve-your-goals-and-objectives.
3 Thompson, L. (n.d.) *What to do with your problem team member.* Northwestern Kellogg. http://www.kellogg.northwestern.edu/news_articles/2014/11072014-bad-team-member.aspx.
4 Resnik, J. (2020). *What is ethics in research & why is it important?* National Institute of Environmental Health Sciences. https://www.niehs.nih.gov/research/resources/bioethics/whatis/.
5 Morris, W. (1980). Plagiarize. *The American Heritage Dictionary.* Houghton Mifflin Company.
6 New Mexico State University. (n.d.). 5.10 – Student academic code of conduct – academic integrity. https://arp.nmsu.edu/5–10/.
7 Geoff. (2017, October 27). William Talley Summerlin and his false claims of transplantation without rejection. *Dr Geoff.* https://drgeoffnutrition.wordpress.com/2017/10/27/william-talley-summerlin-1938-and-his-false-claims-of-transplantation-without-rejection/.
8 Blumsohn, A. (2007, November 5). Memory hole (1 November): the case of Philip Felig. *Scientific Misconduct Blog.* https://scientific-misconduct.blogspot.com/2007/11/memory-hole-1-november-case-of-philip.html.
9 Sathyanarayana Rao, T. & Chittaranjan, A. (2011). The MMR vaccine and autism: sensation, refutation, retraction, and fraud. *Indian Journal of Psychiatry,* 53(2), 95–96. https://www.ncbi.nlm.nih.gov/pmc/articles/PMC3136032/.
10 Maher, B. (2011, April 27). Lab sabotage deemed research misconduct (with exclusive surveillance video). *Nature News Blog.* http://blogs.nature.com/news/2011/04/lab_sabotage_deemed_research_m_1.html.

3 Deliverables

Every formal experiment requires a deliverable, such as a written report or presentation. The deliverable expectations must be confirmed in the earliest planning stages.

Written plans and reports are the primary deliverables for lab experiments. The deliverables for any project must be addressed at the beginning of the endeavor to ensure expectations are understood and teams can begin working toward the goal. While details to improve writing styles and presentation techniques are discussed in later chapters, establishing basic format requirements and deadlines must be firmly understood by everyone at the very outset.

Research proposals may or may not include a preliminary experimental plan. Experimentation often accompanies research, but a proposal document should not be considered sufficient for a properly written experimental lab plan. Proposals are preliminary documents with insufficient detail to execute an experiment. An experiment lab plan should always be developed and written as a complete document, even if it is only part of a larger document or research effort. Well-written plans facilitate both the experimental activities and the final written document.

WRITTEN PLANS

The value of a written plan cannot be emphasized enough. A lab plan prescribes the experimental setup and data collection tools for all activities. In large research efforts or governmental test programs, the lab or test plan may take months or years to develop and involve several organizations and personnel. The plan may outlive the personnel originally involved with the document as assignments change and people move on. The plan must be robust enough to educate new personnel and define activities in sufficient detail.

A written plan is also a justification for resources and personnel. Many experiments require the use of facilities or equipment from outside agencies or alternate sources. Academic examples include a capstone course or experimental research for an advanced degree. It is poor technique to approach another organization or person about the use of equipment without a well-conceived plan. The obvious questions a facility or equipment owner will immediately ask will be the date and time required, personnel involved, intended use, range of parameters and suitability of the requested equipment. A written plan should address these details. The plan also does not need to be perfect and is expected to evolve as resources are secured and details are refined. However, having a hardcopy plan in hand with the basics worked out prior to soliciting a resource demonstrates professionalism and greatly improves the chances of success. Appendix A offers an example lab plan for academic activities.

- The terms *lab*, *test* and *experiment* plan or report are all considered interchangeable

DOI: 10.1201/9781003329237-3

- Lab plans and reports are essentially the same document
  - Both documents follow the same overall format
  - Lab plans are written and submitted prior to the experimental activities
  - Lab reports are written and submitted after the experimental activities
  - Lab plans are used to write the reports, so they are not "extra work"
- Lab plans
  - Plans are the *shell* (outline) of a report
  - Use future tense
  - Outline a systematic approach to conducting the experiment
  - The audience is yourself, team members, outside agencies and an instructor or supervisor
  - Format similar to a lab report
    - Follows a format such as IMRD: Introduction, Method, Results and Discussion
    - The IMRD format is widely accepted as the basis for research report formats
    - Lab plans should include a *methods* section but plans do not have all report-required content

WRITTEN REPORTS

An experiment lab or test report is the continuation of a properly written experimental plan. A final report should be a complete stand-alone document, fully describing the experiment and all results and conclusions. Collaborative documents such as lab reports tend to be divided up into different sections written by different individuals. Without proper editing, the document is often disjointed and repetitive. The report should read like a *story* as a single, cohesive document without forcing the reader to read repetitious material, different verb tenses or varied writing styles. The plot line moves forward without getting bogged down in detailed tables that belong in an appendix. The final report version should have sufficient detail to ensure other researchers can accurately replicate the experiment. Each of the lab manuals in this text is organized and written as examples of completed reports.

- Lab reports
  - Finalized stand-alone document of the experiment
  - Other researchers should be able to recreate the experiment
  - Use present or past tense
  - The audience is other researchers and an instructor or supervisor
  - Should read like *a story*
    - Cohesively written, gets to the point and is interesting
    - Provides enough detail to understand without *boring* the reader
      - Uses concise summary charts in the body of text
      - Place lengthy supporting data tables in the appendix
  - Follows the IMRD format for minimum content and basic organizational structure
- All team members should read and edit each other's written material
  - Improves quality of the document
  - Ensures integrity of document

- Use any available resource to help with grammar, spelling, punctuation and style
- Follow all ethical standards
- For failed experiments
  - Always check with the instructor – usable data might be provided
  - Write the report exactly as it occurred – there is always something to be learned
  - Refer to Chapter 19: Unexpected Results for failures in experimentation
- Do not cite any of the following in lab plans or lab reports
  - Human error (refer to Chapter 6: Managing Experimental Errors)
    - Human error is a failure to meet the objective of the experiment
    - Cite human error only if it truly is a reason to re-accomplish the experiment
  - "*We learned a lot*," or "*The lab was fun*" – stick to the relevancy of the experiment only

ORGANIZATION OF CONTENT

Most requirements for technical reports follow similar outlines, but every agency or organization has its own unique requirements for report documents that should be followed. The IMRD format is a generally accepted approach for research and journal papers. According to one study, the IMRD format, also known as IMRaD (Introduction, Methods, Results and Discussion), was mostly adopted in the 1980s as a standard for scientific writing.[1]

- IMRD section headings for scientific and technical writing
  - Introduction
  - Methods
  - Results
  - Discussion
- Recommended organization and section headings for academic lab plans and reports
  - Title Page (include for both plans and reports)
  - Abstract (include for reports and do not include with plans)
  - Introduction (include for both plans and reports)
  - Theory (include for both plans and reports)
  - Computational methods (include for both plans and reports involving a simulation)
  - Experimental methods (include for both plans and reports)
  - Expected results (include for plans and do not include with reports)
  - Results (include for reports and do not include with plans)
  - Discussion (include for reports and do not include with plans)
  - Conclusions (include for reports and do not include with plans)
  - References (include for both plans and reports as appropriate)
  - Appendices
    - Supporting data (include for both plans and reports as appropriate)
    - Code (include for both plans and reports as appropriate)
    - Deadlines and participation (include for both plans and reports)

Experiments that are limited to computer simulations should only have a computational methods section and no experimental methods section. Experiments with both activities should include both sections. The conclusion section can often be combined with the discussion section. Additional detail for each lab report section follows:

- Title page
  - Provides a presentable appearance to a document
  - Clearly identifies all personnel and organizational information
    - The school, course and instructor
    - All team members, the designated team lead and means of contact
  - Date of document submission
  - Date of experiment
    - Document submission date and experiment date are two separate events
    - Experiment date on the lab plan is important for coordinating activities
    - Experiment date on the lab report is valuable for correlating details after the fact which were not originally recorded, such as barometric pressure on that date

Students who are new to writing often confuse abstracts with introductions. However, an abstract is distinctly different from an introduction. An introduction is primarily a lead-in to the document, supplying context, background information and foresight into the contents of the document. The abstract is a complete and condensed summary of the entire document. After reading an abstract, the reader should understand what the experiment was about and how it turned out. An introduction only informs the reader what they're about to get into with the document. An abstract must be written after the document is complete to properly convey the contents. Referring to Figure 3.1, writing an abstract can almost be a mechanical process by pulling key sentences from each section of the report for the composition.

- Abstract
  - A separate page and stand-alone summary of the entire report in one or two paragraphs
    - Includes key sentences from each section of the report
    - Includes an introductory sentence, hypothesis or objective, and bottom-line results
  - Fundamentally different from an introduction as a complete, concise summary
  - Does not include figures or references
  - Should be written after the report is complete
- Introduction
  - Provides context, relevance, motivation and interest for document
  - Offers a simple concept of the intended experiment
  - Graduate research introductions include a literature review and state-of-the-art information

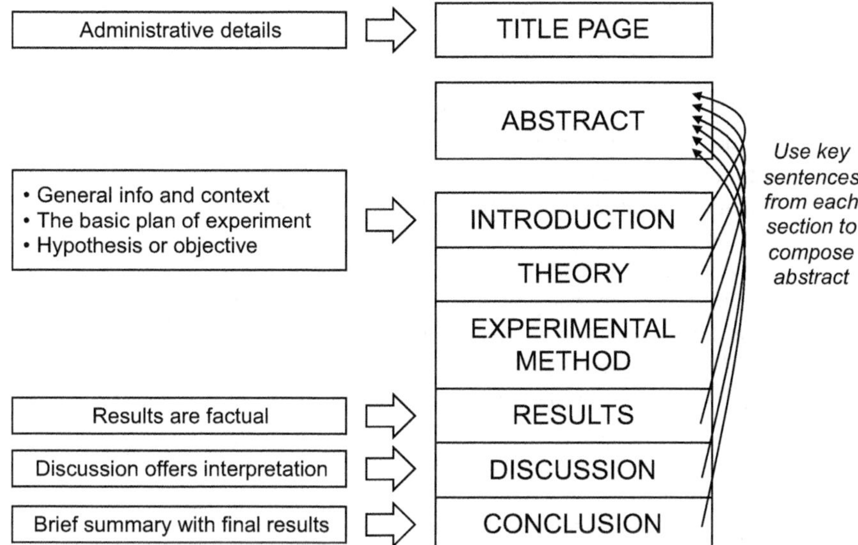

FIGURE 3.1 Graphical organization of a lab report with abstract.

- Undergraduate reports should not include a literature review, references or figures
  - The emphasis should be on composing original thought and introducing the content of the report
- Finishes with the hypothesis, objective or goal the document intends to support

- Theory
  - Also referred to as *Background* section
  - Does not repeat the introductory material
  - Provides theoretical development of relevant concepts and equations in the document
  - Includes assumptions and simplifications in the theoretical model
  - Equations should be developed and presented in the final form used for computations
  - Equations referenced elsewhere in the document require a separate line with enumeration
- Computational Methods
  - Include for computer and numerical simulation experiments
  - Includes the software, computational technique, basic algorithm or approach
  - Includes a theoretical model and assumptions
  - Includes the robustness or limitations of technique
- Experimental methods
  - The "*how to do*" the experiment for other researchers to replicate experiment
  - Describe what *you did*; do not tell the reader what to do

    - Provides detail on how the research or experiment was conducted
    - Includes basic materials, equipment, instruments and resources used in the experiment
        - Detailed lists belong in an appendix
    - Includes sketches or photos of experimental configurations and test sample configurations
    - Includes variables, parameters of investigation and a control reference parameter
    - Provides data collection methods and sample size
    - Outlines any necessary personnel duties and schedule of activities
    - Identifies potential or actual risks, error mitigation plans and contingencies
- Expected results (plans only)
    - Expected results are key to selecting the correct equipment with appropriate sensitivities
    - Essential for evaluating whether an experiment is going well or failing in real time
    - Includes expected values or anticipated ranges of values
    - Includes theoretical, published and estimated values
    - All formulas and values are worked out to the maximum extent possible
    - Includes a reasonable estimation of the unknown values necessary to complete the calculations
    - Includes data collection tables or placeholder summarizing tables
    - Includes hand sketches of anticipated behavior as appropriate
    - Applies only to the lab plan and is replaced by actual results in the report
- Results
    - This is the product of the experiment or research
    - Includes summarizing charts or tables to factually support the hypothesis or objective
    - Includes specific observations and callouts of key data details and anomalies
    - Information is factual and objective without subjective analysis
    - Does not repeat information from the experimental method section
    - Tables and charts should be introduced and described in text
    - Data should be succinct and summarized, with supporting data in appendices
    - Does not include multiple charts when data makes sense combined into a single chart
    - Include results of statistical analysis
    - Does not include inferences or subjective analysis
    - Replaces expected results from the lab plan
- Discussion
    - Provides analysis, interpretation and meaning of results
    - Whereas the results section is strictly factual, the discussion section is interpretive
    - Inferences and discussion should be based on a dispassionate analysis of presented results
    - Indicates potential sources of error and their impact on data

- Conclusion
  - Clearly states bottom-line results with overall error
  - Clearly states how results either support the hypothesis or do not support the hypothesis
  - Very briefly summarizes the experiment and report, as appropriate
  - Offers any considerations for follow-up efforts
- References
  - Includes specific citations or bibliography references used to compose thoughts
  - Includes one specific reference to the lab manual when authorized to transcribe sections
- Appendices
  - Contains all supporting data, detailed equipment lists and procedures
  - Data should be editable to copy and reproduce charts rather than images
  - Contains information secondary to the report's content
  - Not a "*dumping ground*"; tables and figures should be legible, well-formatted, captioned and introduced via text as necessary for any explanations
  - Undergraduate documents should include deadlines and participation information
    - Refer to Appendix B for suggested content

Since academic students should not be expected to recreate the theory and methods of a pre-designed experiment, consideration should be given to the authorized transcription of these specific sections from a comprehensively written lab manual. The goal is to practice writing, compiling and submitting a *complete and comprehensive* report for every experimental session. Authorized transcription of the theory and methods sections avoids forcing students to rewrite material from a lab manual with very limited options for modification. When the theory and methods sections are authorized to transcribe and include in a student-submitted lab plan and report, a single reference should be included to cite the lab manual. Transcribed sections should also be adjusted to flow with the overall student report. The audience is still the instructor and other researchers, so lab manual-specific directives such as "*be sure to wear gloves*" should be omitted unless a unique hazard exists. Reports should inform and not direct the reader to take action. If an academic lab manual is sparsely written, then each section should be fully developed for both the lab plan and lab report. Any transcribed content from an academic lab manual must be specifically authorized and applies only to specific contents of manuals used for academic activities.

While also serving as a report format, the lab manuals in this textbook include a section heading for the experimental objective. These sections are provided only for clarity, and the separate heading should not appear in a lab plan or report. The objective or hypothesis should always be included toward the end of a report introduction.

Specific technical format requirements also vary with each organization. Most format requirements call for a common Unicode computer font at a legible 12-point font size. Technical documents should be double-spaced. Every table and figure must be enumerated, captioned and centered with the caption. Table captions are typically above the table, while figure captions appear below the figure. Failure to adhere to specific format requirements is a ready cause for the rejection of a report. Format errors immediately stand out before even reading the content of the report and provide the first impression of the document overall. If care is not taken to address the technical format issues, then the content of the document is also suspect. Refer to Appendix A and Appendix C, Table C6, for recommended format guidelines.

EXPECTED RESULTS

Simply answer this question to the *fullest* extent: "*What values do I expect to obtain in the lab?*" For experimental research, discussions are often focused on only resolving theoretical values. However, depending on the experiment and available data, expected results may be published as empirical data with no theoretical relationship to work out. An example might include experiments to verify published thermocouple voltages. In this case, expected results are published values rather than theoretical values. Expected results cover the broad range of theoretical values, estimated measurements, published data, empirical data and any available values that contribute to estimating an experimental result in the lab and should be included in the written plan.

Properly completing expected results requires working through all formulations necessary to provide estimations of lab-related measurements and values. Knowing an anticipated value helps select the correct measurement instrument. If an experiment measures the viscosity of a fluid by timing the drop of a sphere in the fluid, expected results should estimate the time of the drop. If time estimates are upwards of a minute, simple observation and a smartphone timer may be sufficient. However, if measuring time for the speed of sound in aluminum, working out the expected time value makes it clear that a high-fidelity instrument such as an oscilloscope is required for time measurements.

Any and all values and computations should be fully worked out in the lab plan and entered into appropriate tables in the expected results section of the lab plan. Reasonable estimates should be used when actual values are not readily available. Aside from choosing the appropriate instrument, the purpose of the expected results is to also fully work out equations to ensure values measured during the lab session actually make sense. All spreadsheet formulas can be verified for accuracy ahead of time and are ready to go for the lab session. Lab session time is both valuable and limited. If formulas are not checked prior to the lab and there is no idea of what to expect for results, lab errors cannot be readily realized and fixed in the available lab time.

As an example of properly completing expected results in a lab plan, consider the simple tensile stress test in Table 3.1. While Table 3.1 is adequately configured for data collection during the lab, without any values, there is no idea what to expect during the lab. Assuming Table 3.1 is associated with a pre-planned academic lab, there may

TABLE 3.1
Simple but Incomplete Expected Results for Tensile Test Data in a Lab Plan

Material	Area (A)	Ultimate Stress (σ_{ULT})	Theoretical Failure Load (P_{ULT})	Experimental Failure Load (P)

not be additional information available prior to the lab. An effort should be made to determine the specimen sample material and cross-sectional area to complete the computations. If this is still not possible, a proxy material and cross section can be entered into the corresponding spreadsheet to test the formulas. Although Table 3.1 is a trivial example, more complex calculations such as wind tunnel data make the issue clear.

Once the lab session is complete, the lab plan's *expected results* develop into the lab report's *results* section. Theoretical calculations dependent on estimated measurements must be updated with actual measurements. A well-written lab plan pays significant dividends in both the lab session and the final report.

- Lab plans must have expected results
- Always know what values to expect to the maximum extent possible prior to a lab session
- Expected results consist of any of the following
  - Theoretical calculations
  - Published data
  - Estimated measurements
    - If the wire diameter is unknown, look up a likely diameter online
  - Sketches of system responses
- Expected results give way to
  - Updated theoretical results based on actual measures versus estimates
  - Experimentally determined results from the lab session

If values are unavailable for the expected results, do not *invent* numbers that could throw off expectations during the lab. Estimates should be reasonable, but if they are highly uncertain, one option is to limit the range of expected results. For example, heat transfer coefficients (HTC) are empirical constants with a very broad published range. If the experimental objective is to determine an HTC and an expected HTC value is required for the lab plan, determine a likely upper and lower bound rather than just guessing at a value. Have the spreadsheet formulations fully worked out and use the likely bounds to evaluate lab results in real time.

A final note of caution with expected results is to be wary of creating bias during experimental data collection. Knowing what to expect is important to properly set up an experiment, but data should be collected faithfully as presented without favoring any presumptions.

EXERCISES

1. TRUE or FALSE. An abstract is similar to an introduction, just with less detail. Explain.
2. Consider Table 3.2 and complete tasks 2a and 2b:
   a. Cross out any column heading that does not belong in a report.
   b. Place an "X" in the appropriate column where the information belongs within the report; more than one section may be appropriate, indicate all appropriate sections.

TABLE 3.2
Report Heading and Content Matrix

	Report Heading									
Report Content	**Abstract**	**Introduction**	**Objective**	**Theory**	**Experimental Method**	**Results**	**Discussion**	**Conclusions**	**Summary**	**Appendices**
Objective										
Figures										
Data collection tables										
Summarizing charts										
Bottom-line results										
Expected results										
Human error										
Assessment about results										

3. Which of the following statements about figures are correct (circle all that are true):
   a. They are left-justified.
   b. The caption can be on the next page from the figure if it does not fit on the same page.
   c. The figure must be introduced in the text before it appears.
   d. Figures should be sequentially numbered.
   e. Figure numbers in appendices should include the appendix letter or number.
4. TRUE or FALSE. The *introduction* of the lab plan or report may be copied from the lab manual.
5. TRUE or FALSE. Estimated *expected results* should be part of the lab *report.*
6. TRUE or FALSE. A lab report can be considered as a *story* that flows with appropriate information to convey the concept and results of an experiment without dwelling on unnecessary detail.
7. TRUE or FALSE. Detailed data are best placed in an appendix rather than the body of a report.

Refer to Appendix A and read the "Tensile Test Experiment" example lab plan. This lab plan outlines the key activities and parameters associated with a planned Capstone course experiment outside of normal classroom activities. Determine if the lab plan is sufficient by answering the following questions:

8. TRUE / FALSE. Always have a basic test plan document (such as in Appendix A) before approaching an organization or individual with an experimentation request.
9. Is the test objective (or hypothesis) clearly identified in the example from Appendix A? If so, what is it?
10. Does the test plan in Appendix A identify the necessary equipment?
11. Are the expected results in Appendix A adequately worked out?
12. Why is it important to include an image of the planned test samples?

NOTE

1 Sollaci, L. & Pereira, M. (2004, July). The introduction, methods, results, and discussion (IMRAD) structure: a fifty-year survey. *Journal of the Medical Library Association*, 92(3), 364–367.

4 Design of Experiments

An experiment to test one condition at a time is not necessarily efficient, especially for large experiments. A properly designed experiment can save time and effort.

Reasons for conducting an experiment with more than a single trial often fall into one of two categories. The first category is narrowly focused, while the second category is far broader:

- *Already having* a specific parameter and causal relationship in mind
- *Looking* for a potential causal relationship among many possible parameters

These two reasons can lead to drastically different requirements when designing an experiment. In the first case, the experimental design is narrowly focused on a specific objective with a single test parameter in an attempt to either support or negate a particular hypothesis. Only a few experimental runs may be necessary to achieve the objective, depending on the statistical requirements.

In the second case, tens, hundreds or maybe even thousands of experimental runs with combinations of parameters may be required to find a possible relationship. This scenario is reflective of the pharmaceutical industry's search for a drug or combination of drugs with a hopeful therapeutic effect. Many experimental lab culture trials may be required to detect a single causal relationship, so a systematic approach is critical to efficiently managing a large number of trials.

Design of Experiments (DOE), as a field of study, offers an opportunity to find a causal relationship from many variables while smartly reducing the number of experimental trials. A significant distinction between engineering experiments and medical study experiments is that medical studies do not necessarily have theory as a guide. Medical interactions have traditionally been based on trial and error, using history and experience as guides. DOE is uniquely important in fields where a great number of experiments are necessary to detect causal relationships.

TERMINOLOGY

A single experiment will have multiple runs (trials) in order to test all the various combinations of a set of changeable parameters, so a single experiment with many runs may yield only *a single data point* of interest or no usable data at all.

- TRIAL or RUN: A single, start-to-finish pass through a complete experimental setup
  - Only one combination of many parameters is tested during a single run
  - Testing another combination of parameters requires another complete run
  - A single run is <u>not</u> a complete experiment when several parameters are investigated

DOI: 10.1201/9781003329237-4

- EXPERIMENT (designed): a collection of runs testing for one or more effects
  - A single experiment requires multiple runs to test multiple parameter combinations
  - A complete experiment fulfills an objective or supports a hypothesis
  - A complete experiment may or may not yield statistically significant results

A single experiment, with all its multiple runs (trials), can be replicated twice (or more) and potentially yield two (or more) data points.

- REPLICATION: repeating an *entire experiment* with its *collection* of individual runs
  - Replication occurs between different experimental sessions
  - Replication can provide a statistically useful data set from an experiment

When considering *how many* experimental events are required, there are two distinct requirements. Both requirements must be considered separately, then combined to know the overall total number of experimental events.

- The number of runs required to test all necessary *parameter combinations*
- The number of replications needed to provide a *statistically convincing* data set

The first is driven by DOE, and the second is driven by statistical requirements.

If a manufacturer seeks to improve product quality, unsatisfactory quality may be due to machine settings, process order, personnel actions, time of day, suppliers or a combination of these variables. DOE techniques avoid trial-and-error approaches and ensure efficiency in the experimental design. In this case, a single experiment might be designed and "run" several times with different combinations of independent variables to find the cause. By definition, multiple runs with changing parameters are *not* multiple experiments.

Continuing with the manufacturing improvement scenario, from all runs of the single experiment (with varying parameters), the resulting data set might show a single combination of parameters (one run, returning a single data point of interest) where both time of day and machine settings together have the largest effect on product quality. This single data point may be sufficient for the manufacturer, especially since obtaining this single data point required multiple runs with different settings, costing time and money. However, a more confident result requires a larger data set and even more effort.

A couple of experimental design options now exist to develop a statistically convincing data set. The first option is to ***replicate*** the entire original experiment by repeating it multiple times. This is the equivalent of saying *all those runs need to be performed again*, multiple times; essentially, a nested loop. This could prove time-consuming and expensive, depending on all the parameter changes, but it makes the case of *that particular combination* of parameters, among many, has the greatest effect. Alternatively, the experiment can be slightly redesigned to narrow the focus only on *that one set* of parameters relative to the control (unchanged) conditions. This option makes the case for improvement *only in that case, relative to the control.*

If the decision is made to test only one condition, then an experiment can be designed with only one run. Now the option exists to replicate the experiment by setting it up, running it, obtaining a data point, tearing it down, setting it up again later, re-running it for another data point, and so on. Alternatively, the experiment could be set up once, run for a data point, then run again for another data point, and so on, and then taken down when enough data is collected. This *single-session* activity is the process of *repetition*.

- REPETITION: repeated samples or measurements of the same item or event
  - Repetition occurs within a single experimental session
  - Repetition can provide a statistically useful data set from an experiment

Overall, the total number of experimental events must consider the number of runs necessary to test all desired parameter combinations, the number of runs or samples necessary to have a statistically significant data set, and the available time, money and resources necessary to perform all activities.

BASELINE CONTROL

A properly designed experiment will always have baseline data, known as a *control*. A control is the *"in reference to"* condition for comparison. The control can be a single value, the mean of a data set, a population mean or null hypothesis in a statistical study. In a statistical study, the *population* usually serves as the reference control compared to the subset *sample*. For an engineering experiment supporting a hypothesis, it is insufficient to perform a single experimental run that appears to support a relationship without also running the same exact conditions with no parameter changes as a control. It is entirely possible the *experimental setup itself* is causing the effect rather than the actual parameter change. The Pons and Fleischmann cold fusion experiment discussed in the chapter on errors is an example where the equipment and setup limitations were confounded with the parameter of interest. Had the original cold fusion experiment also been performed exactly the same way as a control, without the heavy water, the issue may have been identified prior to their press conference on the results.

OVERVIEW OF DOE

DOE offers rational choices for testing every possible combination of parameters or reducing the number of experimental runs to a smaller subset while still retaining the maximum insight into the larger set of possible combinations.

- Definition of DOE
  - A systematic method to determine cause-and-effect
  - Method to *identify and quantify factors with the biggest impact on an output*[1]
  - Method to *quickly screen a large number of factors to determine the most important*[2]
- DOE is usually associated with an objective to discover a causal effect
  - Causal effects may be due to a unique combination of several contributing factors

    - May or may not have statistically significant results (depends on replications)
- Benefits of DOE
    - Potential to reduce the number of experiments compared to a *non-designed* experiment
    - Determine which parameters are responsible for changes in a system
    - Helps in developing models and predicting response
- DOE is particularly useful in industry to[3]
    - Improve process yields
    - Reduce variability in processes
    - Reduce development time
    - Reduce costs
- Fundamental STEPS in DOE[4]
    1. *Set objective(s)*
        - Resolve causal relationships, test hypothesis and validate a process model
    2. *Select the process variable(s)*
        - Decide on independent and dependent variables of interest
        - Establish baseline data, control data or null hypothesis
    3. *Select an experimental design* (discussed in subsequent paragraphs)
        - **One-factor-at-a-time (OFAT)**
        - **Full factorial**
        - **Fractional (partial) factorial**
    4. *Execute the experiment*
        - Follow safety protocols, standard procedures and test plan
        - Calibrate equipment
        - Validate the experimental setup with known samples to establish control
        - Perform a test run to check the expected results
        - Perform experimental run(s) with the desired test parameter changes
        - Record and save the data
        - Record any process steps or deviations from the plan
    5. *Check data consistency with assumptions and expectations*
        - Ensure actual experimental setup is consistent with assumptions
    6. *Analyze and interpret results*
        - Perform data reduction and statistical analysis
        - Draw conclusions based on the data
    7. *Present the results*
        - Write a report or present to the audience as required

EXPERIMENTAL DESIGNS

To illustrate the three basic DOE design options (Step #3), consider a mechanical system where the natural frequency of vibration is a *result*, which might be affected parameter changes of mass, damping, stiffness or a combination thereof. The experimental objective is to determine which parameters or combinations of parameters have the most effect on natural frequency. The independent parameters (variables) are mass, damping and stiffness. The dependent variable (output parameter of interest or result) is natural frequency (or dampened natural frequency). This mechanical vibration example is used in the following definitions and examples.

- **Factorial** experiments
  - DOE term to systematically perform the following types of tests
    - FULL factorial: test ALL possible combinations of all variables
    - FRACTIONAL factorial: test a reduced set of all possible combinations
  - FACTORS: independent variables (test parameters) to change
    - Factor = Independent variable
    - Example of three factors to investigate for natural frequency: mass, damping and stiffness
  - LEVELS: number of changes each factor can have (two is the most common)
    - Two levels: both **more** and **less** of each of mass, damping and stiffness
    - Three levels: usually specific values
      - Three levels of mass: 10, 20 and 30 kg
      - Three levels of damping: damped, undamped and critically damped
  - RESULTS of the factorial test: the dependent variable of interest (natural frequency)
  - TREATMENT: a factor changed by its level; an individual test parameter change
- **DESIGN #1: OFAT**
  - Hold everything constant and vary only *one factor at a time* in each experimental run
  - Runs required: (Factors)*(Treatments)
    - Total runs should also include baseline data run(s)
  - Example: what might cause an increase in the dependent variable of natural frequency
    - Three factors (independent variables or parameters): mass, damping and stiffness
    - Two levels (treatments): more and less of each
    - Runs required: (Factors)*(Treatments) = 3*2 = 6 runs (+baseline)
    - Reference Table 4.1 for an OFAT design
  - Advantages of OFAT
    - OFAT has *fewer* experimental runs than a FULL factorial
    - Simple to implement for small experiments
  - Disadvantages of OFAT
    - No ability to determine if there are <u>combined effects</u>
    - Requires separate experimental runs for each condition
    - OFAT does not consider ***design space*** (discussed in subsequent paragraphs)
    - Is not a *projectable design* into the total design space of an experiment

TABLE 4.1
One-Factor-at-a-Time Design

Run	Factor and Treatment
#0	Baseline parameters (no changes)
#1	Increase mass
#2	Decrease mass
#3	Increase damping
#4	Decrease damping
#5	Increase spring constant
#6	Decrease spring constant

TABLE 4.2
Full Factorial Design

Run	Mass	Damping	Stiffness
#0	Baseline parameters (no changes)		
#1	–	–	–
#2	+	–	–
#3	–	+	–
#4	+	+	–
#5	–	–	+
#6	+	–	+
#7	–	+	+
#8	+	+	+

- **DESIGN #2: FULL FACTORIAL experiment**
  - Test all possible combinations of each factor (independent variables)
  - Runs required: $(\text{LEVELS})^{\text{FACTORS}}$
    - Total runs should also include baseline data run(s)
  - Example:
    - Three factors: mass, damping and stiffness
    - Two levels: more and less of each (where "–" is decrease and "+" is increase)
    - Runs required: $(\text{LEVELS})^{\text{FACTORS}} = 2^3 = 8$ runs (+baseline)
    - Refer to Table 4.2 to create a full factorial design
      - Create a column for each factor
      - Create a row for each run
      - Use *any* scheme of + and – to ensure *all possible combinations* are covered
  - Advantages of full factorial
    - Every possible combination is tested with no ambiguity
  - Disadvantages of full factorial
    - The number of required experiments can become unmanageable

    - Varying four parameters (factors): $2^4 = 16$ experiments
    - Varying five parameters (factors): $2^5 = 32$ experiments
- **DESIGN #3: FRACTIONAL FACTORIAL experiment**
  - Reduces full factorial design to more *manageable* designs
    - A full factorial experiment with one LESS factor: $2^3 \rightarrow 2^2$ (or even less)
  - Provides projectability
    - Testing a subset is representative of the larger set
    - Infer interactions with fewer experimental runs
  - Good initial screening to *weed out* many factors that may not affect results
    - Can switch to full factorial later after screening experiments reduce the number of runs
  - The Pareto principle[5]
    - *Limit experiments to the most important factors* (insight is required)
    - *The 80/20 rule*: testing the most important 20% gives 80% of the necessary info
    - Also known as the "vital few and trivial many"
  - DISADVANTAGES of fractional factorial experiments
    - Factors will be *confounded* or *aliased* (difficult to discern from other combinations)
    - Effects are "lumped together" and can be indistinguishable from others
    - Follow-up experiments may be necessary to improve fidelity

DESIGN SPACE

OFAT and full factorial designs are conceptually straightforward. Tables 4.1 and 4.2 are easy to build. However, a proper fractional factorial design must be an *orthogonal* design to capture as much of the *design space* as possible with fewer experimental runs.

- Definition of design space
  - Design space is the full range of values between the limits of a test parameter
  - Example: the natural frequency of a mechanical system with three test parameters
    - Factors: mass, stiffness and damping
    - In this example, each is set to "5" ($m = 5$ kg, $k = 5$ N/m and $c = 5$ Ns/m)
    - An experiment to test only mass might therefore be set up to test ± 1.5 kg
      - It is not necessary to test every variation between 3.5 and 6.5 kg
      - Testing only the *endpoints* provides insight into the full range
  - The *design space* is the *range of values* captured between the endpoints of a test parameter
- Figure 4.1 provides a graphic depiction of a one-dimensional (1D) design space

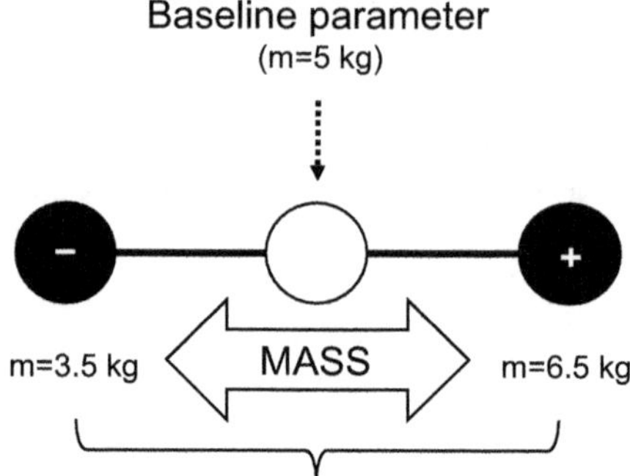

FIGURE 4.1 1D design space.

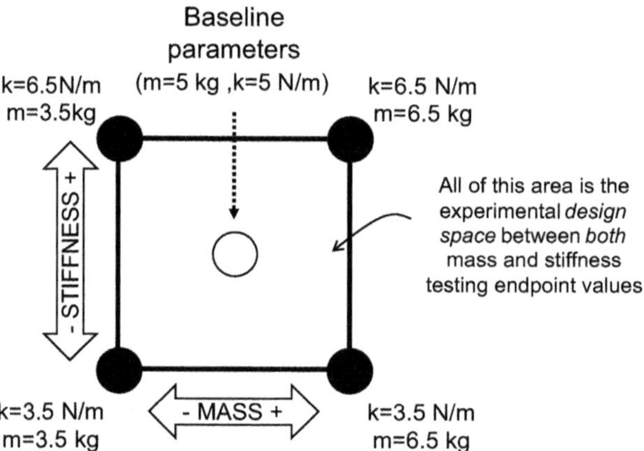

FIGURE 4.2 2D design space.

- In this 1D design space example, the only factor of interest to test is mass (m)
- The design space is the *line* between the two extreme + and – test values for mass
- The test *endpoints* of $m = 3.5$ kg and $m = 6.5$ kg define the full design space

- If <u>only two factors</u> are now tested, say mass (m) and stiffness (k)
    - The *full* design space is an *area* between the test parameter endpoints per Figure 4.2
- Finally, if <u>all three factors</u> are tested, the full design space is a cubic volume, per Figure 4.3
    - A *full factorial* test captures the entire volume (design space) by checking every endpoint

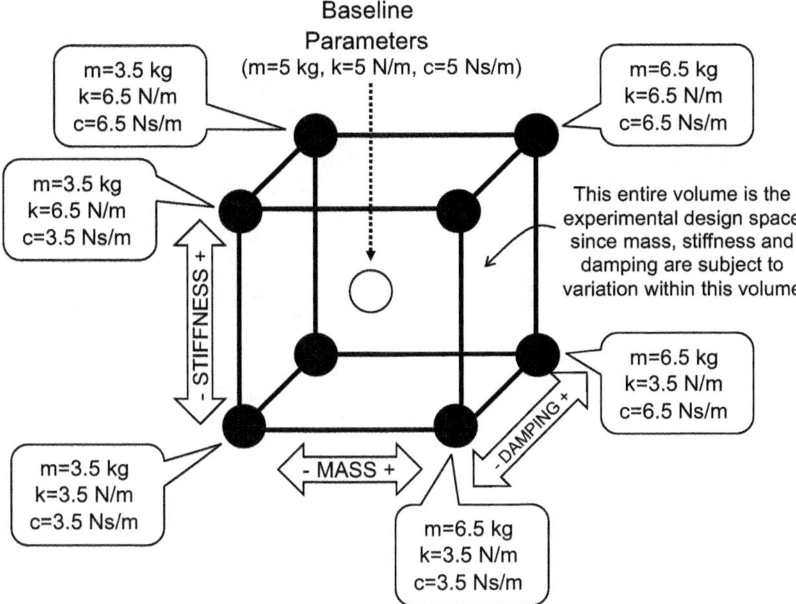

FIGURE 4.3 3D design space.

- $(\text{TWO LEVELS})^{(\text{THREE FACTORS})} = 8$ runs, which are the eight corners of the design space cube

While Figure 4.3 depicts a cube as the complete design space for a full factorial design, a fractional factorial design must therefore be less than a full cube. The key to a *fractional factorial* design is to *still capture as much of the design space as possible without testing every endpoint.* For example, consider a fractional factorial experiment in which the number of runs is reduced from 8 (2^3) to 4 (2^2). Figure 4.4 shows two fractional design space options, both testing only four points. In Figure 4.4a, a poor design results in only a two-dimensional area for the fractional design space, and most of the volume of the full factorial it supposedly represents is completely missed. In Figure 4.4a, this is equivalent to saying the damping parameter was never varied in the experiment. Alternatively, in Figure 4.4b, a proper choice captures most of the full factorial design space volume with only four test points. Figure 4.4b is an *orthogonal design*, while Figure 4.4a is an improperly designed experiment.

- Fractional factorial experiments are a reduced version of a full factorial experiment
- A fractional factorial experiment should *represent* as much of the full factorial experiment as possible; this is called *projectability*
- Creating an orthogonal fractional factorial experiment
    - Determine the total number of *factors* and *levels* to investigate
        - Example: Three factors (mass, damping and stiffness) at two levels (more and less)

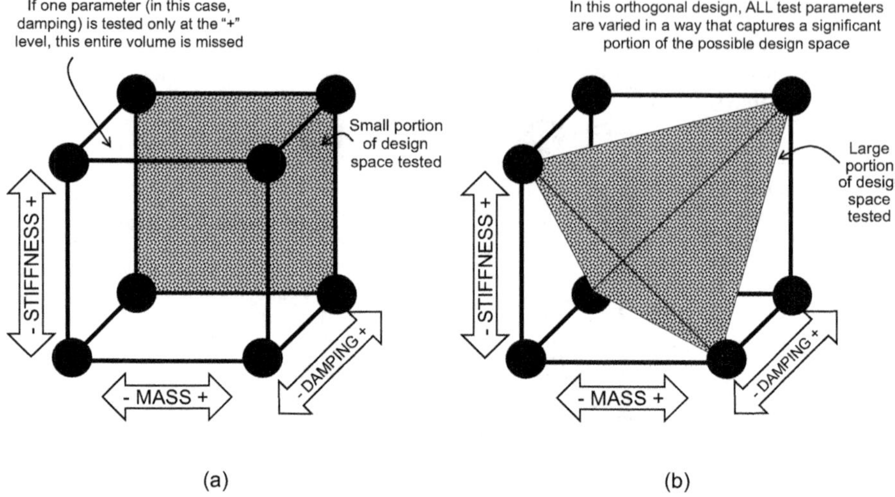

FIGURE 4.4 (a and b) A fractional factorial experiment should capture as much of the full factorial design space as possible with an orthogonal design.

- Calculate the number of *FULL* factorial runs required: $2^3=8$ runs required
- Determine the feasibility of completing all runs in a full factorial experiment
  - Eight experiments may be too costly or time-consuming
- Reduce the number of experimental runs by reducing the exponent
  - $2^3 \rightarrow 2^{3-1}=4$ (instead of 8)
  - Still include all three factors: mass, damping and stiffness
  - Just fewer runs
  - Need to choose the *aliasing* structure
    - Fractional factorial designs cannot uniquely determine every effect or interaction

- Choosing the alias structure
  - Decide (it is a conscious choice) which factor is *confounded* (aliased)
    - If possible, choose the least important factor to confound
      - This retains a full factorial experiment of the most important factors
      - Choosing the least important requires insight, which may or may not be possible
    - If all factors are potentially important, just select one (stiffness in this example)
  - Create a FULL FACTORIAL experiment *without* the aliased factor
  - Still *include* a column for the confounded factor
  - Multiply the first two columns of + and − to get the confounded *treatment* of + or −
    - This scheme creates an orthogonal design
    - Orthogonal: any paired columns have the same number of + and −

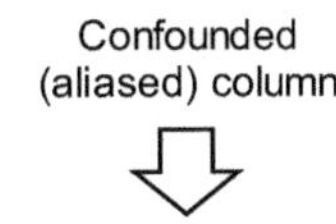

EXPERIMENT RUN	MASS	DAMPING	STIFFNESS	MATH FOR CONFOUNDED COLUMN
#1	-	-	+	← (-)*(-)=(+)
#2	-	+	-	← (-)*(+)=(-)
#3	+	-	-	← (+)*(-)=(-)
#4	+	+	+	← (+)*(+)=(+)

Full factorial here

Orthogonal design requires simple multiplication of non-aliased columns to enter + or - into the confounded column

FIGURE 4.5 Fractional factorial design.

- Example: fractional factorial design with mass, damping and stiffness
  - With three factors, full factorial experiment requires 8 runs ($2^3=8$)
  - A reduced fractional factorial experiment reduces runs to 4 ($2^3 \rightarrow 2^{3-1}=4$)
  - Create a full factorial design for the two non-aliased factors (mass and damping)
  - *Multiply* every + and – in MASS and DAMPING to get confounded STIFFNESS treatment
  - Figure 4.5 shows the resulting fractional factorial design with a column to show the math
  - Any two columns in Figure 4.5 have the same number of + and –
  - This technique *projects* results of a larger full factorial 8 run experiment into 4 runs
  - This technique *captures* the largest portion of the design space for the available runs

DESIGNING AN EXPERIMENT

The following example works through each of the OFAT, full factorial and fractional factorial designs using the same discrete, single-degree-of-freedom vibration study. Damped natural frequency is the dependent variable result of interest. In this case, the experimental objective is to determine if changing mass, stiffness, damping or some combination of these factors can significantly reduce the damped natural frequency. The damped frequency values are real, using the theoretical solution to the second-order vibration equation. An actual DOE experiment could be designed to physically verify these theoretical values.

Vibrational design challenges such as this example are not uncommon, as was the case with the London Millennium Bridge. After construction, the bridge had significant resonant vibrations, which were ultimately reduced by retrofitting dampers.[6] Changing the bridge's stiffness or mass was not considered a viable option.

- Example: DOE design process with actual (theoretical) values
  - This example uses actual mathematical values from a second-order solution equation
  - Consider the damped natural frequency of a discrete vibration experiment
    - Three factors: mass, stiffness and damping (in this case, each have a value of "5")
    - Two levels: more and less of each (±1.5 for each factor)
    - Damped natural frequency, ω_d, is the result of interest
  - An OFAT design is first provided in Table 4.3
    - Each factor is only varied *individually,* low and high
    - The results, ω_d, show only small excursions from the baseline of 0.9 rad/s
    - Combined effects are not considered due to the OFAT design
  - Consider the FULL FACTORIAL test in Table 4.4
    - A full factorial design captures the entire design space
    - Combined effects are captured
    - Full factorial test now shows a *significant combined effect* relationship in run #3
      - Reduced mass, reduced stiffness and increased damping → result of 0.4 rad/s
  - Now consider the FRACTIONAL FACTORIAL test in Table 4.5
    - By capturing a larger portion of the design space, the combined effect *still appears*
    - Run #1 captured the significant combined effect from run #3 in the full factorial
    - Fewer runs were necessary in the fractional factorial with similar results
- A fractional factorial test <u>may not</u> have fully captured the combined effect
  - In this example, the orthogonal design endpoint coincided with the combined effect

TABLE 4.3
OFAT Design

Run #	Mass	Stiffness	Damping	Result (ω_d)
BASELINE→	5.0	5.0	5.0	0.9
VARY BY→	1.5	1.5	1.5	
#1	3.5	5.0	5.0	1.0
#2	6.5	5.0	5.0	0.8
#3	5.0	3.5	5.0	0.7
#4	5.0	6.5	5.0	1.0
#5	5.0	5.0	3.5	0.9
#6	5.0	5.0	6.5	0.8

TABLE 4.4
Full Factorial Design

Run #	Mass	Stiffness	Damping	Result (ω_d)
BASELINE→	5.0	5.0	5.0	0.9
VARY BY→	1.5	1.5	1.5	
#1	3.5	3.5	3.5	0.9
#2	3.5	6.5	3.5	1.3
#3	***3.5***	***3.5***	***6.5***	***0.4***
#4	3.5	6.5	6.5	1.0
#5	6.5	3.5	3.5	0.7
#6	6.5	6.5	3.5	1.0
#7	6.5	3.5	6.5	0.5
#8	6.5	6.5	6.5	0.9

TABLE 4.5
Fractional Factorial Design

Run #	Mass	Stiffness	Damping	Result (ω_d)
BASELINE→	5.0	5.0	5.0	0.9
VARY BY→	1.5	1.5	1.5	
#1	***3.5***	***3.5***	***6.5***	***0.4***
#2	3.5	6.5	3.5	1.3
#3	6.5	3.5	3.5	0.7
#4	6.5	6.5	6.5	0.9

- If the treatment combination did not fully capture the effect, the results still have a high chance of indicating the combined effect
- A further *targeted* experiment or full factorial of fewer factors might be warranted

The results of the designed experiment in both Table 4.4 (full factorial) and Table 4.5 (fractional factorial) show a significant correlation between reduced mass, reduced stiffness and increased damping on the natural frequency. If a mechanical design objective is to reduce the (damped) natural frequency, this combined effect is captured in the design space. However, only *the* single run from the fractional factorial test (4 runs total) and full factorial test (8 runs total) indicates correlation. This is a single statistical data point, despite the multiple runs associated with the experiment. To build a statistically significant case, either the entire designed experiment should be replicated or a subset (or individual) run should be replicated to ensure the apparent correlation is not an anomaly. The number of replications (or runs or samples) should now be chosen to make a statistically significant case (per the next chapter).

EXERCISES

1. Consider an experiment with two levels (more or less from baseline) and five factors. How many experimental runs are required for each of the following three designs:
   a. OFAT?
   b. Full factorial?
   c. Fractional factorial?
2. Explain what an aliased treatment means.
3. What is an orthogonal design?
4. You are designing an experiment to test the failure stress of a material based on three factors and two levels. Complete all areas in the table for a properly designed orthogonal fractional factorial experiment with only four experiments where plus and minus indicate more or less of a level and surface finish is confounded (aliased):

Run	$Area_{xc}$	Temperature	Surface Finish	Fail Stress (Result)
#1	+			TBD
#2	+			TBD
#3	–			TBD
#4	–			TBD

5. Referring to the table in Question #4, what is the likely reason surface finish was chosen as the aliased column?
6. Consider an experiment to determine the tensile stress of 3D-printed sintered aluminum. Five test specimens are checked for an average value. This is an example of:
   a. Repetition
   b. Replication
   c. Duplication
7. If the same experiment to determine the tensile stress of 3D-printed sintered aluminum is performed by another researcher, this is an example of:
   a. Repetition
   b. Replication
   c. Duplication
8. A single, full factorial experiment with eight trials might produce how many usable data points: ____________
9. TRUE or FALSE. The key advantage of a full factorial design compared to an OFAT design is to check for combined effects?
10. TRUE or FALSE. A projectable design may still miss a potential combined effect?

NOTES

1 George, M., Rowlands, D., Price, M., & Maxey, J. (2005). *The lean six sigma pocket toolbook*. McGraw-Hill. p.185.
2 George, M., Rowlands, D., Price, M., & Maxey, J. (2005). *The lean six sigma pocket toolbook*. McGraw-Hill. p.185.
3 Montgomery, D. (2013). *Design and analysis of experiments*. John Wiley & Sons. p.8.
4 NIST Engineering Statistics Handbook. (n.d.). *5.1.3. What are the steps of DOE?* National Institute of Standards and Technology, U.S. Department of Commerce. https://www.itl.nist.gov/div898/handbook/pri/section1/pri13.htm.
5 Coleman, D., Gunter, B. (2014 May 14). A DOE Handbook: A simple approach to basic statistical design of experiments. p.62.
6 Newland, D. (2003) Vibration of the London Millennium Bridge: Cause and cure. *International Journal of Acoustics and Vibration*, 8(1), 9–14.

5 Sample Size

A single experiment with a single result is usually not convincing to support a hypothesis or ensure an accurate value. An appropriate number of samples must be chosen to provide a statistically convincing data set.

The only significant reason to obtain more than one data point by replicating an experiment or testing multiple samples is to be reasonably certain of the results. Multiple experimental activities take more effort, but measurements are inherently inexact and errors creep into every endeavor.

Supporting a hypothesis requires a large enough data set to be statistically convincing. Unfortunately, statistical analysis of data occurs *after* an experiment, but the choice of data set size must occur *before* the experiment. The expected behavior of a planned experimental data set can often be inferred ahead of time, but statistical properties such as data variance remain uncertain ahead of time. If a small data set size is decided upon prior to the experiment and the values turn out to be too varied, the experiment may need to be re-accomplished to obtain a larger data set.

A data set can refer to the number of replications of an experiment, the number of repetitions of a one-run experimental design or the number of specimens or samples subjected to a single experiment. Each of these data set requirements can be considered synonymous. All that matters for this discussion is deciding on the number of data points. Any event or item that can return a single data point may be considered for multiple events or multiple items if there is a statistical need.

- Experiments require multiple data points for two primary reasons
  - To accurately resolve a *truth value* from the mean of sample measurements
  - To make a *statistically convincing case* with the mean from an adequate sample set
- Statistical approaches are essentially the same for both cases, but the goals are different
  - Discerning a truth value implies *narrowing in* on a given value
  - Discerning a hypothesis statement implies *discrimination from* a given value
  - Both situations involve multiple data points, resulting in a distribution with a mean
- The number of experimental data point samples (replications or runs) is a function of
  - Resources available: time, funding, materials, facilities and equipment
    - Too few samples: inaccurate
    - Too many samples: expensive
  - Requirements and nature of experiment
    - Published report: the number of samples should be statistically significant

DOI: 10.1201/9781003329237-5

    - Academic curiosity or industrial results: only one or few samples may be sufficient
- Common options to choose a large enough data set to be statistically significant
  1. Use an established *rule of thumb*
  2. Use a t-score confidence interval plot
  3. Use a z-score calculation
- Statistically driven sample size selection should also consider
  - Level of significance: increased significance in results requires more samples
  - Power of the study: higher power studies require more samples
  - Effect size: detecting small effect sizes requires more samples
  - Variance: greater variability in data requires more samples
  - Precision: improving precision may require more samples
- Data sample set size selection assumes
  - Random, independent samples or events
  - Normal distribution of the resulting data set
- Central Limit Theorem for normal distributions states
  - Independent, random samples will tend toward a central mean
  - Sample sets should be *large enough* to have a central tendency
  - No definition of *large*, but 30 samples are generally accepted as the minimum cutoff value

Statistical analysis underlies the choice of sample size selection discussed here, but a more complete discussion of statistics and associated math is not presented until later chapters on statistics. The current goal is to choose a data set size while including enough statistics and associated math to complete the effort.

RULE OF THUMB

Rule of thumb offered by Hines and Montgomery[1] for sample size selection is based on statistical approaches but avoids the math. Sometimes the only thing needed is a rough number of samples for a convincing result without having to make a more rigorous case.

- Where the data behavior is _______ then number of samples is $n \geq$ ______:
  - "Well-behaved": $n \geq 4$ (data behaves very much like a normal distribution)
  - "Fairly behaved": $n \geq 12$ (data has "*no prominent mode*" and appears more uniform)
  - "Ill-behaved": $n \geq 100$ (data distribution is unclear or with substantial outliers)
  - "Reliable estimates": $n \geq 20$
  - Often quoted: $n \geq 30$ (the cutoff between a z-test and t-test)

CONFIDENCE INTERVAL PLOT

A *t-value* (*t*-variable or *t*-score) and *t-distribution* (student distribution) are useful statistical tools for small data sets. A *t*-distribution has greater variance than a normal distribution but converges to a normal distribution as sample size increases. As the number of samples increases, confidence in the mean increases and the *t*-score *confidence interval* decreases (tightens). The payoff in effort becomes less and less since the relationship is non-linear. Figure 5.1 depicts this non-linear relationship and offers a guide for sample size selection. Figure 5.1 does not necessarily suggest choosing a specific confidence interval (vertical axis) so much as it suggests a rapidly improving confidence toward resolving a truth value based on the number of samples or size of the data set.

- Sample sizes can be chosen based on Figure 5.1
  - Four or five samples may be sufficient for determining a suitable mean value
  - Four or five samples may be sufficient if they are expensive or time-consuming
  - 10 to 12 samples[2] provide a reasonable tradeoff between accuracy and effort
- Figure 5.1 assumes the following:
  - A well-behaved sample set
  - A standard 95% confidence level
  - A small data set
  - The distribution variance is a fixed unknown and removed from consideration
- Different confidence levels (other than 95%) will change the axis values of the plot

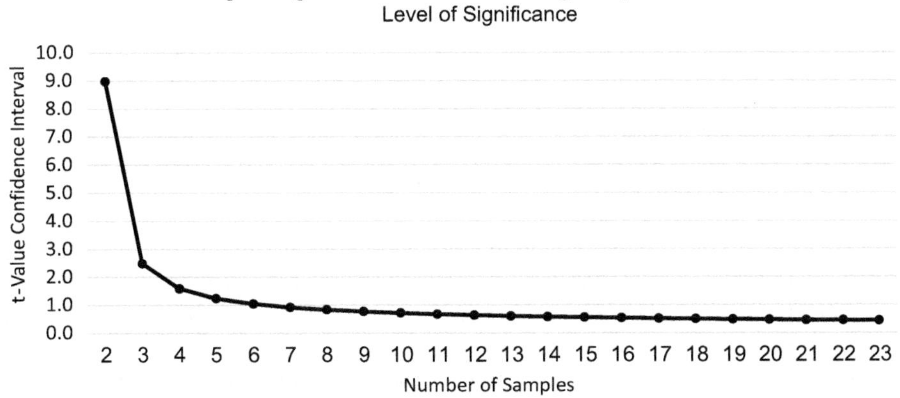

FIGURE 5.1 Reducing uncertainty by increasing sample size. Plot generated from *t*-value function at 95% confidence and *t*-values at $\alpha/2$ for every $n - 1$ samples. (Plot based on concept derived by Montgomery.) (Montgomery, D. (2013). *Design and analysis of experiments*. John Wiley & Sons. p.45.)

- The plot's non-linear *shape* (square-root) remains the same
- Confidence interval values will change
- Sample sizes suggested by the plot remain *relatively unchanged*
- Figure 5.1 assumes an experimenter needs to choose a sample size, n, to
    - Resolve a single truth value from the mean of a single normally distributed data set
    - Compare a sample mean to a single baseline or control value
- Figure 5.1 does *not* assume discerning between *two* data sets

Figure 5.1 and its derivation conceptually follow an approach presented by Montgomery (*Design and Analysis of Experiments*).[3] Whereas Montgomery develops his plot based on a *two independent sample t-test* formula and half confidence interval, Figure 5.1 presented here uses the simpler *one sample t-test* formula. Regardless of the particular derivation, the results of Figure 5.1 are essentially the same and suggest similar sample sizes due to the rapid non-linear fall-off.

DERIVATION of the confidence interval plot. The goal of this derivation is to create a relationship between sample size, n, and a confidence interval as a justification for Figure 5.1. This derivation is based on Montgomery's approach, but the direction and presentation here uniquely attempt to build an understanding of the underlying statistical concepts. The plot in Figure 5.1 is sufficient for use without knowing the derivation. However, there is value-added in discussing the statistical principles used in the derivation. The emphasis here is on *discussion* rather than statistical and mathematical symbology.

A rough derivation for Figure 5.1 begins with the standard statistical *one independent sample t-test* formula in Equation 5.1.

$$t = \frac{x - \mu}{S/\sqrt{n}} \tag{5.1}$$

where: t = t-value
x = test sample mean
μ = truth value
n = number of samples
S = sample variance

The numerator in Equation 5.1 merely establishes a point value in a distribution. It is the *error* of the experimental sample set mean, x, from the truth value, μ. For the experimenter, the sample mean, x, may or may not accurately reflect the truth value, μ. For example, 10 measurements of the gravitational constant, g, might return a sample mean of $x = 9.75$ m/s^2, compared to the nominally accepted standard value (truth value) of $\mu = 9.81$ m/s^2. The numerator error in this example computes to 0.06 m/s^2 due to the experimental sample mean *miss*. The denominator in Equation 5.1 *standardizes* the numerator based on the actual (non-standard) sample variance, S, and the number of samples, n. Therefore, a t-score formula returns *the experimental sample-mean error, from a truth value, standardized for sample size and variance.* Another way of restating this is a *t*-score is the sample mean's *error, standardized* for table use. Having a standardized *t*-score allows using standardized *t*-distribution tables to predict where data is expected.

Using one interpretation, entering a *t*-table with a *t*-score returns a probability, *P*, of data falling to one side (or the other) of the *t*-score, based on the *t*-distribution function (student's *t*-distribution). The *t*-distribution is essentially a normal distribution with greater variance to account for the small sample sizes typical of experimentation.

The next step in the derivation of sample size selection for Figure 5.1 is to set a desired confidence level. This choice is a statistical decision that is made ahead of time and is generally separate from any experimental details. The standard confidence level is 95%, which is exactly the same as setting a 5% *level of significance*, denoted as alpha, α (for a 95% confidence, $\alpha = 0.05$). Any confidence level can be chosen, but 95% is the standard and is used in this derivation.

Normally, the number of samples is required along with the confidence level to enter the *t*-score table. However, the number of samples is the result of interest, and we need to develop a relationship between the confidence level of a *t*-score and the number of samples. This sets up the next relationship of the derivation with Equation 5.2.

$$P(t_L \leq t \leq t_U) = 0.95 \tag{5.2}$$

where: $P(\ldots)$ = probability
t = *t*-value from Equation 5.1
t_L = lower *t*-value limit
t_U = upper *t*-value limit
0.95 = 95% confidence

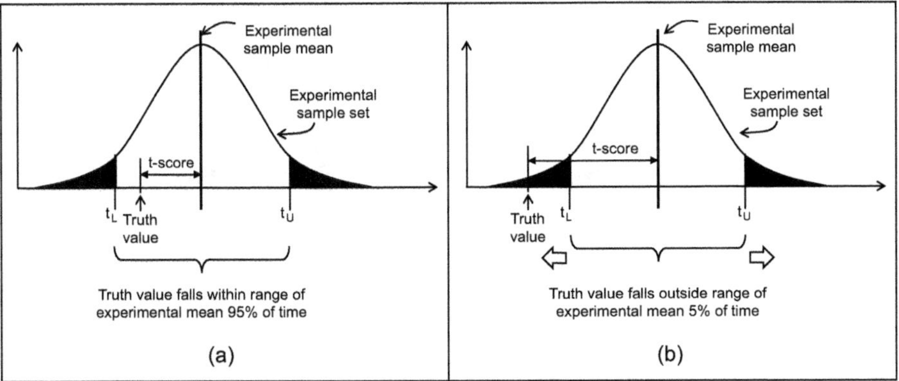

FIGURE 5.2 (a and b) Developing the confidence interval with a *t*-distribution.

Equation 5.2 merely states we want our standardized error, the t-value (from Equation 5.1), to fall between a lower and upper limit 95% of the time. This is a *two-tailed* test since we are considering results anywhere between the lower and upper bounds.

Figure 5.2 is a visual depiction of Equation 5.2. Figure 5.2 shows the planned sample data set will return a normal distribution with a mean. The experimental mean is the result of *however many* samples are decided upon. We have chosen a 95% confidence interval as a matter of convention, which sets the upper and lower bounds of the distribution at t_U and t_L (returned from a t-table). This establishes our goal of choosing a sample size large enough to ensure the sample set distribution will capture the truth value within the upper and lower bounds 95% of the time.

As the number of samples increases, the t-distribution *sharpens* (less variance), as shown in Figure 5.3. The tightening distribution drives down the t-values (as returned from a t-value table), which requires a more accurate mean to maintain the 95% confidence level, as shown by the shift in Figure 5.3b. The more samples, the tighter the distribution, and the closer we can expect our mean to be the truth value if we hope to maintain 95% confidence.

It may help to recognize in both Figures 5.2 and 5.3 that the truth value does not change; the truth value is the truth value, regardless of results. We have no control over the truth value, only our experimental sample set and sample mean. The confidence interval is based on the experimental data set distribution, and with a 95% confidence level, the experimental sample set should *capture* the truth value within the upper and lower bounds 95% of the time.

Technically, entering a t-table requires the *degrees of freedom* and *not* the number of samples. Degrees of freedom are merely the number of samples, n, minus one ($n - 1$). The significance level is divided by two. As shown in Figure 5.2, a 95% confidence range means 5% ($\alpha = 0.05$) of samples fall outside either the upper or lower bound. Therefore, the bounds are located at 0.05÷2 for both the lower and upper bounds on the symmetric profile. The bounds, t_L and t_U, are functions of $\alpha/2$ and $n - 1$. We've chosen to fix $\alpha/2$ at 0.025 (95% confidence or 5% significance divided by the

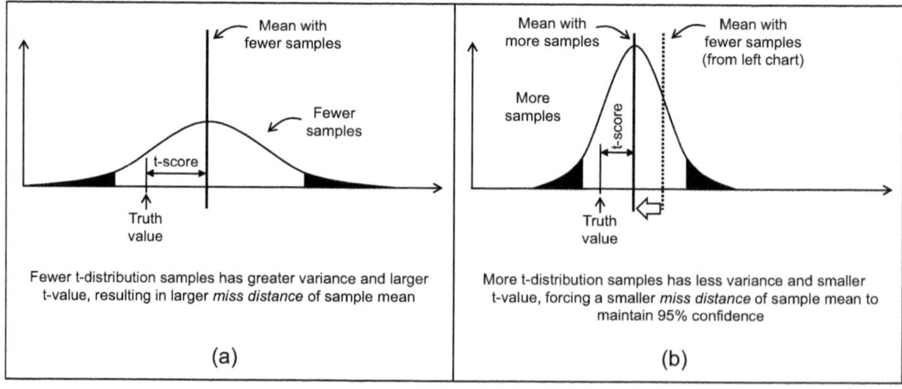

FIGURE 5.3 (a and b) Increasing sample size effect on t-distribution.

two tails in the distribution) while $n - 1$ varies. This can be written as $t_L = -t_{0.025,\, n-1}$ and $t_U = +t_{0.025,\, n-1}$. Equation 5.2 can now be expanded into Equation 5.3.

$$P\left(-t_{0.025,\, n-1} \leq \frac{x-\mu}{S/\sqrt{n}} \leq +t_{0.025,\, n-1}\right) = 0.95 \tag{5.3}$$

Slightly modify inside the parentheses of Equation 5.3 so the unknown but fixed constants (x, μ and S) remain within the upper and lower bounds (inside the inequalities), but the number of samples (n) move outside the inequalities with the bound limits as shown in Equation 5.4.

$$P\left\{\left(-t_{0.025,\, n-1}\right)/\sqrt{n} \leq \frac{x-\mu}{S} \leq \left(+t_{0.025,\, n-1}\right)/\sqrt{n}\right\} = 0.95 \tag{5.4}$$

Equation 5.4 now states our *normalized error*, $(x - \mu)/S$, should fall between the bounds (confidence interval) of t_L and t_U 95% of the time. The t-distribution is symmetric, so either bound in Equation 5.4 can be used to enter a t-table. Care must be taken to use a two-tailed t-table with Equation 5.4 to capture the full 95% interval. If a one-tailed table is used, the value must be doubled. All that remains is to pull either the upper or lower bound from Equation 5.4 and plot the results. We're only interested in how the bounds (confidence interval) change with increasing samples. As shown in Figure 5.3b, we expect additional samples to tighten up the results and improve our sample set mean relative to the truth value. Equation 5.5 is the absolute value of either bound of the confidence interval in Equation 5.4.

$$\left(t_{0.025,\, n-1}\right)/\sqrt{n} \tag{5.5}$$

To generate the plot in Figure 5.1, consider that 12 samples have 11 degrees of freedom ($DOF = n - 1$). From a standard two-tailed t-table, the t-score lookup value at 95% confidence and 11-DOF ($t_{0.025,11}$) is 2.201. The confidence interval bound in Equation 5.5 evaluates to 0.6354 as shown in Equation 5.6 and plotted in Figure 5.1. This process is repeated from $n = 2$ upward as desired.

$$(2.201)/\sqrt{12} = 0.6354 \tag{5.6}$$

HYPOTHESIS TESTING

Sample data set size Options #1 and #2 are *qualitative* approaches to choosing a data set size by using a rule of thumb or a plot to decide. The z-score test (Option #3) involves a more *quantitative* approach by solving an equation. This approach assumes an experimenter needs enough sample data points to statistically satisfy a hypothesis test, so the sample sets are large enough to approach a normal distribution and use z-table numbers instead of t-tables. Equation 5.7 prescribes the sample size formula[4]:

$$n = 0.25\left(\frac{z}{E_{PCT}}\right)^2 \tag{5.7}$$

where: n = number of samples required
z = z-table look up for desired confidence level
E_{PCT} = desired % difference (error) from established value

As with many statistical hypothesis tests, the normal application of Equation 5.7 is to answer the question of *how many samples* are required *from a population* to discern a difference. In other words, there is only one relatively large data set, and we need to pull a certain number of samples from this single pool of data to test an idea about the pool of data with our small sample set.

A quality process control engineer would fall into the above scenario where they might need to choose a sample size from a large production batch to discern quality or attributes. However, the typical engineering experiment needs to generate a data set for later analysis. In this application, Equation 5.7 is not about choosing a sample size from a population to test a hypothesis; it is about generating a sufficient data set to meet a hypothesis test. Either way, the sample set generates a mean with the goal of discernment, where some *distance from another value* (population mean or a criteria value) is necessary to feel confident in the results.

- Application of Equation 5.7 assumes the following:
  - Experimental sample data sets will be normally distributed
  - Variance (and standard deviation) of the data set is unknown ahead of time
    - The probability distribution variance is assumed to be at a maximum, such that $\sigma^2 = 0.25$
    - *Refer to the derivation at the end of this section for rationale*
  - The experimental results seek to demonstrate a discernible difference *from* a given value
    - Enough experimental samples or replications are necessary for discernment
- The error, E, in Equation 5.7
  - Is a percentage
  - Is the error between the likely sample set mean and another value of interest
  - Alternatively, the desired error can be chosen to decide the number of samples

Example 5.1

A team of aerospace engineering students is interested in knowing the first natural frequency mode of vibration of a model aircraft wing being designed and built for an intercollegiate competition. Competition rules require a wing vibration frequency greater than 90 Hz. Unfortunately, the wing design is not a homogeneous cantilever beam, so experimentation rather than theory is necessary to ensure the requirement is met. The students want to provide a statistically convincing case that their aircraft wing does in fact meet or exceed the 90 Hz minimum requirement, so more than one experimental test run is planned. How many experiments (data points via experimental replications) are necessary to be confident the wing's natural frequency is actually greater than 90 Hz?

Solution

A first test is performed to get some idea of the unknown natural frequency and appears to return a value of 110 Hz. The aircraft appears to meet the requirement, but students do not know how much variance to expect in the measurement process, and a single data point is not convincing. The worst-case scenario would be to accept the single 110 Hz measurement and have the competition judges measure 88 Hz and disqualify the aircraft.

This example happens to be a *one-tailed test* where only a lower limit of at least 90 Hz is required. A 5% significance level is chosen as a typical measure of statistical confidence (95%). The standard *z*-score at 5% for a one-tailed test is 1.645. (The more common situation is a *two-tailed test* where data can be anywhere within an upper and lower bound at 5% significance, and the *z*-score would be 1.96 for a two-tailed test.) The *z*-scores are discussed more in the chapter on statistics.

To solve this problem now, simply consider the normal distribution plot in Figure 5.4 with a single marker (one-tailed test) placed at the lower 90 Hz point of interest. *By definition,* 100% of *all data* is in the *area* under this entire normal curve. At 95% confidence for a one-tailed test, we can accept 5% of all our data points falling into the tail, to the left of the 90 Hz marker cutoff. This also means 95% of all our data points will fall to the right of the marker in the rest of the distribution. It may be better to state that *we want this to happen*; *we want* 95% of our measurements to show up *in the green,* greater than the 90 Hz cutoff. *If we can make this happen,* we have a 95% chance that the contest judge will also measure a passing result. The only question to resolve is *where* does this marker occur *according to the normal distribution*? Yes, the marker is at 90 Hz, but that value actually says *nothing* about the normal distribution and the 5% cutoff.

A *z*-score correlates the 90 Hz marker position with the percentages of the normal distribution. Simply use any *z*-score calculator, *z*-score table or normal distribution table and figure out where the 0.05 and 0.95 area cutoffs occur. Most *z*-score tables show a graphic with a shaded area to help indicate how much of the curve area is being returned by the table values. In statistical notation, this problem is phrased as $P(x > z) = 0.95$, where the *probability* that any *x-data value* will occur, greater than *z*, is 95%. The *z*-score that makes this statement true is $z = -1.645$. All normal distributions are the same, so *z* always equals ± 1.645 for a 5% and 95% cutoff in a one-tailed test. Negative or positive values are mostly irrelevant since the curve is symmetric, but a negative *z*-value is left of the anticipated mean. We assume every measurement taken will vary somewhat, and it will vary according to the normal distribution.

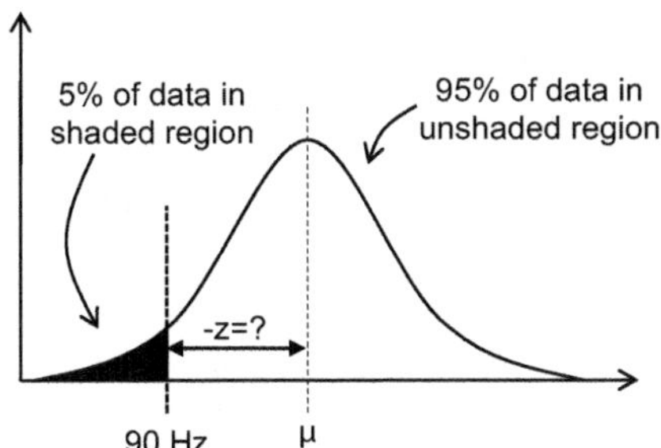

FIGURE 5.4 The *z*-scores identify area percentage cutoff points for the normal distribution.

The students enter the numbers into Equation 5.7, with the expected sample mean of 110 Hz as shown in Equation 5.8.

$$n = (0.25)\left[\frac{1.645}{(110-90)/90}\right]^2 \tag{5.8}$$

A total of n = 13.6 sample runs are required, which rounds up to n = 14. Without knowing the variance in the samples ahead of time, the formula assumes a maximized distribution variance, and 14 samples will help ensure 95% of all measurements will be at least 90 Hz <u>if</u> the sample set mean does in fact result in 110 Hz. Assuming the contest measurement techniques are the same, the students can be 95% certain the contest judges will not measure less than 90 Hz on their aircraft and disqualify it. If the mean of all samples turns out to be less than 110 Hz, then the students may need to accept a lower confidence.

In the next example, an alternative application of Equation 5.7 uses a *t*-score value instead of a *z*-score value due to the small sample sizes. However, with this approach, unlike the *z*-score test, a *t*-score requires knowing the sample size, which is what we are solving in the first place. One or two iterations might be sufficient to utilize a *t*-score, and the sample size requirement should be greater due to the greater variance in the *t*-distribution.

Example 5.2

Equation 5.8 returned a sample size of n = 14, which does not reach the cutoff of $n \geq 30$ for a normal distribution. However, n = *14* is a reasonable first iteration to enter the *t*-table. Using the same one-tailed test at 5% significance from the preceding example, a *t*-score table returns a *t*-score of t = 1.7709 for *DOF* = 13 (n – 1). Re-entering Equation 5.8 with a numerator of t = 1.7709 instead of z = 1.645, results in $n \approx 16$. As expected, a few more samples are required if a *t*-table is used, which is a more conservative choice given the small sample size.

DERIVATION. The *z*-score and *t*-score formulas are essentially the same, but the *z*-score assumes a larger sample size and complete information about the data set. The *z*-score formula is given in Equation 5.9.

$$z = \frac{x-\mu}{\sigma/\sqrt{n}} \tag{5.9}$$

where: z = *z*-value
x = test sample mean
μ = truth value
n = number of samples
σ = standard deviation of entire distribution

An alternate formulation of the *z*-score is also available based on population *proportions*, as shown in Equation 5.10.

$$z = \frac{(p-p_0)}{\sqrt{p_0(1-p_0)/n}} \tag{5.10}$$

where: z = z-value
p = sample proportion (%)
p_0 = hypothesis proportion (%)
n = number of samples

Both Equations 5.9 and 5.10 are equally valid and return the same result. The choice of formula is a matter of available information. Since the sample set standard deviation, σ, is unknown ahead of time in Equation 5.9, Equation 5.10 is instead selected for this technique. Solve Equation 5.10 for the number of samples, n, as shown in Equation 5.11.

$$n = p(1 - p_0)\frac{z^2}{(p - p_0)^2} \tag{5.11}$$

Seemingly, our situation in Equation 5.11 is no better than that in Equation 5.9 since we do not know the sample proportion, p, nor the hypothesis proportion, p_0. However, we can set up useful equivalencies.

Hypothesis testing is about discernment between a given value (null hypothesis) and a sample set mean. The denominator, $(p - p_0)$, in Equation 5–11 is the difference in proportions (percentages) of the sample mean compared to a given mean, and this is simply (percentage) *error*, E. Equation 5.11 can be updated in Equation 5.12.

$$n = p(1 - p_0)\frac{z^2}{E^2} \tag{5.12}$$

We still don't know p or p_0 in Equation 5.12, but we can maximize the function, $p(1 - p_0)$, just to be sure. This function can be plotted or a derivative taken to show it is maximized at $p = p_0 = 0.5$. Maximizing this value assumes the worst-case scenario and returns the maximum possible samples. Equation 5.7 is the result since $0.5(1 - 0.5) = 0.25$.

If the variance can be estimated from similar results, previous studies or other published data, Equation 5.13 can be used instead of Equation 5.7. The numerator from Equation 5.9 <u>is</u> the error, <u>not</u> the percentage error. The number of samples can therefore be calculated from Equation 5.9 to arrive at the result in Equation 5.13.

$$n = \left(\frac{z\sigma_e}{E_{ABS}}\right)^2 \tag{5.13}$$

where: n = number of samples
z = z-value
σ_e = estimated standard deviation
E_{ABS} = desired error value from established value

EXERCISES

1. List three methods to help determine a sample data set size for an experiment.
2. Assuming a random distribution, how many samples are necessary to be considered a large enough sample set to approximate a normal distribution?

3. Assume an experiment requires determining the yield stress of 3D-printed PLA test specimens. What is the minimum number of test specimens for a statistically significant mean value?
4. Assuming a normal distribution, an experiment requires determining the yield stress of 3D-printed PLA test specimens. The results must show less than 7% error with a 95% confidence level. How many test specimens are required (use Equations 5.5–5.7)?
5. Assume the result from Question #4 is unreasonable, so further investigation is required. Why is the required number of samples so high?
6. Assume the published yield stress for PLA is 60 MPa. Based on Question #4, what is the plus or minus acceptable error value?
7. Based on Question #6, what range of yield stress values should we expect 95% of the time?
8. Consider that a previous study indicated a standard deviation for PLA yield stress of ± 10 MPa. Use Equations 5.5–5.9 to determine the number of samples for Question #4 with the additional information of the standard deviation.

NOTES

1 Hines, W. & Montgomery, D. (1980). *Probability and statistics in engineering and management science*. John Wiley & Sons. p.183.
2 Montgomery, D. (2013). *Design and analysis of experiments*. John Wiley & Sons. p.45.
3 Montgomery, D. (2013). *Design and analysis of experiments*. John Wiley & Sons. p.45.
4 Sullivan, L. (n.d.). *Power and sample size determination*. Boston University School of Public Health. https://sphweb.bumc.bu.edu/otlt/mph-modules/bs/bs704_power/bs704_power_print.html.

6 Managing Experimental Errors

Uncontrolled errors negatively impact experimental results. While some errors are unavoidable, most can be controlled within reasonable limits by implementing a proper strategy.

Errors cannot be completely eliminated in an experiment, but they can be minimized to an acceptable level. Understanding errors requires understanding where they enter the experimental process and show up in the results.

VALUES

Values are the hard numbers. Experiments begin and end with values. Every measurement and physical parameter is an input value into an experiment. When a final data set is returned as an experimental output, the data points are values. Since outputs depend on inputs, care must always be taken to minimize error during initial measurements. Unfortunately, measurements are also inherently inexact values and always subject to some degree of error.

- Values are numbers: measurements and quantities (real numbers and integers)
- Values have[1]
  - True value: the *exact* value of a parameter, which may be precisely unknowable
  - Measured value: the value of a parameter as returned by an instrument
- Measurements are inexact values and are always subject to some degree of error
  - Inherent measurement errors always affect experimental results
- Measurements have
  - Accuracy: measurements that return as close to the truth value as possible
  - Precision: measurements with very consistent results (may or may not be accurate)
  - Error: the difference between *measured* value and *truth* value
  - Uncertainty: the *plus or minus* range of possible values in a result

VARIABLES

An experimental variable has a couple different definitions, depending on context. On one hand, a variable is the mathematical placeholder for a value. A variable is also *anything* that can change and impact an experiment. Both definitions identify changeability. The first definition considers the mathematical relationships *within* the experiment.

DOI: 10.1201/9781003329237-6

Input measurement errors get propagated through an experiment based on the relationships between these variables. Once an error has found its way into this definition of variable, there is little to be done. In this case, it is important to take precise measurements before entering values into an equation. Examining the mathematical relationships also helps to identify areas where special attention is required. Any formula in an experiment with higher order terms will have a greater impact on results.

The second definition implies all of the influences on an experiment. These variables surround every experiment and may or may not be part of, or become part of the experiment. Environmental variables that could negatively impact an experiment, such as temperature, wind and noise, can be mostly mitigated in a controlled lab environment.

- Types of variables
    - Independent: do not influence each other
    - Dependent: values change based on other variables
    - Extraneous: not purposely controlled but may affect result
- Independent variables are either[2]
    - Controlled: recognized and held fixed so they are not a factor
    - Test parameter(s): allowed to vary as the point of the experiment to see an effect
- Test parameter(s) are
    - The independent variable(s) of primary interest
    - Changed during the experiment to test causal relationships
    - A single variable or multiple variables (parameters)
    - Definable, measurable, configurable and changeable
    - Defines the nature of the test
        - The independent variable (test parameter) is changed under controlled conditions
        - Other variables are held fixed to test the result of the dependent variable(s)

ERROR

Error is the difference between two values of the same item or event where one value is considered correct or true. However, the measure of error is better defined in relative terms of a *value of investigation* compared to a *reference value*. In most scientific or engineering experimentation, we expect an actual truth value to exist for physical parameters. For example, Earth's gravitational constant might be considered true when designing an experiment to verify the value of $g = 9.81\ \text{m/s}^2$. However, this is an empirically determined value that varies from location to location on earth, so it can be difficult to discern whether a high-fidelity gravitational experiment returns the true value, which is different from $g = 9.81\ \text{m/s}^2$ or whether the $g = 9.81\ \text{m/s}^2$ value is more correct and the experiment is wrong. Equation 6.1a provides the measure of error in broadest terms, while Equation 6.1b is the most common application and interpretation of error.

$$Error = |Value\ of\ Investigation - Reference\ \ value| \tag{6.1a}$$

$$Error = |Measured\ value - Truth\ value| \tag{6.1b}$$

Either the *value of investigation* or the *reference value* could be a theoretical value, a derived value, a measured value, a published value or any other generally accepted value. When computing a percentage error, as given in Equation 6.2, it is important to specify the reference value in the denominator. Usually, when comparing experimental results to theoretical results, theoretical values are considered the reference, but this scheme depends only on the objective of the experiment. For example, a pedestrian bridge might have an experimentally determined natural frequency at very high fidelity. If a predictive mathematical model is being developed for the bridge, the *experimental value is the reference truth value*, while the theoretical model is subject to error.

$$\%Error = \frac{|Value\ of\ Investigation - Reference\ value|}{Reference\ value} \tag{6.2}$$

Acceptable levels of error depend on the experimental objective. Experimental errors compared to theory are generally considered reasonable within about 10%. However, a 10% error on a fuel-critical rocket test could prove disastrously excessive.

Sources of error can form a never-ending list, from measurement inaccuracies to equipment limitations to noise or observational limits. However, despite a great variety of sources, errors fall into only three basic *types*. Types of error are classified as *random*, *systematic* and *human*, and these three types influence results in three distinct ways, as seen in Figure 6.1. Random errors create a wide normal distribution of data, while systematic errors bias results away from the truth value. Human

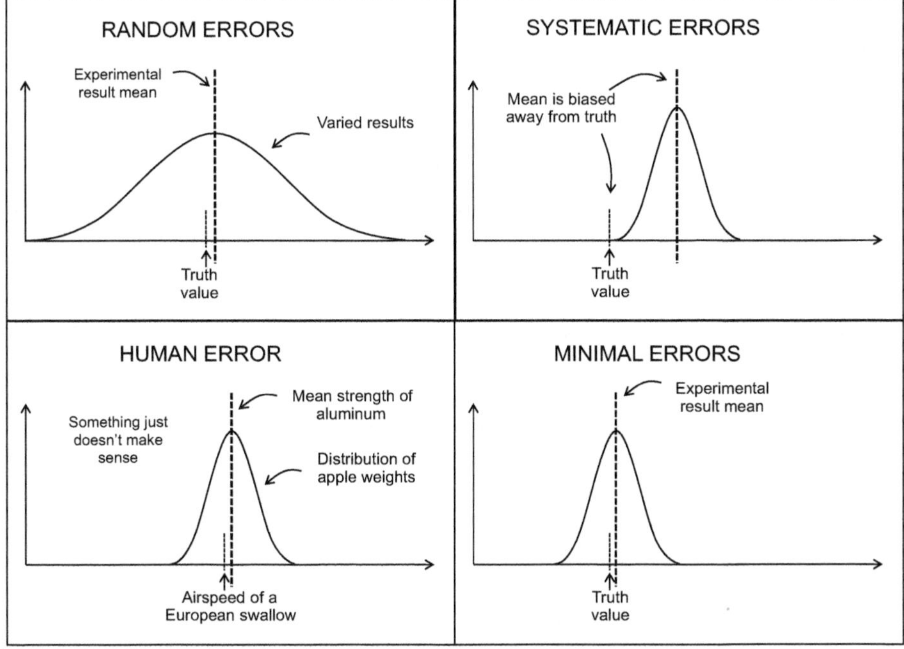

FIGURE 6.1 Types of error and their influence on a data set.

errors are mistakes that should be caught and fixed since they can call into question the entire experiment.

- Errors are divided into TYPES and SOURCES
- TYPES of error
  - SYSTEMATIC
    - Fixed or biased, trends, consistently too high or too low
  - RANDOM
    - Fluctuations in values with no particular trend
  - HUMAN (or operator)
    - *Blunders* or mistakes
    - Misread values or wrong units
    - Do not confuse observational random errors from humans *as* human error

HUMAN ERROR

Human error is sometimes used as a catch-all for measurement variances or as an unknown reason why experimental results differ from expected results. However, human error should seldom, if ever, appear as an attributed error within a report. By definition, human error is a mistake, and if it is known, it should be corrected. If a mistake is unknown, there is simply no knowledge to even discuss it within a report. A banal (not banana) example of human error is an experimental objective to measure the weight of ten oranges, but ten apples were mistakenly weighed instead. Why publish a report with the weight of apples when an objective calls for the weight of oranges? The most professional option in this situation is to rerun the experiment with oranges to match the objective, so there is no need to include human error. Alternatively, the report could be published with an objective of oranges but data from apples, but *also* include an important discussion of human error, where apples and oranges are so similar, a mistake was made, and we can all learn from this published human error report. Reports citing human error then become a knowledge base for others to avoid similar mistakes. Finally, it is important to note when people take measurements that slightly vary between individuals, these are random TYPES of errors due to an observational human SOURCE but not *human error.*

The Pons and Fleischmann experiment outlined below is an infamous example of human error.[3] If this experiment was known to have human error, it would never have been publicized.

- The Pons and Fleischmann experiment (example of error TYPE)
  - In 1989, Stanley Pons and Martin Fleischmann claimed to have created *cold fusion*
  - They claimed energy creation from deuterium-deuterium fusion through electrolysis
  - There was competition between BYU and University of Utah
    - Pons and Fleischmann were from UU
    - The competitors had an agreement for joint publication if there was a breakthrough

- Pons and Fleischmann had the first results
  - They measured neutron signatures thought to be from a fusion reaction
  - They broke the cooperation agreement and announced it at a press conference
  - They chose not to first submit a publication for peer review (normal protocol)
- There was a worldwide rush to replicate the experiment after the news reports
  - Replicating was difficult without an available report (the purpose of a report)
- When initially replicated, the experiment *did* appear to create the neutron signature
  - Neutron counters were placed near the experiment and then behind shielding
  - The counters all showed neutron action from the experiment
- The neutron counter tubes were temperature sensitive
  - The indicator of fusion (neutron count) was found to be flawed
- TYPEs of error in Pons and Fleischmann experiment
  - Systematic: high neutron counts created bias in the results
  - Random: noise due to the presence of water molecules[4]
    - *"The hole in the glass insulator grows (due to temperature) very slightly... There is a strong ionizing chemical reaction (with) the water molecule... It looks, to the electronics, exactly as if a neutron was just captured..."*
  - HUMAN: It was an error *not* to create a control experiment with everything exactly the same except for the active parameter. It was an error *not* to scrutinize the measurement equipment for such a monumental claim. It was an error *not* to submit the results for peer review. The conclusion was wrong. Obviously, there were significant personality issues, but personal agendas are not the point here. Errors were made in the rigor of the process, such that the conclusion wrongly supported the hypothesis. As human error, it simply passed by the experimenters without recognition. Had they recognized the mistakes, the results would not have been publicized. Or, they could've written a report and cite human error in failing to verify equipment limitations leading to the wrong conclusion, so others would not make the same mistake. Mistakes are part of the experimental process, even human error mistakes. Ethical issues such as broken agreements and the non-standard publicity approach placed this situation outside forgiveness of the scientific method.
- Pons and Fleischmann *retreated* to other countries

- SOURCES of error
  - Theoretical (TYPE: systematic)
    - Over-simplification of model or equations can create consistently incorrect results

- Observational (TYPE: systemic, random and human)
  - Individual differences in reading results, particularly analog values
  - Parallax: not being visually aligned for accurate analog meter values
  - Human error would be a major mistake, such as recording the wrong units
- Selection bias: picking and choosing data only to support a hypothesis
- Sampling bias: all data are included, but biased in the initial plan and design
  - Choosing sample sets that do not represent the actual true population
- Noise (TYPE: random): variations in measured signals and quantities
  - Johnson-Nyquist (white noise, thermal activity and frequency-independent)
  - Shot noise (low-level signals and discrete nature of signal carriers)
- Interference (TYPE: systemic and random)
  - Bias due to voltage offsets, ambient pressures and contaminants
  - Random, occasional interferences (vibrations from passing vehicles)
- Instrumentation (TYPE: systemic)
  - Static calibration: failure to zeroize or refer to an accepted standard
  - Dynamic calibration: measurements drift over time (components heat up)
  - Lag or warm-up: measurements begin before equipment is ready
  - Resolution: finite limits in ability to resolve value (may also be TYPE random)
  - Loading: act of measurement affects value
- Incomplete or improper procedures (TYPE: systemic and random)
  - Improper use or storage causes unrealized instrument damage
- Hysteresis (TYPE: systemic)
  - Values that change over time due to testing (magnetization and stretch)
  - System may not rebound to previous condition after excitation
  - Ropes, springs and cords may stretch and rebound in different ways
- Test sample preparation and consistency
  - All samples should be at same temperature and consistent
- Environmental (TYPE: systemic and random)
  - Vibrations, air drafts and electromagnetic interference
- Settling (TYPE: systemic and random)
  - Recording data prior to steady-state conditions when required
  - Switch bounce interference for high-speed events

STRATEGIES

Errors in an experiment can come from a nearly endless variety of sources. These sources influence measurements either randomly or systematically. Since the goal of an experiment is to ascertain a truth value as accurately as possible, it is important to eliminate or minimize the impact of error sources. Error analysis and reducing the impact begin during the planning stage of an experiment since sources cannot be deliberately controlled if they are not first identified. Error control continues through the entire process of planning, conducting and reporting the experiment. Several established strategies are available to control errors.

- Reduce error by controlling *independent variables* in an experiment
    - Independent variables propagate to results (dependent variable)
    - Fundamental first step: *identify* sources of error
        - Cannot control errors if unknown and unaccounted
    - During test design, identify and list all possible sources and impact on test variables
- Strategies to control errors[5]
    1. Control
    2. Repetition
    3. Replication
    4. Randomization
    5. Blocking

1. Control
    - Establish a *controlled environment*
        - The lab: general control over temperature, wind, noise and light
        - Barometric chamber: precise control over pressure
        - Faraday cage: control over electromagnetic interference
    - Constantly measure and maintain influential variables
        - Example: record and monitor ambient temperature and pressure
        - Ensures ability to correct in real-time or compensate in results after the fact
    - Establish procedures to ensure consistency
        - Ensures mistakes are not made during a complex set-up
        - Avoids introducing undesirable changes during an experiment

The next two strategies, repetition and replication, are often confused. Repetition typically involves several test sample measurements during a single experiment to obtain an average value. Replication means performing an entire experiment again to ensure validity. Both techniques provide a more comprehensive data set for statistical analysis of the results.

2. Repetition: multiple measurement events during a single lab session
    - Repeated measurements of a test parameter during a single test
    - Multiple runs or trials of a single parameter during a single session
    - Recording an event more than once before changing parameters
3. Replication: returning to the lab to *rerun* an entire experiment
    - Re-accomplishing an entire experiment at a different time
    - Reproducing an experiment by another person or at another location
    - Starting over entirely from the beginning

The next strategy of randomization seems counter-intuitive to purposely introduce random error into an experiment. However, this strategy recognizes that following the same exact process over and over might lead to some unforeseen systematic bias in the results. Some steps or activities during an experiment *should have no impact* on the results, so they can be randomized to ensure no undue bias is introduced. A simple example of randomization is to have more than one person take measurements. The measurements *should be* independent of the person using the equipment, but each person may read the results slightly differently. Changing personnel to take measurements would randomize the potential bias of the same person always reading values the same way. Randomizing in this manner changes the observational source error from *type systematic* to *type random*, and the resultant mean will be statistically closer to the truth value.

4. Randomization
    - Helps neutralize bias not specifically accounted for in test design
    - Purposefully changes an independent variable that should have no impact on results
    - Randomizes the order in which an inconsequential test sequence is performed
    - Uses different test operators or different equipment

A block is a group of experiments under one set of conditions. If experiments are also run under a different set of conditions, they are grouped into a different block, even though the experiments are exactly the same. An example of three blocks is to run five thermocouple tests on Monday, five thermocouple tests on Tuesday and five more thermocouple tests on Wednesday. Ideally, all 15 tests are exactly the same and can be compared together in a single statistical analysis. However, the three different days may unintentionally introduce variations in the results due to the different ambient pressure and temperature on each day. A *blocking* strategy acknowledges potentially uncontrolled variables between groupings (blocks) of the same experiment due to time, location, personnel or industrial batches. The term (*blocking*) can be construed as *blocking* the unwanted effects between groups of experiments.

5. Blocking
    - A *block* is a grouping of tests within an experiment, either intentionally or unintentionally
        - *Two* blocks could be running a single experiment multiple times over *two* days
    - Experimental blocks (groups) can create undesired biases in data (systematic error)
        - Ambient temperature and pressure might affect results between different days
        - Any bias present in the data is the *blocking factor*
        - Blocking factor (bias) can *confound* the results
        - Statistical results might be due to either the treatment <u>or</u> the blocking factor
    - Blocking strategies typically apply when
        - Extraneous variables are known but not directly controllable

  - Multiple runs of an experiment cannot be completed under homogeneous conditions
  - Different days, suppliers or locations create differences in conditions or materials
- Strategy to account for the blocking factor (bias)
  - Divide the treatment plan evenly between blocks
  - Randomize treatments within each block
- Blocking example
  - Plan an experiment with 20 trials testing two factors (A and B) over two days
  - A poor plan without a blocking strategy
    - Day #1: run all ten trials of parameter A
    - Day #2: run all ten trials of parameter B
    - Differences in results might be due to *either* day 1 or 2 *or* parameter A or B
  - A better blocking strategy is to *block* the effects of different days
    - Day #1: run five trials of parameter A and five trials of parameter B
    - Day #2: run five trials of parameter A and five trials of parameter B
    - Effects of non-homogeneous days will not create undue bias between A and B

LOADING

All experiments require measurements to obtain data. Unfortunately, the very act of measurement can introduce an error known as *loading*. Proper instrumentation and measurement technique can minimize loading to the point of being insignificant. A poor measurement scheme can confound the results of the experiment. A clumsy agricultural worker haphazardly walking through a test field of grain to measure average stalk height is a form of loading, as some stalks are damaged and skew the measurement process. Electrical resistance measurements are an easily quantifiable example of loading.

- Loading is the "difference between the value of the measurand and the indicated value as a result of the act of measurement"[6]
- An instrument extracts some energy from the measurand
- The measurand is disturbed by the act of measurement
- Good measurement systems minimize the device's influence on the signal
- Example: 95°C water temperature measured with a glass thermometer
  - Glass thermometer is at room temp of 22°C
  - Water temperature drops to 94.9°C during the measurement
- Example: A tire at 32 psi is measured with a pressure gauge

  Air escapes from the pressurized tire into the non-pressurized gauge

  The tire pressure drops to 31.9 psi during the measurement

Example 6.1

Consider an experiment to measure a 5 Ω resistor. The first experimental design uses a poor ohmmeter with an internal resistance of 1000 Ω. The second experimental design measures the same 5 Ω resistor with an ohmmeter of 1E6 Ω. Each ohmmeter *loads* the circuit by creating a parallel resistor circuit in the measurement process. What effect does loading have on the results between the 1E3 Ω ohmmeter and the 1E6 Ω ohmmeter?

- Electrical resistance formulas:
    - Resistors in series: $R_{TOTAL} = R_1 + R_2 + \ldots + R_n$
    - Resistors in parallel: $1/R_{TOTAL} = 1/R_1 + 1/R_2 + \ldots + 1/R_n$
- High loading due to electrical resistance measurement of *poor ohmmeter*
    - True resistance of individual resistor: $R_1 = 5\ \Omega$
    - Internal resistance of measurement device (ohmmeter): $R_2 = 1000\ \Omega$
    - Act of measurement creates a parallel circuit as shown in Figure 6.2
        - $1/R_T = 1/R_1 + 1/R_2$
        - $1/R_T = 1/5 + 1/1000$
        - $R_T = 4.975\ \Omega$
    - R_{TOTAL} is the equivalent resistance of the circuit
    - Loading effect of instrument is *indistinguishable* from actual resistor value
- Low loading due to electrical resistance measurement of a *better ohmmeter*
    - True resistance of individual resistor: $R_1 = 5\ \Omega$
    - Internal resistance of measurement device (ohmmeter): $R_2 = 1{,}000{,}000\ \Omega$
    - Act of measurement creates a parallel circuit as shown in Figure 6.2
        - $1/R_T = 1/R_1 + 1/R_2$
        - $1/R_T = 1/5 + 1/1{,}000{,}000$
        - $R_T = 4.99998\ \Omega \approx 5.0\ \Omega$
    - Loading effect is significantly reduced with proper measurement device

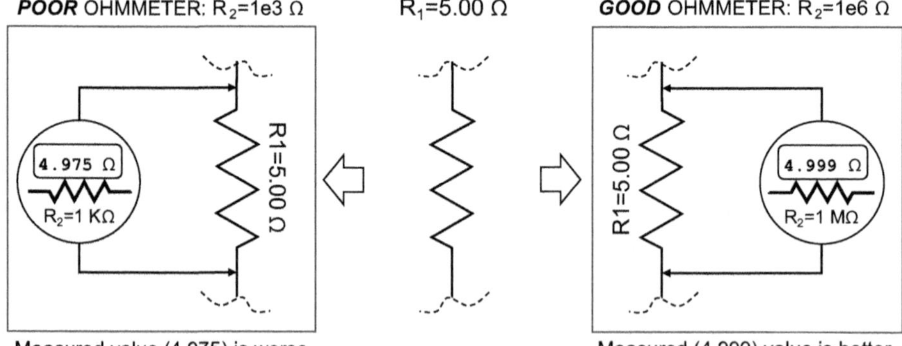

FIGURE 6.2 Loading effects are indistinguishable from actual measurement but can be minimized with proper devices and techniques.

EXERCISES

1. A theory is postulated that alternating current reactance measurements can predict the length of a crack in metal. An experiment is performed, returning a robust data set of reactance values associated with a crack. The theoretical model is checked against the data set. What value belongs in the denominator for the error calculation?
2. An experiment is planned to measure airspeed values using a water u-tube manometer Pitot tube, similar to Lab 13. The airspeed is calculated based on the height difference in the u-tube columns of water. The water column heights appear to fluctuate and change every time the height is measured. Answer the following questions.
   a. What is this called when measurement affects the result?
   b. What is the type of error?
   c. What is the source of error?
   d. A strategy is chosen to minimize the airflow disturbance by using a camera to record reference values with the height difference and measure from the camera instead of using calipers directly. What is this strategy called?
   e. Assume this experiment requires 10 sets of measurements for one airspeed value and 10 measurements for a different airspeed value. The experiment cannot be completed in one session and must occur over two sessions. Explain how best to configure the experiment to avoid errors and the name of the strategy.
3. TRUE or FALSE. A measurement can return an exact value.
4. TRUE or FALSE. A count can return an exact value.
5. A u-tube manometer experiment called for isopropyl alcohol as the fluid, but water was used. Calculations were based on the alcohol requirement. Answer the following questions:
   a. What type of error is this?
   b. What is the correct course of action?
   c. If the error is not noticed, what can be done?
6. A strain gauge with a resistance of 120 Ω is checked with an inexpensive ohmmeter having an internal resistance of 1E5 Ω. What value will be displayed on the ohmmeter due to loading (to hundredths of an ohm)?

NOTES

1 Figliola, R. & Beasley, D. (2015). *Theory and design for mechanical measurements.* Jon Wiley & Sons. p.17.
2 Figliola, R. & Beasley, D. (2015). *Theory and design for mechanical measurements.* Jon Wiley & Sons. p.6.
3 Mahaffey, J. (2017). *Atomic Adventures: Secret islands, forgotten n-rays, and isotopic murder- a journey into the wild world of nuclear science.* Pegasus Books.
4 Mahaffey, J. (2017). *Atomic Adventures: Secret islands, forgotten n-rays, and isotopic murder- a journey into the wild world of nuclear science.* Pegasus Books. p.128.
5 Montgomery, D. (2013). *Design and analysis of experiments.* John Wiley & Sons. p.12.
6 Figliola, R. & Beasley, D. (2015). *Theory and design for mechanical measurements.* Jon Wiley & Sons. p.235.

7 System Modeling and First-Order Systems

Mathematical models describe a theoretical relationship between parameters. Once a model is verified, it becomes predictive of responses to experiments. Temperature decay follows a first-order model. Lab #11 explores the first-order nature of temperature decay and the cooling of fins.

Experimentation is generally about revealing or demonstrating causal relationships in the physical world with real phenomena. Mathematical modeling provides conceptual relationships where an equation with one or more independent variables dictate the result of a dependent variable. If the variables and constants in the mathematical equation correlate to physical properties, the equation can serve as a theoretical construct, ready for experimental validation. A good mathematical model, validated by experimentation, becomes predictive and useful for engineering applications.

- Physical systems are generally modeled mathematically to
  - Understand the observed phenomenon
  - Validate a theoretical hypothesis
  - Develop a predictive relationship
- Experiments use mathematical models to
  - Determine theoretical and expected results
  - Select necessary measurement instrumentation based on
    - Response speed, settling time, sensitivity and measurement range
- Approach to modeling #1: begin with specific existing relationships
  - Categorize the experimental topic of interest: thermal, mechanical, chemical, electrical
  - Use an existing relationship: Newton's law of cooling, Hooke's law, Ohm's law
  - Derive a variation of an existing relationship suitable to the parameters of study
- Approach to modeling #2: work from general principles to a specific application
  - Begin with a conservation law: first law of thermodynamics, mechanical conservation of energy, electrical conservation of charge
  - Add in terms relevant to the topic of interest, such as deciding whether to include a radiant heat transfer term in a conservation of thermal energy equation
  - Neglect smaller effects in the interest of simplification

DOI: 10.1201/9781003329237-7

- Approach to modeling #3: work from specific parameters and units to a general solution
  - Perform a "parameter round-up" by first identifying all the variables of interest
  - Analyze how units combine and cancel for consistency
  - Research existing or postulate new relationships involving key parameters and units
- Approach to modeling #4: begin with experimental data
  - Make observations and look for correlations in parameters and behaviors
  - Develop a hypothetical relationship based on observed phenomenon
  - Perform curve fits of the data and check for generally predictive behavior
- Approach to modeling #5: develop equations of motion unique to a particular system
  - Use a Newtonian, Lagrangian or Hamiltonian approach
  - Account for all forms of energy and mass
  - Define system boundaries, along with initial and boundary conditions

Mathematical models can be developed using a variety of approaches. The process can be more challenging when hypothesizing new relationships, compared to applying tried-and-true techniques to a particular implementation. Since many models already exist for physical phenomena, the *parameter round-up* approach can help point to an available model by reviewing the parameters and units of interest. Experimental models involving forces and objects with mass will likely lead to a variation of Newton's second law, of which there are several variations. Experiments involving temperature and heat transfer may lead to Newton's law of cooling, Fourier's law of heat conduction, the first law of thermodynamics or possibly all three laws combined. When the theoretical problem is not well understood, simply writing down all of the parameters of interest *and units* can give insight into available mathematical models or new ideas about how parameters may be related.

- Mechanical system theoretical models include
  - Conservation of mechanical (potential and kinetic) energy
  - Newton's second law of motion
  - Elastic or inelastic collisions
- Thermal system theoretical models include
  - Conservation of energy (first law of thermodynamics)
  - Newton's law of cooling
  - Fourier's law of heat conduction
- Fluid system theoretical models include
  - Bernoulli's principle
  - Navier-Stokes equations

MODEL FIDELITY

Some mathematical models are better than others. An initial model may prove inadequate. A (super?) model's fidelity is the qualitative measure of how faithfully the model predicts actual performance. The difference between theoretical behavior and actual behavior is the error, as described in the previous chapter. Modeling a physical system offers the opportunity to test inputs and changes in a virtual environment, but the model needs to satisfactorily represent reality. Experiments can be performed on the physical system, and the model can be compared to these actual values. If the error is high, then the model requires improvement. Sometimes this can be achieved with higher order terms or more degrees of freedom.

SYSTEM ORDER

The system order refers to the highest derivative or exponent power of the mathematical model. When referring to a dynamic model, these derivatives or powers are with respect to time as the independent variable. A typical point of interest in a dynamic system is the time required for a system to return to or arrive at a steady-state value after an input change. Many time-based dynamic systems are reasonably modeled as zero-, first- or second-order systems. Complex systems may require higher order models but can often be simplified to a second-order system under certain assumptions or restrictions.

As a matter of definition, consider the simple form of Ohm's law, $V = IR$. This is a zero-order system because there are no derivatives and voltage is directly proportional to current for a given resistance. However, resistance can and does change with time due to temperature effects. Even though resistance can drift with time, in this particular model of Ohm's law, voltage is still directly proportional to current without a lag to reach equilibrium. This is a zeroth-order, time-independent system.

- Systems can often be modeled as zero-, first- or second-order systems, as shown in Figure 7.1
    - The *order* identifies the highest level of derivative or exponent in the model
    - Higher order models are sometimes required for improved fidelity
- Overview of the mathematical forms for each model order
    - Zero-order general form: $y(x) = k\,F(x)$
        - The result exactly equals input times any constant of proportionality
        - Example: $y = 3x$, where "y" (result) exactly equals "x" (input) times a factor of 3
        - *No derivatives, no time dependency*
    - First-order general form: $a\dot{y} + by = F(t)$
        - Similar to zero-order, but involves a delay in getting to the final result
        - Example: $\dot{y} + y = 0$, where "t" is the implicit independent variable
        - *Includes a first-order derivative term*
    - Second-order general form: $m\ddot{y} + c\dot{y} + ky = F(t)$
        - Allows for, but does not require, oscillatory behavior
        - Example form: $\ddot{y} + \dot{y} + y = 0$
        - *Includes a second-order derivative term*
    - All basic zero-, first- and second-order differential equations have *standard solutions*

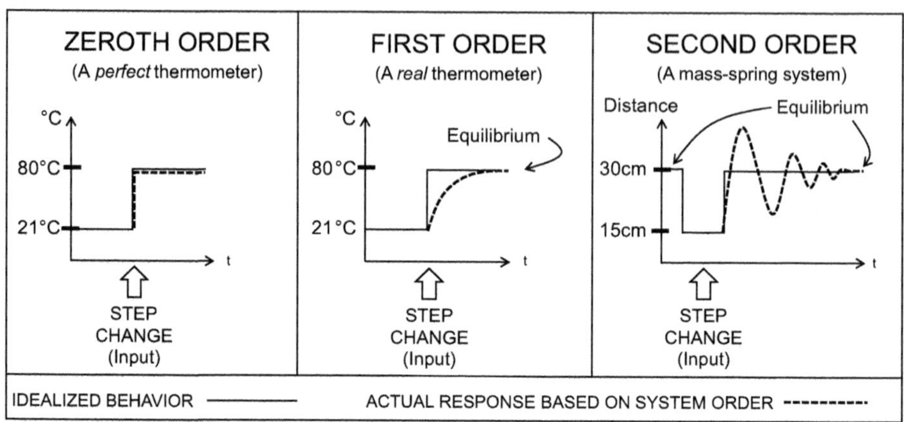

FIGURE 7.1 System response based on order.

Zero-order system characteristics

- A static system with no time dependency
- The system may change over time, but any modeled response is immediate
- Output exactly follows input, to include any scaling factors
- Measurements do not require settling time
- No lag, no derivatives, no conditions on linearity or non-linearity
- Zeroth order is somewhat trivial as an idealized system with a dynamically insignificant response

First-order systems characteristics

- A system that changes with time (dynamic)
- Response to change (input) is delayed
- Typically need to consider settling time to take a measurement
- Applies to systems with *storage* or dissipative capability
- Does not oscillate: oscillation requires two (or more) energy forms to swap between (hence the oscillation); first order has only one energy component to source or sink
    - A first-order system can still oscillate due to an *external forcing function*
    - It is important to distinguish between *internal* system response orders and *external* inputs that may initially appear (incorrectly) to be part of the *internal* response
- No ability to overshoot
- Examples of the first-order system
    - Cart sliding on friction surface
        - Dissipating energy by friction; no exchange of energy between states
        - Energy is dissipated in one direction only

    - A hot object cooling to room temperature
        - Heat energy dissipates in one direction only, to the surrounding environment
- First-order mathematical model
    - $a\dot{y} + by = K\ F(t) \leftarrow$ General ODE form
        - $F(t)$ is an external forcing function which may or may not be present
            - If $F(t)$ term is present, the ODE is non-homogeneous
            - If $F(t)$ term is not present, the ODE is homogeneous
    - Slightly re-arrange ODE into a more useful form by collecting generalized constants
    - $\dot{y} + ky = 0 \leftarrow$ General ODE form for modeling many real-world first-order systems
        - $\dot{y}$ is the rate (change or velocity)
        - y is position or the function's value at any particular time, t
        - Model is a first-order, homogeneous, ordinary differential equation (ODE)
            - Solutions to homogeneous first-order ODEs are known
- The standard solution for a first-order ODE takes on variations of the following form
    - $y(t) = Ce^{-\lambda t} + D$, where
    - $Ce^{-\lambda t}$ is a decaying exponential term for stable systems that depend on initial conditions
    - λ is the decay time constant
    - D is the steady-state term, independent of initial conditions
- Example of a first-order system: a thermometer placed in hot water, per Figure 7.2
    - The thermometer is slow to respond and takes a few moments to display the correct temperature
    - Response lags from input but has no ability to overshoot in a first-order model

Heating and cooling are good examples of first-order systems. These situations have one form of system energy, heat, which either transfers into or out of the system with no ability to return without an external forcing function.

Sir Isaac Newton recognized that a hot object cools to the surrounding temperature at a rate proportional to the temperature differential. This makes sense since a very hot object will initially shed heat quickly, but stabilizing at room temperature still takes a fair amount of time. This relationship is an exponential decay rate, governed by Newton's law of cooling in the differential form of Equation 7.1.

$$\frac{dT}{dt} = k(T_\infty - T) \tag{7.1}$$

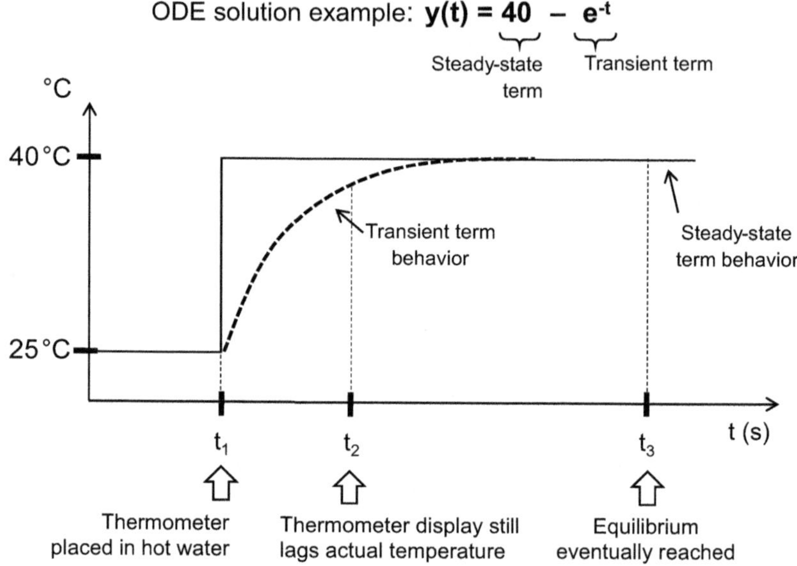

FIGURE 7.2 Example of how first-order model solution terms model actual response.

- Newton's law of cooling in Equation 7.1 states the following:
  - An object's temperature, T, changes with time, $\frac{dT}{dt}$
  - That object's temperature is changing at a rate proportional, k, to the difference between the object's current temperature, T, and the surrounding environmental temperature, T_∞.
  - Environment temperature is equally referred to as *surrounding* or *ambient* temperature
    - The T_∞ notation implies steady-state temperature as time approaches infinity
    - T_∞ is a numerical constant

When Equation 7.1 is solved for the cooling case, the result is the decaying temperature function, Equation 7.2a. Equation 7.2a is written with the decay constant, k, and re-written in Equation 7.2b with the time constant, τ, which is the inverse of the decay constant ($1/k$).

$$T(t) = T_\infty + (T_0 - T_\infty)e^{-kt} \tag{7.2a}$$

$$T(t) = T_\infty + (T_0 - T_\infty)e^{-\left(t/\tau\right)} \tag{7.2b}$$

where: $T(t)$ = temperature at any time, t

T_∞ = ambient, environmental or steady-state temperature

T_o = initial temperature of hot object
k = decay constant
τ = time constant, tau
$k = 1/\tau$

- Equation 7.2 can be used to predict when an object has reached the desired temperature
    - Initial and ambient temperatures are usually known
    - Decay rate or time constant is still required to complete the mathematical model

If an object has a homogeneous temperature throughout, the temperature decay rate, and therefore the time constant, is given by Equation 7.3.[1]

$$k = \frac{\rho VC}{hA_S} \tag{7.3}$$

where: k = decay constant
ρ = material density
V = volume of the object
C = specific heat constant of material
h = heat transfer coefficient
A_S = surface area of the object subject to cooling

For non-homogeneous substances, complex objects or non-uniform temperature distributions, Equation 7.3 is difficult to implement. In these cases, the decay rate, k, can be experimentally determined and used to predict future behavior. Experimental data is usually a plot of units such as temperature versus time. Since the decay constant, k, and time constant, τ, are inverses as shown in Equations 7.2a and 7.2b, it is easiest to select the time *when* the quantity (temperature) is within 36.8% of the equilibrium value as given in Equation 7.4. Since first-order systems can increase or decrease exponentially, the time constant occurs at a point 36.8% of the equilibrium value, or an overall change of 63.2%.

$$\tau = t, \textit{ only } @ \left(\frac{1}{e}\right) \textit{ of equilibrium value} \tag{7.4}$$

Equation 7.4 states that *tau equals time*, but only at the unique moment of *1/e*, which is 0.368. If the data is insufficient to include the 36.8% data point, an alternative to Equation 7.4 for any value along the exponential curve is given in Equation 7.5.

$$\tau = -\frac{\Delta t}{\ln\left[\dfrac{T_C - T_\infty}{T_H - T_\infty}\right]} \tag{7.5}$$

where: τ = time constant, tau
Δt = elapsed time between any two temperature points
T_∞ = ambient, equilibrium steady-state temperature

T_H = any hotter temperature point to define Δt
T_C = any cooler temperature point to define Δt
$k = 1/\tau$ (exponential decay constant)

Mathematically, an exponential function never actually reaches a steady-state value, so steady state is considered to occur after five time constants.

Example 7.1

Consider the following arbitrary, simple exponential decay function in Figure 7.3. This function has the most basic form of $y(t) = 100e^{-kt}$. Use the data in Figure 7.3 to write the complete mathematical expression of the function.

- GIVEN: $y(t) = 100e^{-kt}$ and data in Figure 7.3.
- FIND: Determine "*k*"
  - The maximum value of the function is 100
  - The time constant occurs at 36.8% of the overall amplitude range
  - The time associated with 36.8 is already provided in the figure
    - At this unique point, $t = 22$ seconds
    - Only for this unique point, tau also equals time: $\tau = 22$ seconds
  - The decay function is $k = 1/\tau$: $k = 0.046$
  - The function is: $y(t) = 100e^{-0.046t}$
  - This function can be checked for validity at $t = 0$ and $t = 22$

While Example 7.1 is the simplest example, temperature decay and distribution functions are slightly more involved. Working in SI units, all temperature calculations should be performed in Kelvin rather than Celsius to ensure consistent units. Also, temperature functions always require an offset since they do not end at absolute zero. Finally, Newton's law of cooling is based on temperature *differences*. The actual temperature does not matter. An object can cool from 1200K to a

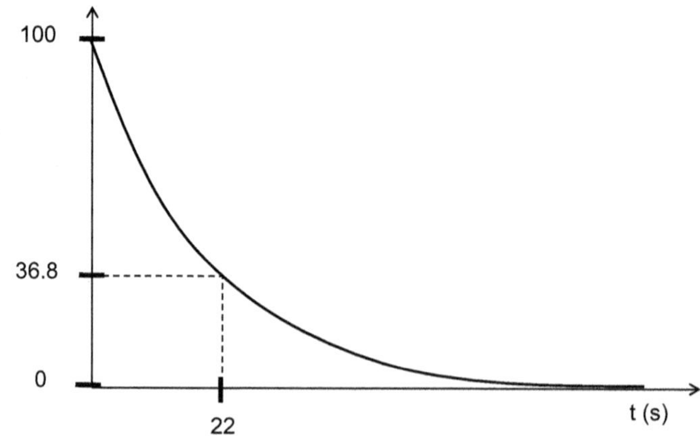

FIGURE 7.3 Example of modeling first-order function data.

steady state of 1000K, just as an object can cool from 400K to a steady state of 200K. Both situations see a 200°C difference, but the first case is very hot while the second case is below freezing. The exponential curve is therefore based on the temperature difference, which is the $(T_0 - T_\infty)$ in Equation 7.2. The temperature *difference function* will eventually decay to zero as an object's temperature reaches ambient. As the difference function dies out, the ambient offset, T_∞, remains.

Example 7.2

Compute and plot the theoretical temperature decay for a small cylinder of AL6061T6 aluminum, initially at 150°C (423K) in a room at 26°C (299K)

- GIVEN: Properties of aluminum for use in Equation 7.3
  - $\rho = 2700\,kg/m^3$
  - $V = 3.45E{-}5\,m^3$ (an arbitrary cylindrical volume)
  - $C = 900$ J/kg-K (specific heat capacity of aluminum)
  - $A_S = 6.33E{-}3\,m^2$ (surface area of the arbitrary cylinder)
  - $h = 11\,W/(m^2K)$
  - Note: Heat transfer coefficients, h, vary significantly from their published values. For free-air convection, at temperatures near 200°C, $h \approx 10\,W/m^2K$.
- FIND: The theoretical time constant and plot the temperature curve
  - $k = \dfrac{(2700)(3.45E-5)(900)}{(11)(6.33E-3)}$
  - $\tau = 1/k = 1203\ s$
  - Using Equation 7.2b and converting all temperatures to Kelvin for consistent units
    - $T(t) = T_\infty + (T_0 - T_\infty)e^{-(t/\tau)}$
    - $T(t) = 299 + (423 - 299)e^{-(t/1203)}$
    - $T(t) = 299 + (124)e^{-(t/1203)}$
  - Figure 7.4 provides a plot of the function along with the time constant

Figure 7.4 indicates the added complexity of working with time constants for temperature distributions. It is important to note that the *curve* is *only* the exponential term (temperature difference function) without the offset. This means the time constant must be taken only between the limits of the curve, excluding the offset. Next, subtracting two Kelvin temperatures results in a Celsius temperature, which is used to identify the 36.8% point between the limits on the curve. This value must be added to the offset, either in Celsius or Kelvin, to arrive at the time constant temperature.

A summary interpretation of Figure 7.4 states that a small cylinder of aluminum at 150°C will cool according to the curve in Figure 7.4. At 1203 seconds, or approximately 20 minutes, the cylinder will have cooled to about 72°C. A *sanity check* confirms that a hot 150°C chunk of aluminum placed in a 26°C room will still be rather warm, somewhere between the two temperatures and probably closer to the lower temperature. The result makes sense.

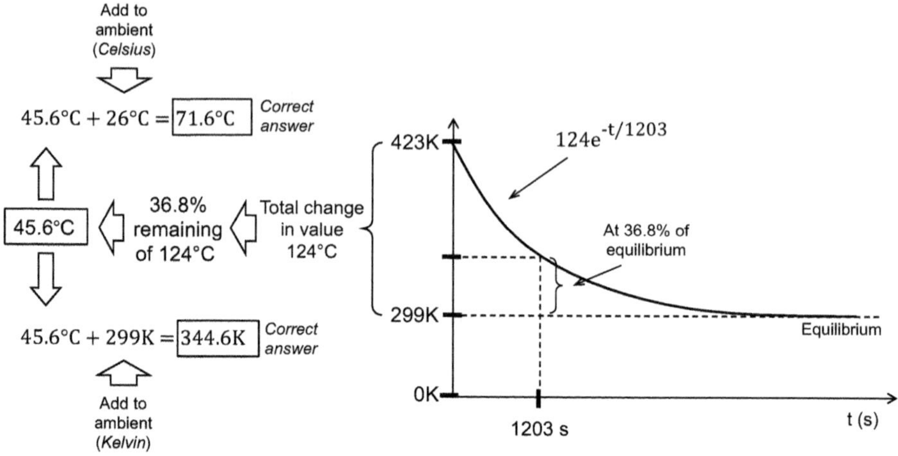

FIGURE 7.4 Temperature distribution function.

Example 7.3

Consider the same exact situation in Example 7.2 from an experimental perspective. The cylinder of aluminum is heated to 150°C then allowed to cool in the room. Unfortunately, data recording was interrupted prior to 20 minutes, and the aluminum only cooled to 86°C when data recording ceased. Ideally, given the 150°C starting temperature and the 26°C ambient temperature, cooling would've been recorded at 36.8% between these two temperatures, which corresponds to 71.6°C. This should've happened 1203 seconds after cooling began, and the data would have been sufficient to write the equation. Since cooling data only exists up to 86°C, Equation 7.5 can be used to complete the necessary information.

- GIVEN: Same physical parameters as in Example 7.2 and an experimental data set
- FIND: The time constant from an arbitrary point on the exponential curve
  - Choose arbitrary data collection point, say at 3 minutes, where the temperature is recorded at 107.2°C
  - Use Equation 7.5
    - $\tau = -\dfrac{\Delta t}{\ln\left[\dfrac{T_C - T_\infty}{T_H - T_\infty}\right]}$
    - $\tau = -\dfrac{180}{\ln\left[\dfrac{107.2 - 299}{423 - 299}\right]}$
    - $\tau = 1163\ s$

In this case, the experimental time constant is 1163 seconds, compared to the theoretical time constant of 1203 seconds. When plotted, these two curves are very close, and it is reasonable to expect variation between theoretical values and experimental values. This is particularly true for the theoretical calculation since the *heat transfer coefficient* value is difficult to know with precision.

DERIVATION. The obvious question is, "why 36.8%" for a time constant. The answer is only a matter of convenience. Equation 7.5 provides the correct formula to compute the time constant for any point on an exponentially decaying curve. Equation 7.5 is Equation 7.2b solved for τ. However, a complete data set will have values (temperature) versus time (seconds) between an upper and lower limit. When a convenient "time constant" data point is chosen, Equation 7.5 can be skipped. The exponential curve decays at a rate of $e^{-t/\tau}$. If $t=\tau$, then finding t returns τ. Setting the two equal, results in e^{-1}, which equals 0.368. Since this *is* the curve relating temperature values to time, choosing a value with 36.8% remaining on the curve provides the time where $t=\tau$. The time constant is the inverse of the decay constant, and the time constant is a measure of the exponential function for comparing between functions.

EXERCISES

1. Consider an experiment to test the effectiveness of propulsive force acting on a moving object. For example, a model rocket is launched on a trajectory, and a single-use maneuvering thruster is fired to re-target the rocket's trajectory. Consider the five approaches to theoretical modeling discussed in this chapter. What general approach would you use to model the maneuvering thruster? Explain
2. Provide any example of a first-order system.
3. Why is temperature decay modeled as an exponential function?
4. How are the time constant and decay constant related in an exponential function?
5. At what time does the following temperature function decay to a steady state of 300K: $T(t)=300+125e^{-0.02t}$ where time is in seconds.
6. If a function decays exponentially from 50 at $t=0$ to 20 at $t=10$ minutes, what is the time constant?
7. A hot object is removed from a heat source and placed in a room at 27°C ambient temperature. Temperature is not initially recorded, so some time goes by, and then the object is recorded at 240°C and 5 minutes later, it is recorded again at 180°C. Answer the following questions:
   a. Can the temperature decay function be determined without the initial conditions? Explain.
   b. If possible, what is the temperature decay function?

NOTE

1 Cengel, Y. (2003). *Heat transfer: a practical approach.* McGraw-Hill. p.211.

8 Discrete Vibrations and Second-Order Systems

Many real-world oscillatory systems can be effectively modeled as second-order, one-degree-of-freedom systems. Lab #1 explores the classic mass-spring-damper 1-DOF oscillatory system.

A classic example of a one degree-of-freedom (1-DOF) system is the wheel suspension of a vehicle. The vehicle has mass, which is suspended on the spring connected to the wheel. While traveling along a very smooth road, the vehicle's motion is stable and in equilibrium. As soon as the vehicle hits a momentary bump in the road, the force is transmitted through the spring, and the vehicle is displaced from equilibrium. The vehicle seeks to return to equilibrium, but the energy imparted into the system now trades off between the kinetic energy of movement and the potential energy stored in the spring. With no dissipation, the imparted energy remains in the system, continuously alternating between the two forms, and the system oscillates. If a damper is added to the system, energy is dissipated through heat, and the oscillating vehicle can settle back to equilibrium.

SECOND-ORDER SYSTEMS

The three primary elements of a mechanical 1-DOF oscillatory system are the mass, spring and damper depicted in Figure 8.1. The spring is the system stiffness where "spring" and "stiffness," are interchangeable terms. The system is discrete because the components are idealized individual elements. It is 1-DOF because it can only move back and forth along a single path described by a single independent variable (time). It is a second-order system because it can be described by a second-order ordinary differential equation (ODE).

- Figure 8.1 depicts the physical model of a discrete, 1-DOF second-order system with mass, spring and damper components
- Dampers are like walking in a chest-high pool of water
  - Walking (velocity) in any direction is met with resistance
  - Stand still (no velocity), and there is no effect
  - Damping only shows up on velocity term of equation
- Second-order systems consist of the following
  - Dynamic (change with time)
  - Oscillate (or are sufficiently damped)
  - Have inertia
  - Described by second-order ODE

DOI: 10.1201/9781003329237-8

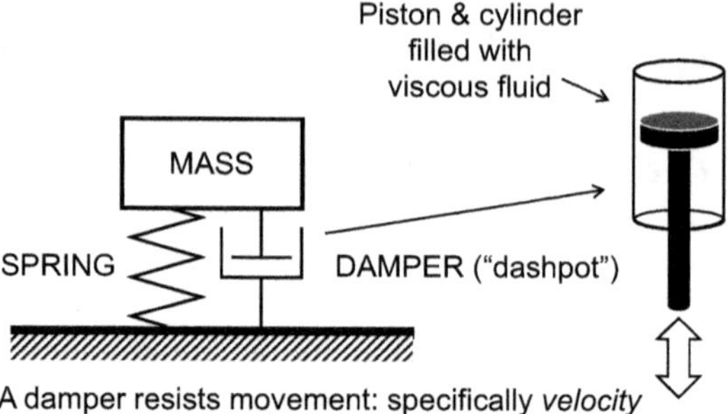

FIGURE 8.1 Physical model of second-order system.

MATHEMATICAL MODEL

It is important to distinguish between internal system response and external forcing functions in the model. A system in equilibrium will remain in equilibrium unless acted upon by an external force. The external force may be a momentary step input or a continuous function. The simplest case is a step input that displaces the system from equilibrium but then allows the system to respond without further external influence. The return behavior is the internal system response due only to the parameters of mass, stiffness and damping. A time domain plot and analysis of a freely oscillating response generally begin at the moment the system is allowed to respond from maximum displacement, without having to consider the nature of the initial input itself. The standard second-order mathematical model is analyzed in Figure 8.2.

- Second-order mathematical model: $m\ddot{y} + c\dot{y} + ky = F(t)$
- Analyzing the model in Figure 8.2, there is no weight (mg) term
  - Gravitational constant, g, does not exist in the equation
  - Internal system response is independent of weight (independent of gravity but not mass-independent)
  - The frequency of the system is unaffected by whether it is vertical or horizontal

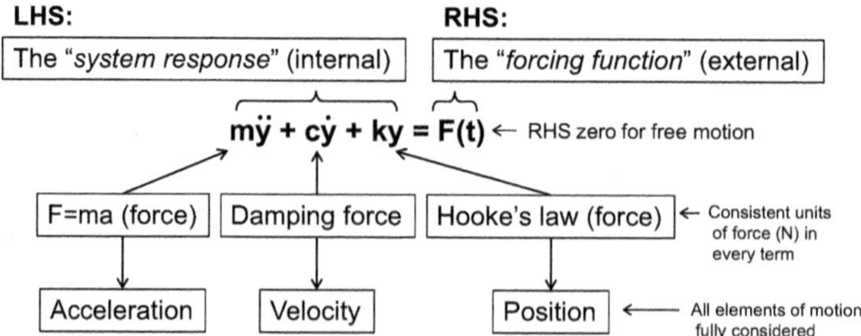

FIGURE 8.2 Term-wise analysis of second-order system mathematical model.

- The LHS of equation, sorted with only dependent variable terms has the following characteristics
  - It models the inherent system itself
  - It describes how the system responds after an input displacement from equilibrium
  - LHS does not include input or external functions
  - LHS results in the homogeneous solution to the ODE
- The RHS of equation, sorted with only independent variable terms and constants has the following characteristics
  - RHS is zero (no terms) for a freely oscillating system
  - RHS is non-zero, if an external forcing function is present
  - A RHS non-zero term forcing function could be a sinusoidal (unbalanced motor) or any other input driving system excitation
  - RHS results in a particular solution to the ODE
  - (Forcing functions and particular solutions are not considered further)
- There are two different forms to write a second-order mathematical model
  - Using physical *parameters* of mass, damping and stiffness (m, c and k) in Equation 8.1

$$m\ddot{y} + c\dot{y} + ky = 0 \tag{8.1}$$

  - Using physical *response* of frequency and damping ratio (ω, ζ) in Equation 8.2

$$\ddot{y} + 2\zeta\omega_n\dot{y} + \omega_n{}^2 y = 0 \tag{8.2}$$

  - Both equations are the same, requiring only algebra to change forms
- The terms of Equation 8.2 must now be defined as follows for algebra to change between Equations 8.1 and 8.2
  - Define undamped natural frequency as ω_n with Equation 8.3

$$\omega_n = \sqrt{k/m} \tag{8.3}$$

Equation 8.3 is characteristic of both frequency and velocity. Equation 8.3 has the general form of the square root of stiffness divided by mass or density. This basic relationship is true of the vibration frequencies of a continuous system discussed in the next chapter. This stiffness-density relationship is also true of the velocity of solids, liquids and gases discussed in later chapters. No derivation is suggested here to demonstrate that Equation 8.3 is in fact a natural frequency. Instead, by manipulating Equation 8.1 and creating a term with the square root of stiffness-to-mass, the result *chosen* in the form of Equation 8.3 is uniquely characteristic of natural frequency.

- In general, (undamped) natural frequency is a function of stiffness and mass (density)
- Equation 8.3 is characteristic of the natural frequency form for other systems
- Velocity and frequency are related by wavelength, so velocity also shares this form

- Define a damping ratio (zeta): $\zeta = \frac{c}{2\sqrt{km}}$
- Define the critical damping coefficient: $c_{cr} = 2\sqrt{km}$
- Equating above two definitions, zeta (ζ) also equals c/c_{cr}
- Define damped natural frequency: $\omega_d = \omega_n\sqrt{1-\zeta^2}$
  - Note that the above damped natural frequency is a simplification for *only* $\zeta < 1$
  - The quadratic equation *actually* solves to $\omega_d = \zeta\omega_n \pm \sqrt{1-\zeta^2\omega_n - \omega_n{}^2}$
  - However, there is no natural frequency for $\zeta \geq 1$ since there is *no oscillation*
  - With $\zeta \geq 1$, the system is over-damped and does not oscillate
- Using newly defined terms in the "response" form of Equation 8.2 tells us more
  - ω_n is the *natural frequency* of the system; the oscillation frequency
  - ω_d is the *damped natural frequency* of the system
    - The system has a natural frequency, ω_n, but it oscillates differently at a damped natural frequency depending on the amount of damping
  - Damping, c, (also the damping ratio, c/c_{cr}) changes response of the system
    - Like a swinging door
      - 1) Under-damped: swings back and forth several times before settling
      - 2) Critically damped: closes quickly without any swinging
      - 3) Over-damped: takes "forever" to close and does not swing
- Second-order mathematical solution
  - Systems without external forcing functions are homogeneous
    - $\ddot{y} + a\dot{y} + by = 0 \leftarrow$ Standard solution form (highest term has "1" coefficient: $1*\ddot{y}$)
    - $\lambda^2 + (c/m)\lambda + (k/m) = 0 \leftarrow$ Characteristic equation for ODE solution
  - Second-order ODE has two roots with two solutions: $y_h = y_1 + y_2$
  - Solution of characteristic equation is roots given by the standard quadratic Equation 8.4

$$\lambda = \frac{-c \pm \sqrt{c^2 - 4mk}}{2m} \tag{8.4}$$

- The general solution is a standard known technique where the two solutions are
  - $y_1 = c_1 e^{\lambda 1 t}$
  - $y_2 = c_2 e^{\lambda 2 t}$
- The roots, λ_1 and λ_2, determine how the system behaves for damping

RESPONSE PLOTS

The roots of the mathematical solution dictate the system response, which can be plotted in the time domain for greater clarification and analysis.

- Analyze the roots of solution for three damping cases (under-, critical- and over-damped)
  - Three system elements: mass (m), stiffness (k) and damping (c)
  - Consider the magnitudes of each system element relative to one another
  - Consider only the bold terms under the radical in the root solutions of Equation 8.5

$$\lambda = \frac{-c \pm \sqrt{\mathbf{c}^2 - 4\mathbf{mk}}}{2m} \tag{8.5}$$

  - Under-damped system as depicted in Figure 8.3
    - Damping, c, is *low* compared to mass, m, and spring force, k (stiffness): $c^2 < 4mk$
    - The radical term contains a "−1" and its root becomes an imaginary term
      - *Oscillatory systems have an imaginary term*
    - Damping ratio is low (less than critical): $0 < \zeta < 1$

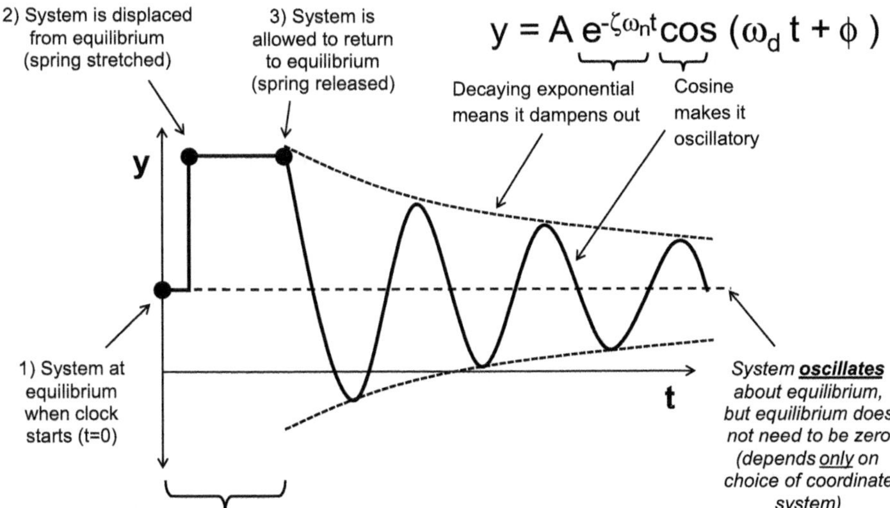

FIGURE 8.3 Analysis of second-order system mathematical model of under-damped case.

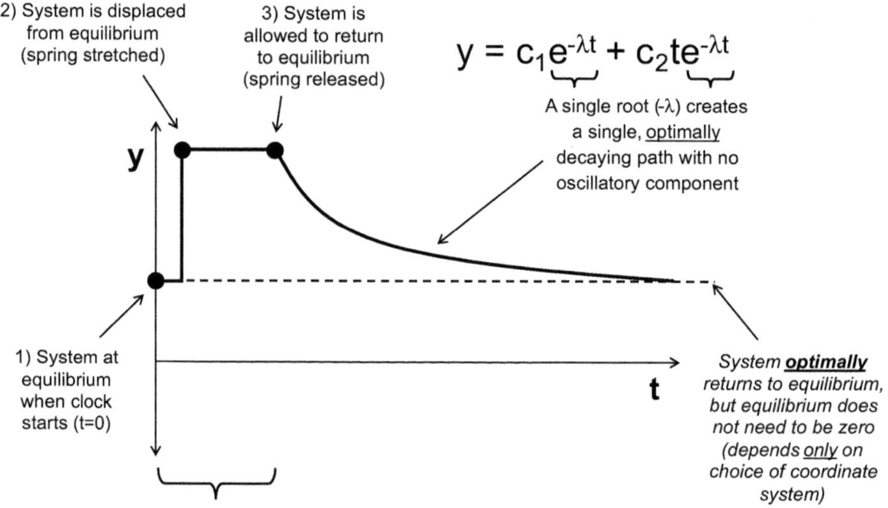

FIGURE 8.4 Analysis of second-order system mathematical model of critically damped case.

  - Example: a door that swings back and forth but still eventually closes
  - Typical ODE solution form: $y = Ae^{-\alpha t}\,[\cos(\omega t + \phi)]$
- Critically damped system as depicted in Figure 8.4
  - Damping, c, is *comparable to* mass, m, and spring force, k (stiffness): $c^2 = 4mk$
  - The radical term is zero, and the root becomes a single real number: $\lambda_1 = \lambda_2 = \lambda$
    - *System does* <u>*not*</u> *have an imaginary term and is therefore non-oscillatory*
  - Damping ratio is *just right* (critical): $\zeta = 1$
  - Example: a door that swings shut fastest with no oscillation
  - Typical ODE solution form: $y = c_1e^{-\lambda t} + c_2t\,e^{-\lambda t}$
- Over-damped system as depicted in Figure 8.5
  - Damping, c, is *high* compared to mass, m, and spring force, k (stiffness): $c^2 > 4mk$
  - The radical is real, and the roots are two real numbers (+ and − radical term)
    - *System does* not *have an imaginary term and is therefore non-oscillatory*
  - Damping ratio is high (higher than critical): $\zeta > 1$

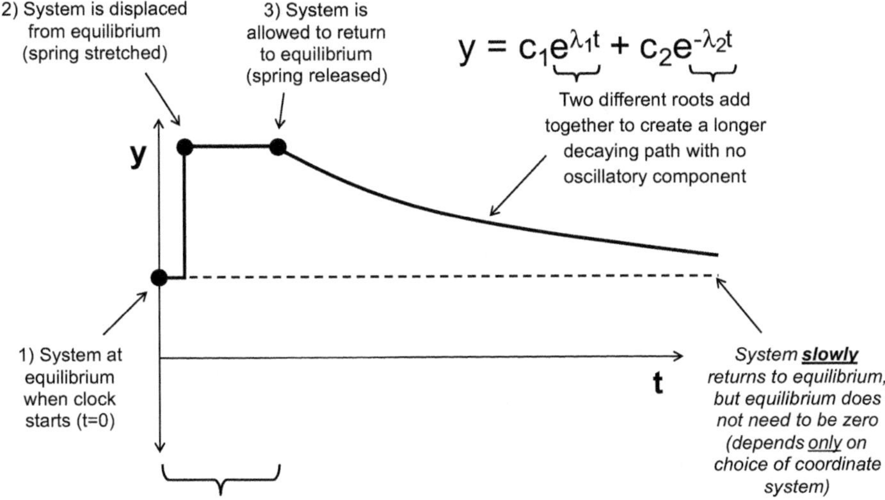

FIGURE 8.5 Analysis of second-order system mathematical model of over-damped case.

- Example: a stiff door that takes a while to close with no oscillation
- Typical ODE solution form: $y = c_1 e^{\lambda 1t} + c_2 e^{-\lambda 2t}$

- No Damping (undamped): *special case* of general second-order system
  - $\ddot{y} + a\dot{y} + by = 0 \leftarrow$ Damped second-order homogeneous equation (general form)
    - $\dot{y}$ term is velocity, generic coefficient "a" (usually "c") is damping coefficient
  - $\ddot{y} + \cancel{a\dot{y}} + by = 0 \leftarrow$ Undamped second-order equation with no velocity term
  - Solution changes to sinusoidal-only for no damping: $y = A\cos(\omega_n t + \phi)$
  - With *any* damping, the system eventually settles into equilibrium
  - *No damping* results in a continuous sinusoid about equilibrium
- Analyzing oscillatory waveforms
  - Applies both to *under-damped* and *undamped* oscillatory waveforms
  - Period, T: one FULL wavelength as shown in Figure 8.6
    - Period can be measured from peak-to-peak, trough-to-trough or between zero-crossings
    - $T = \frac{1}{f}$, where f = frequency (in Hertz, 1/s or "cycles per seconds")
    - $T = \frac{2\pi}{\omega}$, where ω is the angular frequency (in "radians per second")

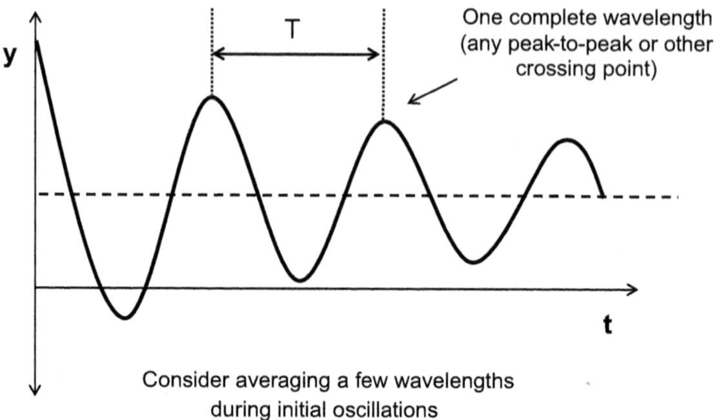

FIGURE 8.6 Determining the period of an oscillatory waveform.

- ω: use radians (not degrees) in the "*number crunching*" → $\sin(\omega t)$, $e\omega^t$
  - Never mix degrees with numbers; only radians can be mixed with real numbers
- Hertz, f: Always the physical *counts* of an event
  - $\omega = 2\pi f$ ←Conversion from Hertz to radians

Example 8.1

A mountain bike is designed with a single-point suspension spring carrying about 40 kg of mass at the suspension point. The spring constant is 20 kN/m and the damping constant is adjustable. Determine the oscillatory attributes of the bike.

- FIND: The undamped natural frequency and cycles per second
  - Use Equation 8.3: $\omega_n = \sqrt{20000/40} = 22 \text{ rad/s}$
  - Cycles per second: $\omega = 2\pi f \rightarrow f = \omega/2\pi \rightarrow f = 22/(2\pi) \rightarrow f = 3.6\,\text{Hz}$
- FIND: The damping constant to ensure quick equilibrium with no oscillation
  - Critical damping: $c_{cr} = 2\sqrt{km} \rightarrow c_{cr} = 2\sqrt{(20000)(40)} \rightarrow c_{cr} = 1789$ kN-s/m
- FIND: The second-order mathematical model for the under-damped system if damping constant is set to $c = 500$ kN-s/m, and plot the solution
  - Use Equation 8.1: $m\ddot{y} + c\dot{y} + ky = 0 \rightarrow 40\ddot{y} + 500\dot{y} + 20000y = 0$
  - Standard ODE form: $\ddot{y} + 12.5\dot{y} + 500y = 0$
  - Characteristic equation: $\lambda^2 + 12.5\lambda + 500 = 0$
  - Roots from quadratic equation, Equation 8.4: $\lambda = -6.25 \pm 21.5i$
    - The system is oscillatory with the imaginary roots
  - Damping ratio, zeta: $\zeta = \dfrac{c}{2\sqrt{km}} \rightarrow \zeta = \dfrac{500}{2\sqrt{(20000)(40)}} \rightarrow \zeta = 0.28$
  - Damped natural frequency: $\omega_d = \omega_n\sqrt{1-\zeta^2} \rightarrow \omega_d = 22\sqrt{1-0.28^2} \rightarrow \omega_d = 21.5$ rad/s

    - The damped natural frequency is the imaginary root term
    - The damped natural frequency is less than the undamped frequency
- Solution from Figure 8.3: $y = Ae^{-\zeta\omega_n t}\left(\cos\omega_d t + \phi\right) \rightarrow y = Ae^{-6.25t}$ $(\cos 21t)$
    - The amplitude, *A*, is the initial condition of how far the spring is initially displaced
    - The phase, ϕ, is zero since the cosine timing begins at maximum displacement
- The solution can be plotted with a graphing calculator or online resource

EXERCISES

1. A second-order system has the following parameters: $m = 5\,\text{kg}$, $k = 4\,\text{N/m}$ and $c = 10$ N-s/m. Complete the following questions.
    a. Is the system oscillatory?
    b. Write the ODE for this system in terms of physical parameters (Equation 8.1).
    c. Is there a damped natural frequency?
    d. What is the time domain solution to the ODE?
2. Consider a mass-spring-damper cart system on an inclined track. Which one of the following is a true statement (ignore any friction-related effects):
    a. The system has the same natural frequency on the moon with 16% of Earth's gravity.
    b. A greater track inclination angle would have no effect on the equilibrium position.
    c. A greater track inclination angle would reduce the system's natural frequency.
    d. Changing the mass of the cart would have no effect on the system's natural frequency.
3. Consider a system described by the ODE: $\ddot{y} + 5y = 0$. Answer the following.
    a. Is the system damped?
    b. What is the natural frequency?
    c. What is the period of oscillation?
    d. What is the critical damping for this system?
    e. What is the time domain solution?
4. Assume the time domain solution of a system is: $y = A\cos(5t + \phi)$, and the system is displaced 15 cm from equilibrium. What are the values for A and ϕ?

9 Continuous System Vibrations

Similar to discrete systems, solid structures also have vibratory behavior. The vibration of a non-discrete system is more complex, but mathematical models can still be developed. Lab #2 explores the vibratory modes and frequencies of a cantilever beam subjected to both free and forced vibrations.

NATURAL FREQUENCIES

Knowing a continuous structure's natural frequencies can be important to avoid destructive vibration-induced deformations due to the large amplitude excursions associated with these natural frequencies. One of the most famous engineering examples of natural frequency excitations was the newly constructed Tacoma Narrows bridge in Washington state. In 1940, a combination of wind and bridge aerodynamics provided the external forcing function to excite the bridge at its natural frequency. The maximum amplitude excursions of the natural frequency resulted in structural failure of the bridge.

- Continuous structure vibrations share characteristics with discrete systems
  - Natural frequencies are a function of mass, stiffness and damping
  - If a continuous structure is displaced from equilibrium, it will vibrate at its natural frequency as it seeks to return to equilibrium
- Continuous structures have more than one vibration frequency
  - The frequencies are interchangeably called *natural, resonant* or *harmonic* frequencies
  - The lowest vibration frequency is the *fundamental* frequency
  - The fundamental frequency is associated with either free or forced vibrations
  - Higher harmonic frequencies require excitation with an external forcing function
- A cantilever beam is an example of a continuous structure
  - The natural frequencies of a homogeneous cantilever beam can be calculated theoretically
  - Experimentation may be required to resolve the natural frequencies of complex structures
- The first three natural frequencies of a cantilever beam are given in Equations 9.1(a–c)

$$f_1 = \frac{1}{2\pi}(1.875)^2\sqrt{\frac{EI}{mL^3}} \tag{9.1a}$$

DOI: 10.1201/9781003329237-9

$$f_2 = \frac{1}{2\pi}(4.694)^2 \sqrt{\frac{EI}{mL^3}} \tag{9.1b}$$

$$f_3 = \frac{1}{2\pi}(7.854)^2 \sqrt{\frac{EI}{mL^3}} \tag{9.1c}$$

where: f_n = natural frequency
E = modulus of elasticity
I = moment of inertia
m = mass
L = cantilever length

- *Note: Many references indicate the frequency formulations with L^4 instead of L^3 per Equations 9.1(a–c). To avoid the discrepancy, mass should be considered a function of integration and expressed as mass per unit length or in terms of density. (Refer to the derivation that follows to arrive at the L^3 formulation.)*
- As noted in the previous chapter, natural frequencies in Equations 9.1(a–c) follow the form of the square root of material stiffness relative to material density (mass)

- Mode shapes, as shown in Figure 9.1, provide insight into the natural frequencies of a cantilever beam
  - *Mode shapes* are physical bending deformations of a beam due to vibration
  - Mode shapes correspond to harmonic frequencies of a vibrating beam ($f_1, f_2, f_3, \ldots$)

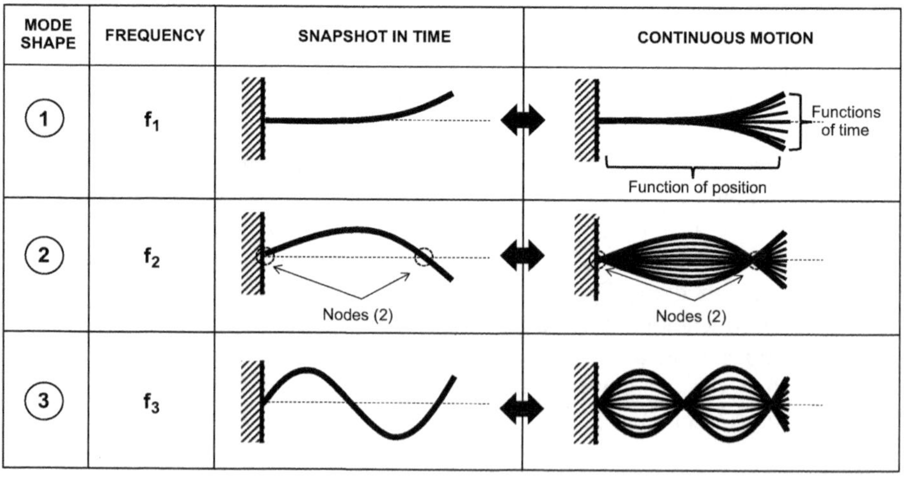

MODE SHAPE	FREQUENCY	SNAPSHOT IN TIME	CONTINUOUS MOTION
1	f_1		Functions of time; Function of position
2	f_2	Nodes (2)	Nodes (2)
3	f_3		

FIGURE 9.1 Beam bending mode shapes.

- Mode shape vibratory excursions due to harmonic frequencies have
  - Anti-nodes: maximum amplitudes
  - Nodes: minimum (zero) amplitudes
  - Amplitude excursions of a cantilever beam, y, are a function of
    - Position along the length of the beam, x
    - A particular moment in time, t
- Mode shape *nodes*
  - Nodes are points of zero amplitude excursions
  - The *number* of nodes corresponds to the mode shape and frequency
  - The first mode shape at f_1 has one node, the second mode shape has two nodes, the third mode shape has three nodes

Example 9.1

An aluminum diving springboard is observed to vibrate at about 1.2 Hz after a diver takes off from the board. If the board is modeled as a solid, homogeneous cantilever beam, how thick is the board? In SI units, regulations require a width of 50.8 cm and a length of 4.87 m.

- GIVEN: Aluminum: $E = 70$ GPa, $\rho = 2700$ kg/m^3
- FIND: Thickness based on first mode of vibration
  - Use Equation 9.1a: $f_1 = \frac{1}{2\pi}(1.875)^2\sqrt{\frac{EI}{mL^3}}$
  - Using consistent units: $1.2 = \frac{1}{2\pi}(1.875)^2\sqrt{\frac{(70e9)\left(\frac{1}{12}\right)(0.508)(t^3)}{(2700)(0.508)(4.87)(t)(4.87)^3}}$
  - Solve for t: $t = 3.5$ cm
- The moment of inertia for a rectangular beam is based on its cross-section and should be memorized
- The springboard vibrates in the first mode shape, with a single node at the cantilever attach point

DERIVATION: Each structure has its own set of natural vibration frequencies based primarily on its unique characteristics of mass, stiffness and damping. A non-homogeneous structure is more complex, with possible variations in mass distribution, stiffness, damping, materials or geometric cross-sections. In many cases, simplified models of complex structures can provide theoretical insight into the natural vibration frequencies. A cantilever beam is one of the most basic continuous structures to model for vibrations, which requires a combination of static bending analysis and dynamic beam theory. An overview of the theoretical approach to derive Equations 9.1(a–c) follows. The detailed partial differential equation solution is not offered here. Several good sources are available for the detailed solution.[1]

- Motivation for derivation
  - Avoids pulling frequency equations (f_1, f_2, f_3) from *"thin air"*
  - Provides insight into the nature of continuous structure vibrations
  - Provides a detailed review of shear-moment diagrams

- Determining the natural frequencies and mode shapes of a cantilever beam has three main tasks
  1. Identify **relationships** among load, shear, moment, slope and deflection
     - Defines beam parameters in terms of the same dependent variable: y (position)
  2. Develop **static beam bending** theory
     - Determine static bending characteristics of a cantilever beam
     - Governing *statics* equation: $\Sigma F_y = 0$
  3. Develop **dynamic beam** theory
     - Accounts for time in vibrating beam
     - Governing *dynamics* equation: $\Sigma F_y = ma$
     - The above three tasks will result in the following **goal**:
  - Beam bending equation solution: $y(x) = A\cos(\beta x) + B\sin(\beta x) + C\cosh(\beta x) + D\sinh(\beta x)$
     - Where β is convenient constant, helpful to find solutions (roots) to the equation above
     - β contains the beam's physical and material properties as a constant
     - As a trigonometric argument, β also contains a repeating (harmonic) frequency value
     - $\beta^4 = \dfrac{\omega_n^2 m}{EI}$
     - Once β is resolved, the above equation can be solved for ω_n
     - $\omega_n = \beta^2 \sqrt{\dfrac{EI}{m}}$, a root ratio of stiffness to mass (density), characteristic of frequency
  - The solution equation describes the *bending of a beam at any point*
     - Describes vibration amplitude excursion, y, as a function of position, x, along the beam
     - Several roots (eigenvalues) satisfy the solution; these roots are the frequencies
  - Derive equation → Solve roots → Get natural frequencies: $f_1, f_2, f_3, \ldots$
     - Where $f = \omega_n / 2\pi$

SHEAR AND MOMENT EQUATIONS

The three tasks to develop the natural frequencies of the cantilever beam begin by creating a shear and moment diagram. The following development works through a detailed internal load analysis to create the diagram while also establishing the fundamental relationships identified in Task 1. Figure 9.2 provides the beam geometry for the desired relationships. By memorizing *the Task 1* relationships between derivatives, accurate shear and moment diagrams can also be constructed without a detailed internal load analysis. While the more comprehensive internal load analysis is used here, the simplified approach to shear and moment diagrams is discussed in the chapter on stress and strain.

- **TASK 1: Relationships of dependent variable, *y*, as a function of position, *x***
  - Deflection: y
  - Slope (first derivative): y'

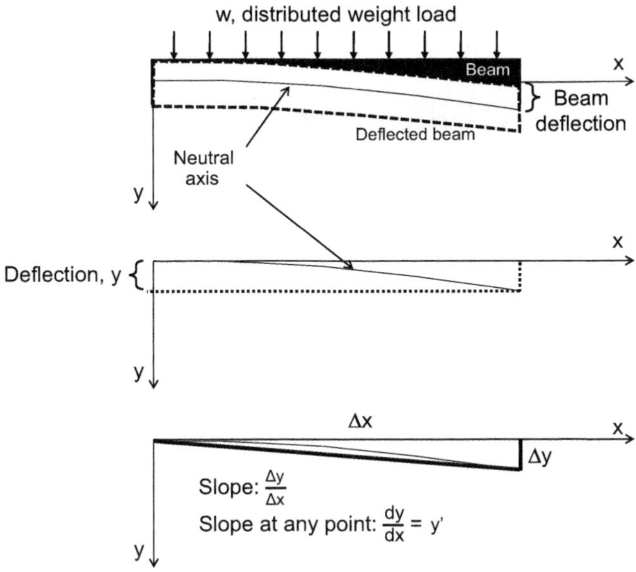

FIGURE 9.2 Cantilever beam bending geometry.

  - Moment (second derivative): y''
  - Shear (third derivative): y'''
  - Load (fourth derivative): y''''
- Begin with an x–y coordinate system and the simple bending geometry in Figure 9.2
  - Determine first two (easy) relationships: y and y'
  - Choose $+y$ in down direction since the cantilever beam will deflect down from the weight
- The first two relations are straight from the geometry of Figure 9.2
  - Deflection: y
  - Slope: y'
- Internal load analysis is necessary for the next three relationships
  - Moment: y''
  - Shear: y'''
  - Load: y''''
- *Any* beam configuration analysis is valid for developing relationships
  - Cantilever, simply supported, …
  - The analysis results in relationships between moment and shear that are common to any beam
- Resolve the reaction forces from the FBD per Figure 9.3
- Create shear and moment diagrams
- With reaction forces resolved, create a virtual cut to analyze forces at any point in Figure 9.4
  - This will resolve the shear and moment equations for any point within the beam

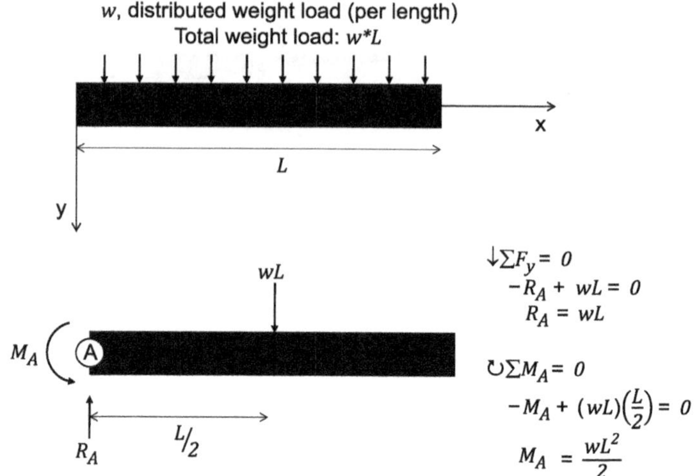

FIGURE 9.3 Cantilever beam reaction forces.

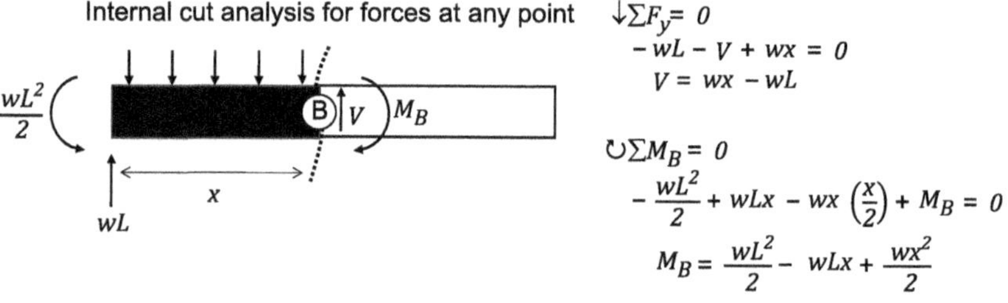

FIGURE 9.4 Cantilever beam shear and moment equations.

- Generalized shear and moment equations have now been developed from Figures 9.3 and 9.4
  - Shear: $V = wx - wL$
  - Moment: $M = \dfrac{wL^2}{2} - wLx + \dfrac{wx^2}{2}$
  - These equations can be used to plot shear–moment (V–M) diagrams from $x = 0$ to $x = L$
    - V–M diagram is not offered here since it is not relevant to derivation
- Take derivative of shear equation, V, with respect to "x"
  - $\dfrac{dV}{dx} = \dfrac{d}{dx}(wx - wL) = w \leftarrow$ Result is the load
  - **Therefore V′ = load**
- Take derivative of moment equation, M, with respect to "x"
  - $\dfrac{dM}{dx} = \dfrac{d}{dx}\left(\dfrac{wL^2}{2} - wLx + \dfrac{wx^2}{2}\right) = wx - wL \leftarrow$ Result is the shear equation

TABLE 9.1
Cantilever Beam Relationships Established up to This Point

Physical Property	Relation	Derivatives of Y
Deflection	y	y
Slope	y'	y'
Moment		
Shear	M'	
Load	V' or M''	

- **Therefore $M' = V$**
- *Note that the second derivative of moment also equals load:* M'' **= Load**
- Relationships thus far are independent of material properties and are entered into Table 9.1
- **TASK 2: Static beam bending**
  - It is necessary to make the connection between M and y'' in Table 9.1
  - This development accounts for material and physical properties
  - There are two general approaches
    - Euler–Bernoulli (E–B) beam theory: no rotational effects, theoretically stiffer beam used in this development
    - Timoshenko beam theory: includes rotational effects, more difficult but more accurate (not used for this development)
- E–B beam theory provides the following
  - A static, analytical model of how a beam bends under load
  - Relates the second-order derivative of deflection, y, to moments, M
  - The E–B beam equation is only *presented* in Equation 9.1
    - Refer to companion Lab #2 for the E-B derivation

$$EI\left(\frac{d^2y}{dx^2}\right) = M \leftrightarrow EI\left(y''\right) = M \tag{9.1}$$

- Equation 9.1 equates y'' to M as a function of material properties (E and I)
- Table 9.1 can be completed by relating moment to deflection per Table 9.2
- The relationships are now complete
  - Moment: $y'' \leftarrow$ As given by the E–B equation
- Since $M = y''$ (E–B) and M' = shear
  - Take the derivative of moment to get to shear such that shear equals y'''
    - Shear: $V = y''' \leftarrow$ Since $M = y''$, shear is M' and $M' = y'''$ = shear
    - Load: $y'''' \leftarrow$ The derivative of shear is load
- Since the E–B equation relates moment to load, it picks up a negative sign since the bending load creates a negative moment by sign convention
- When integrating, the *constant of integration is the reaction force* constant

TABLE 9.2
Cantilever Beam Relationships

Physical Property	Relation	Derivatives of *Y*	
Deflection	y	y	
Slope	y'	y'	
Moment	**M**	$\mathbf{y''}$	**← Euler–Bernoulli**
Shear	M'	y'''	
Load	V' or M''	y''''	

- **TASK 3: Dynamic beam bending**
  - Statics: $\Sigma F = 0$
  - Dynamics: $\Sigma F = ma$
    - Decompose acceleration into its second time derivative
    - $\Sigma F = m\left(\frac{d^2y}{dt^2}\right)$
- Qualitative analysis using all of the previous information
  - Note that **TASK 2** gave ***LOAD as the second derivative of MOMENT***
  - Therefore, take the second derivative of the E–B equation to get load (ΣF)
    - E–B: $M = EIy''$
    - E–B second derivative: $\frac{d^2}{dx^2}(M) = \frac{d^2}{dx^2}(EIy'')$
    - $M'' = \frac{d^2}{dx^2}(EIy'')$ ← RHS of equation *must* be load since M'' *is* load
  - Substitute load (M'') into dynamics equation (summation of forces is load)
    - $\Sigma F = m\left(\frac{d^2y}{dt^2}\right)$
    - $\frac{\mathbf{d^2}}{\mathbf{dx^2}}(\mathbf{EIy''}) = m\left(\frac{d^2y}{dt^2}\right)$
    - $\left[\frac{EI}{m}\right]\frac{d^2}{dx^2}(y'') = \frac{d^2y}{dt^2}$ ← Re-arrange above to pull material properties over to LHS
    - $\left[\frac{EI}{m}\right]\frac{d^4}{dx^4}(y) = \frac{d^2}{dt^2}(y)$ ← Clean up the differential notation
    - Note that "y" is a function of both *x*-position, *x*, *and* time, *t*; make this clear…
    - $\left[\frac{EI}{m}\right]\frac{d^4}{dx^4}[y(x,t)] = \frac{d^2}{dt^2}[y(x,t)]$ ← Partial differential equation of two variables
    - The above form is a PDE, and *separation of variables* can be used

- $\left[\frac{EI}{m}\right]\frac{d^4}{dx^4}[Y(x)T(t)]=\frac{d^2}{dt^2}[Y(x)T(t)]$ ← Two separate functions substituted
- Use separation of variables to pull out one function as constant, then other as constant
- $T(t)\left[\frac{EI}{m}\right]\frac{d^4}{dx^4}Y(x)=Y(x)\frac{d^2}{dt^2}T(t)$ ← Derivatives of x on LHS and t on RHS
- $\frac{\left[\frac{EI}{m}\right]\frac{d^4}{dx^4}Y(x)}{Y(x)}=\frac{\frac{d^2}{dt^2}T(t)}{T(t)}$ ← Divide for like terms on LHS and RHS
- Use the separation of variables technique to evaluate both terms to a constant, c
- Consider only the separated function of position (LHS) set to a constant
- $\frac{\left[\frac{EI}{m}\right]\frac{d^4}{dx^4}Y(x)}{Y(x)}=c$
- Re-arrange as a fourth-order ODE
- $\frac{d^4}{dx^4}Y(x)-c\frac{m}{EI}Y(x)=0$
- The above equation is not completely set up for integration
  - The most overlooked fact in derivations is that *mass* is <u>not</u> constant during integration
  - Integration occurs in the x-direction and at $x=0$, $m=0$
  - Mass must therefore be *per unit length* and subject to integration
  - Both E and I and the natural frequency remain constant along the length
  - Since $m=\rho V$ and $V=Ax$, mass per unit length is: $m=\rho A$
- $\frac{d^4}{dx^4}Y(x)-c\frac{\rho A}{EI}Y(x)=0$ ← Mass per unit length substitution
- With some foresight into the solution technique, choose
  - c as a function of natural frequency, ω_n
  - Combine all constants into a convenient variable, $\beta^4=\frac{\omega_n^{\,2}\rho A}{EI}$
- Clean up the result for integration and boundary condition evaluations
- $y''''(x)-\beta^4 y(x)=0$

• The above equation is solved as: $y(x)=A\cos(\beta x)+B\sin(\beta x)+C\cosh(\beta x)+D\sinh(\beta x)$
  - The constants are evaluated by the beam boundary conditions at $x=0$ and $x=L$
  - Moment and shear are zero at the free end; slope and deflection are zero at the fixed end

- The following condition is a result of applying the above boundary conditions
- $\cos(\beta L)\cosh(\beta L) = -1$ (length, L, shows up when solved at $x = L$)
- The above equation is transcendental, requiring numerical or other methods to solve
- The following first three periodic values will satisfy the transcendental equation
  - $\beta L = 1.8751, 4.6941, 7.8548, \ldots$
- Given the earlier convenient definition: $\beta^4 = \frac{\omega_n^{\,2}\rho A}{EI}$, re-arrange: $\omega_n = \sqrt{\frac{\beta^4 EI}{\rho A}}$
- Substitute $\beta = \frac{1.8751}{L}$ for the first periodic solution, along with $\omega = 2\pi f$
- $f_1 = \frac{1}{2\pi}(1.8751)^2\sqrt{\frac{EI}{\rho AL^4}}$
- Since integration is complete, the mass term can be recovered above, if desired
  - Since $m = \rho AL$, an L from L^4 can be "*borrowed*" to achieve the final form
  - $f_1 = \frac{1}{2\pi}(1.8751)^2\sqrt{\frac{EI}{mL^3}}$

• Above approach yields the harmonic frequencies of the cantilever beam in Equations 9.1(a–c)

EXERCISES

1. Circle each of the following to make the relationships correct:
   a. Shear is the (first, second or third) (DERIVATIVE or INTEGRAL) of load.
   b. Shear is the (first, second or third) (DERIVATIVE or INTEGRAL) of moment.
   c. Deflection is the (first, second or third) (DERIVATIVE or INTEGRAL) of moment.
   d. Load is the (first, second or third) (DERIVATIVE or INTEGRAL) of moment.
2. A simply supported beam with a distributed weight load has a shear function of $V = 16 - 4x$, where x is position along the beam to a maximum length of L. What is the moment function?
3. TRUE or FALSE. A beam requires a forcing function to vibrate at a frequency higher than the first fundamental frequency?
4. Refer to the beam in Figure 9.4. Use the given shear and moment functions to draw a shear-moment diagram.

5. A polycarbonate beam has the following dimensions: $L = 18\,\text{cm}$, $W = 2\,\text{cm}$, $t = 2\,\text{mm}$. What frequency is required to excite the beam at its third natural frequency?
6. A 15 cm aluminum cantilever beam is desired to resonate at a frequency of 256 Hz. Assuming a square beam, what dimensions are required?

NOTE

1 Irvine, T. (2019). *An introduction to shock and vibration response spectra.* http://www.vibrationdata.com/tutorials2/Tom_book_12_1_19.pdf.

10 Waves

Waves surround us. Information is carried in waves, and understanding wave behavior is key to understanding many interconnected properties in our world. Lab #3 explores mechanical wave properties carried in both a gas (air) and a solid (a cord).

Waves are constantly in motion and carry both energy and information. To utilize waves in design applications from musical instruments to mobile phone communications, it is necessary to understand the fundamental relationships between frequency, velocity and wavelength. Since waves are dynamic, one technique to study behavior is to utilize the constructive and destructive interference patterns of waves to create a standing wave that has the characteristics of the source waves.

WAVE TYPES

While electromagnetic waves do not require a medium for propagation, mechanical waves do. Solids, liquids and gases propagate waves as energy is transferred between the bonds of atomic and molecular structures.

- Three main types of waves (there are other less common types)
  - Transverse
  - Longitudinal
  - Surface (not discussed here)
- Transverse waves are depicted in Figure 10.1
  - Wave travels away from disturbance
  - Particles of medium are displaced perpendicular to the direction of the wave
  - Example: water waves rippling from rock thrown into a calm pond
- Longitudinal waves are depicted in Figure 10.2
  - Wave travels away from disturbance (same as transverse)
  - Particles of medium are displaced parallel to the direction of the wave
  - Example: sound traveling through air from sharp noise
- Standing "waves" are depicted in Figure 10.3
  - Not a *type* of wave
  - Result of interference patterns of either transverse or longitudinal *source waves*
  - Appears motionless to an observer (fixed nodes and anti-nodes)
  - Source waves travel until they run out of energy or hit an obstruction
  - If waves hit obstructions, they reflect back in opposite direction
  - Oncoming and reflected waves interact with each other, creating an interference pattern

DOI: 10.1201/9781003329237-10

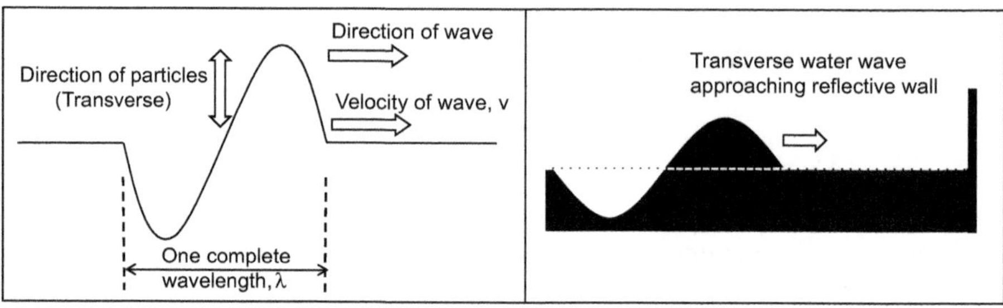

FIGURE 10.1 Transverse wave.

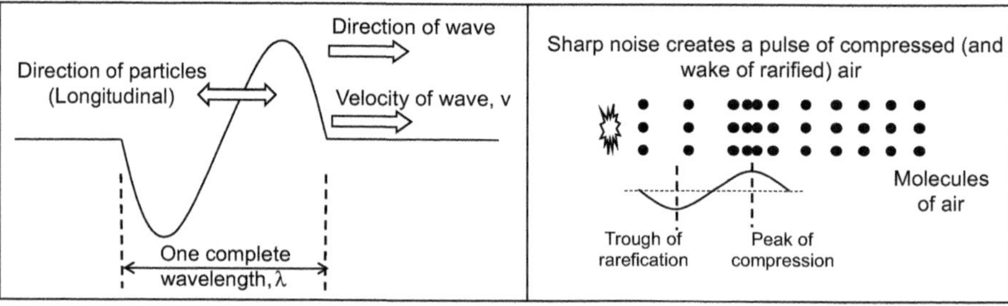

FIGURE 10.2 Longitudinal wave.

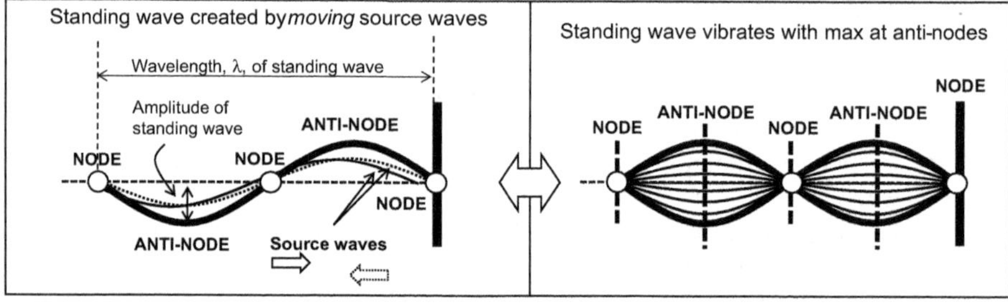

FIGURE 10.3 Standing wave nodes and anti-nodes created by moving transverse or longitudinal source waves reflected upon themselves after hitting a bound.

- Conditions for a standing wave
  - Source waves are continuous and constantly move back and forth (sourced and reflected)
  - Source wavelength and reflective endpoint lengths must be correct intervals of each other
  - Standing wave *appears* stationary
    - Amplitude oscillates between maximums at anti-nodes
    - Nodes appear as fixed points of zero amplitude on the stationary wave

  - All waves are functions of time, and usually just *snapshots* in time are depicted
    - Figure 10.3 shows a typical wave snapshot
    - Figure 10.3 also shows oscillatory movement as nodes and anti-nodes form
- If source wavelength and bounding space are not at correct intervals from each other
  - Interference pattern is irregular
  - A stationary wave cannot develop
- Standing waves are of interest because
  - A standing wave has all the properties of the source waves
  - Analyzing a source wave is easier if a stationary wave is created
  - For analysis purposes, standing waves and source waves are synonymous
  - Standing waves create sound amplifications with anti-nodes

BOUNDARY CONDITIONS

Boundary conditions are how the endpoints of a wavelength may or may not be constrained.

- Consider a single wave constrained with fixed-fixed endpoints in Figure 10.4
  - To create a standing wave, a wavelength must *neatly fit* within the bounds
  - The min or max amplitude must occur at a bound (fixed or free)
  - These min and max points occur at ¼, ½, ¾, or one full wavelength
  - Min and max points also occur at higher wavelength multiples as the wavelength repeats
  - These key wavelength bounds (fractions or multiples) permit reflected waves to exactly interfere with traveling inbound waves for constructive and destructive interference
- Bounds
  - Either end can be fixed or free
    - Fixed-fixed (also, closed-closed)

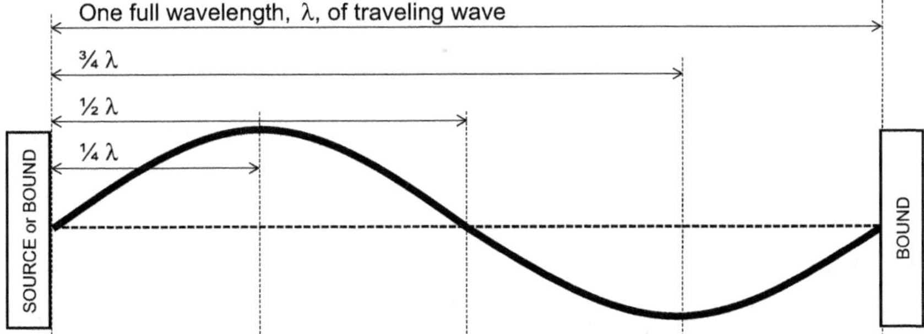

FIGURE 10.4 Intervals of a wavelength within a bound.

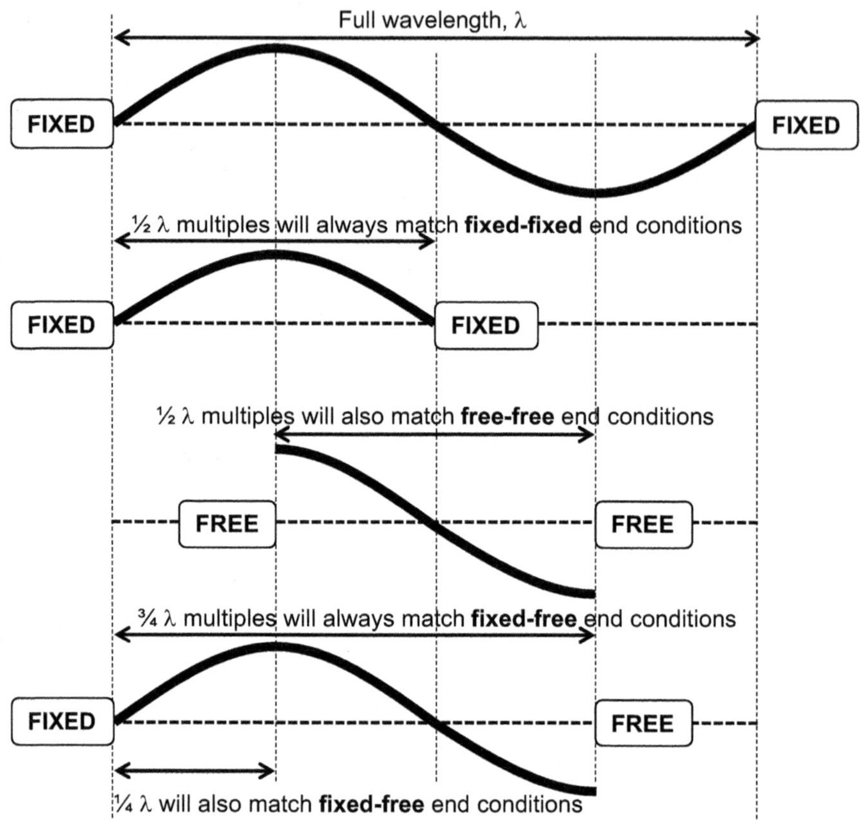

FIGURE 10.5 Boundary conditions to create a standing wave.

  - Fixed-free (also, closed-open)
  - Free-free (also, open-open)
- A node or anti-node must be allowed to form at a bound to create a clean standing wave
  - For a fixed end → only a node (not anti-node) can occur
  - For a free end → only an anti-node (not node) can occur
  - These occurrences can happen at different portions of the wavelength per Figure 10.5
    - These portions are specific fractions or multiples of the wavelength
    - Not just integer multiples

WAVE EQUATIONS

Two fundamental equations are available to describe wave behavior: the *properties* equation to relate velocity, frequency and wavelength, and the *displacement* equation to plot a wave solution in the time domain.

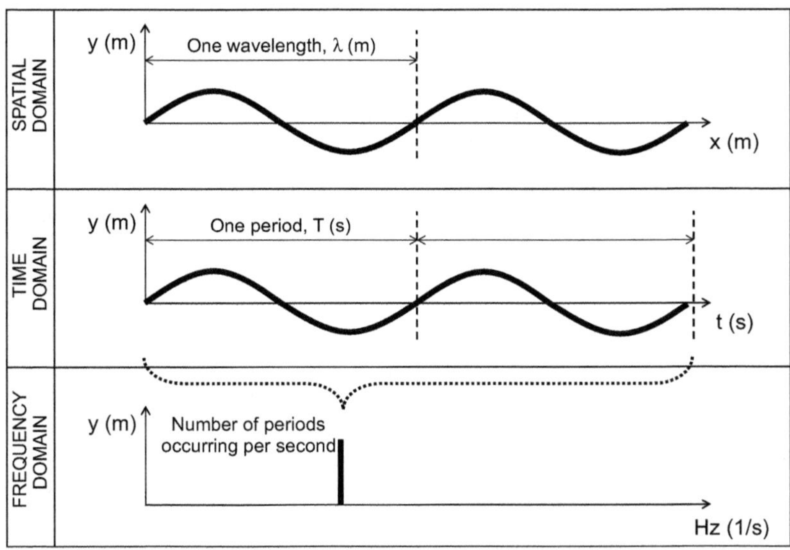

FIGURE 10.6 Plotting a wave in different domains.

- *Properties* (two properties must be known to solve for the third property)

$$v = f\lambda \tag{10.1}$$

where: v = velocity
f = frequency
λ = wavelength

- *Displacement*

$$y = A\sin(kx - \omega t) \tag{10.2}$$

where: y = vertical displacement at a given time and position
A = maximum vertical amplitude
k = wave number, $2\pi/\lambda$
ω = angular frequency

  - Wave velocity, v, is wave number ÷ angular frequency (k/ω)
  - Wave number is radians per wavelength, $2\pi/\lambda$
- Wave frequency, wavelength and velocity are all interrelated
- Waves can be represented in different domains shown in Figure 10.6
  - Spatial domain: displacement (meters) versus length or distance (meters)
    - Used to measure or depict wavelength, λ
  - Time domain: displacement (meters) versus time (seconds)
    - Used to measure or depict period, T
    - One cycle is equivalent to one period
    - Number of periods can be counted to determine frequency, $f = 1/T$

- Frequency domain: displacement (meters) versus (Hertz)
  - Frequency, $f = 1/T$ has units of s^{-1}
  - Angular frequency, $\omega = 2\pi f$ has units of *radians per second*
  - Always ensure unit consistency in equations
    - Use frequency, f, with *counts* or *cycles*
    - Use frequency, f, directly in the relationship of Equation 10.1
    - Convert f to ω for unitless values when combining with numbers

Example 10.1

A flute is a wind instrument open at both ends. If designing a flute, what is the shortest length of the flute to play the first fundamental frequency of middle-C at 262 Hz?

- GIVEN: Velocity of sound in air is approximately 340 m/s
- FIND: Length of wavelength for $f = 262$ Hz and shortest length for open-open BCs
  - Use Equation 10.1: $v = f\lambda$
  - Solve for wavelength: $340 = 262\lambda \rightarrow \lambda = 1.3\ m$
  - 1.3 m is a single wavelength for the given frequency and speed of sound
  - Figure 10.5 does not specifically show one full wavelength fitting into the free-free boundary conditions, but it is possible to see this by shifting the full wavelength by one-quarter
  - The shortest possible portion of a wavelength matching the free-free BCs is the third example in Figure 10.5 which is half a wavelength
  - Multiply the wavelength by one-half to get the shortest match to the necessary BCs
  - Shortest length is $0.5(\lambda = 1.3\text{ m}) = 0.65$ m
- A correction factor is required for a complete answer (discussed in companion Lab #3)

DERIVATIONS. Wave properties in Equation 10.1 is a simple analysis of relationships.

- Properties equation: $v = f\lambda$
  - Consider the wave in Figure 10.7 plotted as snapshots in time in the spatial domain
  - A point of interest moves with velocity from one position at t_0 to another position at t_1
    - Velocity is m/s
    - One wavelength, λ (m), is always traveling at a rate of one period by definition
      - Velocity is one wavelength per period: $v = \lambda/T$,
      - where $T = 1/f$
      - Therefore, $v = f\lambda$

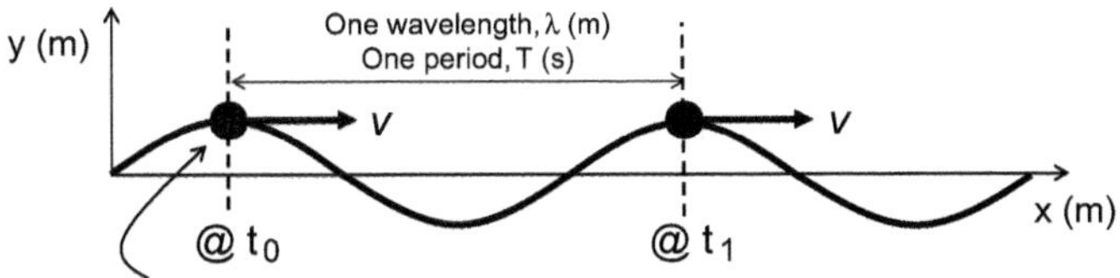

FIGURE 10.7 Velocity of a point on a wave at different time snaps.

The wave displacement equation derivation can be approached in a few different ways. The most common approach is the one-dimensional partial differential equation (PDE) modeled from a string under tension. The PDE development and solution are beyond the scope of this discussion, but the basic results and relationships are relevant.

- Displacement equation: $y = A\sin(kx - \omega t)$
  - The sine argument $(kx - wt)$ must resolve as radians (cannot take the *sine* of meters)
  - Where k is defined as *wave number*: $k = {}^{2\pi}\!/_{\lambda}$
  - k merely provides the repeating 2π sinusoidal constant based on wavelength, λ (m)
  - x is position (m) along the *x-axis* at any time, t, of interest
  - The combined term, kx, cancels units of meters for a dimensionless sine argument
  - Where ω is the angular frequency: $\omega = 2\pi f$
    - Since $f = {}^{v}\!/_{\lambda}$, angular frequency can be written in terms of v and $\lambda : \omega = {}^{2\pi v}\!/_{\lambda}$
  - t is the time (s) at any time of interest
  - The combined term, ωt, cancels units of seconds for a dimensionless sine argument
- Displacement equation is a solution to the 1D wave equation
  - Wave equation is PDE of two variables, position and time
    - 1D PDE wave equation: $\frac{\partial^2 y}{\partial x^2} = \left(\frac{T}{\mu}\right)\frac{\partial^2 y}{\partial t^2}$, where
    - T is the string tension; μ is the string mass per unit length
  - Displacement equation can be verified as a solution to wave PDE by taking derivatives of the displacement equation and substituting derivatives into PDE
    - $y = A\sin(kx - \omega t)$
    - Derivative of y with respect to x:
    - ${}^{\partial y}\!/_{\partial x} = kA\cos(kx - \omega t)$
    - ${}^{\partial^2 y}\!/_{\partial x^2} = -k^2 A\sin(kx - \omega t)$
    - Derivative of y with respect to t:
    - ${}^{\partial y}\!/_{\partial t} = -\omega A\cos(kx - \omega t)$

- $\partial^2 y / \partial x^2 = -\omega^2 A \sin(kx - \omega t)$
- Make these substitutions into original PDE
- $-k^2 A \sin(kx - \omega t) = -\left(\frac{T}{\mu}\right)\omega^2 A \sin(kx - \omega t)$
- $-k^2 = -\left(\frac{T}{\mu}\right)\omega^2$
- $\frac{T}{\mu} = \left(\frac{\omega}{k}\right)^2$

- Previous definitions of ω and k can be substituted into equation
  - $\frac{T}{\mu} = \left(\frac{2\pi v/\lambda}{2\pi/\lambda}\right)^2$
  - Simplifying the above set of substitutions results in wave velocity in Equation 10.3

$$v = \sqrt{T/\mu} \tag{10.3}$$

where: v = wave velocity
T = string or cord tension (not period)
μ = mass per unit length

Equation 10.3 is significant because it shares the form of other velocity equations

- In general terms: $v = \sqrt{Rigidity\ (or\ Elasticity) / Density}$
- Velocity of wave in solids: $v = \sqrt{B/\rho}$, where B is the elastic bulk modulus
- Velocity of wave in air: $v = \sqrt{\gamma RT/M}$, where γRT is the elastic bulk modulus

EXERCISES

1. The speed of sound in metal travels about 6000 m/s compared to about 340 m/s in air. For a frequency of middle-C at $f = 262$ Hz, answer the following.
   a. What is the wavelength in metal?
   b. What is the wavelength in air?
   c. Why is the wavelength longer in metal than in air?
2. A string instrument has a length of 0.81 m. The string has a mass per unit length of 5.8*e*–3 kg/m. To achieve a frequency of 82 Hz, how much tension is required?

3. Assume all the relationships and results from Question #2. If the string is plucked with a displacement of 1 cm, write the wave displacement function, Equation 10.2, for the string.
4. Remove time dependency from the result in Question #3 by setting $t = 0$ and plot the function. Determine the wavelength from the function plot.
5. A string has a mass per unit length of 4.8 kg/m and is stretched with a tension of 40 N between two fixed points 2 m apart. The string is vibrating where a standing wave has developed with four nodes. Answer the following.
   a. How many wavelengths are visible?
   b. What is the wavelength?
   c. What is the vibration frequency?

11 Stress and Strain

Engineers design structures, and these structures should not fail in service. Several tools are available to the engineer, such as stress-strain curves and load cell analysis, to ensure materials are chosen correctly and perform as expected. Lab #4 employs these tools to validate material properties in tension and bending.

Structural materials such as steel and aluminum are stiff and strong, but only up to a point. It is important to design structures within limits of material capabilities so structural materials do not bend or break during service. Yield stress and elastic modulus are two of the most fundamental considerations when choosing a design material. Both yield stress and elastic modulus are material properties that can be verified through experimentation. Several techniques are available to assess these strength-related material properties, but the most fundamental approach is the stress-strain plot.

Tension. Stress is defined by tensile loading. Although bending loads are different from tensile loads, tension is still the defining method for stress.

- Stress and strain definitions
  - Stress (σ): force applied per cross-sectional area (units of *pressure* on a material)
  - Strain (ε): change in length over original length (*stretch* of a material)
- Hooke's law for linear-elastic behavior
  - Solid materials and coiled springs behave similarly
  - A solid material can deform and then spring back to its original condition
  - Linear relationships
    - Standard equation of a line $y = mx + b$, where m is the slope
    - $F_{SPRING} = k*x$ (Force = Proportional constant * Displacement)
    - $\sigma = E*\varepsilon$ (Stress = Proportional constant * Strain displacement)
    - Not valid beyond a material's proportional limit stress
- Stress-strain diagrams
  - Plot of a material's stress on vertical axis against strain on horizontal axis
    - Plots describe a material's deformation due to applied stress
    - Plots are different (unique) for different materials
    - Figure 11.1 shows stress-strain plots typical of steel and aluminum
  - Two general regions
    - Elastic: material will return to its original shape after strain deformation
    - Plastic: material becomes permanently deformed
  - Particular points of interest
    - Pt A: linear-elastic material behavior with slope E (elastic modulus)
    - Pt B: proportional limit where linear relationship begins to break down

DOI: 10.1201/9781003329237-11

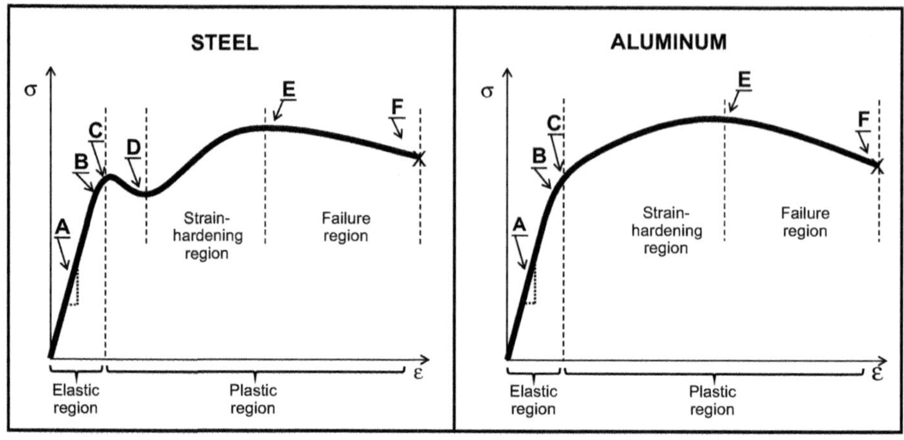

FIGURE 11.1 Stress-strain behavior of steel and aluminum.

- Pt C: yield stress (σ_{YS}) typical of maximum design limit stress (no safety margin)
- Pt D: "*Dip*" unique to steel
  - Imperfections in the molecular lattice (dislocations) begin to slip as the material stretches
  - Material loses strength and stress dips from the maximum yield stress point
  - Aluminum does not exhibit dislocation movement dip due to grain structure
- Pt E: ultimate stress (σ_{ULT}) is the maximum stress limit of a material
  - Past Pt D, dislocations *pile up* as they run out of options for continued movement
  - Strain hardening begins as dislocations stop slipping
  - Material begins strengthening again to its maximum at Pt E
- D: Breaking stress (σ_F) where material fails completely

- Stress-strain curves vary based on various material properties, as shown in Figure 11.2
- Two common load conditions
  - Tension: defines stress and is easy to analyze material properties using load cells
  - Bending: common load condition for beams
  - Other common load conditions: compression, buckling, shear and torsion
- Fundamental relationships of tensile and bending stress and strain in Equations 11.1–11.3

$$TENSILE\ STRESS\text{: } \sigma = \frac{F}{A}\ or\ \sigma = \frac{P}{A} \tag{11.1}$$

$$BENDING\ STRESS\text{: } \sigma = \frac{Mc}{I} \tag{11.2}$$

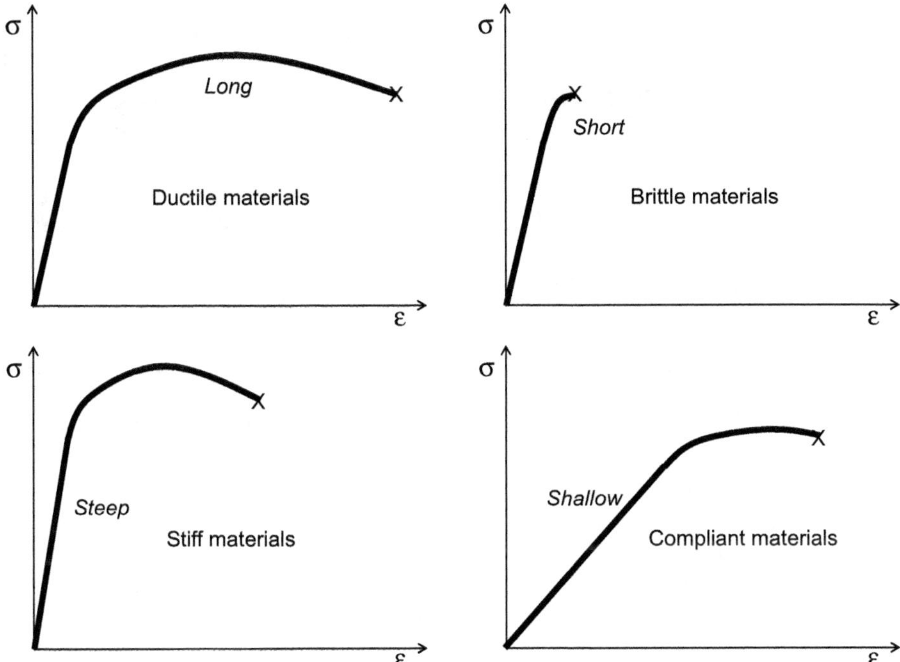

FIGURE 11.2 Typical stress-strain curves based on various material properties.

$$STRAIN\text{: } \varepsilon = \frac{\Delta L}{L_0} \text{ also } \varepsilon = \frac{\sigma}{E} \quad (11.3)$$

where: σ = stress
F or P = applied load
A = cross-sectional area
M = applied moment
c = half-thickness of beam (neutral axis to edge)
I = cross-section moment of inertia
ε = strain
ΔL = change in length
L_0 = original length

- DERIVATION. The bending stress equation derivation is presented in companion Lab #4
  - Stress is *defined* as $\sigma = \frac{Force}{Area}$
  - This stress definition equation is used in the analysis of beam bending
  - Refer to Lab #4 to show how $\sigma = \frac{P}{A}$ is rewritten for bending as $\sigma = \frac{Mc}{I}$

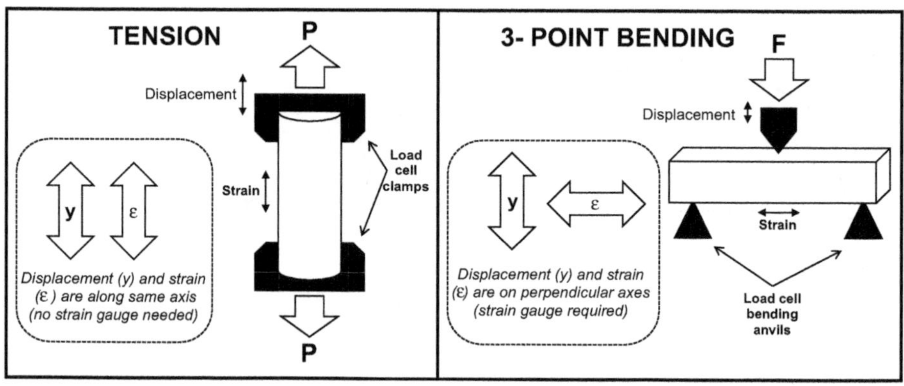

FIGURE 11.3 Typical load cell configurations to measure tensile and bending stress and strain.

- Measuring stress and strain to create a stress-strain plot
  - Determine goal of measurement
    - Complete stress-strain plot → destructive test of sample to failure
    - Verifying elastic modulus → non-destructive test within the elastic region
  - Material test specimens are placed into a test load cell as shown in Figure 11.3
  - Load cell applies a force to the test specimen and measures displacement

Example 11.1

Consider a non-destructive tensile test to determine elastic modulus, E, on a cylindrical aluminum test specimen. A standard load cell is used with the ability to plot applied load versus displacement. What is the maximum allowable load to ensure a non-destructive test?

- EXPERIMENTAL OBJECTIVE: Create stress-strain plot to determine elastic modulus, *E*, without exceeding yield stress
- GIVEN: A 6061T6 aluminum rod, 5 mm in diameter, 5 cm in length
- FIND: Calculate maximum applied load to remain within elastic region
  - Look up yield stress for 6061T6 aluminum: 276 MPa
  - Calculate maximum load cell *force* based on material yield stress
    - Area (5 mm diameter): $A = \pi\left(\frac{5\times10^{-3}}{2}\right)^2 = 1.96\times10^{-5}\ m^2$
    - Stress equation:
      $$\sigma = \frac{F}{A} \rightarrow 276\times10^6\ Pa = \frac{F}{1.96\times10^{-5}\ m^2} \rightarrow F = 5.4\ \text{kN}$$
  - Maximum load cell force should be less than 5.4 kN to remain in elastic region

- Load cells measure both force and displacement, which are aligned with tensile strain
- Load cells provide a direct measure of *tensile* strain via displacement along the same axis
- Tensile stress distribution is essentially constant throughout the material
- To complete the experimental objective:
    - Record load cell applied force and displacement at desired intervals *below max load*
    - Convert load cell applied force to stress with tensile stress equation, $\sigma = \frac{P}{A}$
    - Convert load cell displacement to strain with strain equation, $\varepsilon = \frac{\Delta L}{L_0}$

BENDING

Using a load cell for bending is less straightforward compared to tensile tests. Since load cell displacement and strain are not along the same axis, bending stress must be rewritten in terms of the applied load.

- Two primary issues to resolve for bending tests with a basic load cell
    1. Rewrite the bending stress equation in terms of force instead of moment
    2. Measuring strain since load cells do not offer a direct bending strain measure
- LOAD CELL ISSUE #1: Load cells typically apply force and not moments
    - 3-point bending through force application is a common load cell bending configuration
    - Bending stress equation must be rewritten in terms of applied force (not moment)
    - Rewriting bending stress equation in terms of force is *unique to the load* condition
    - A shear-moment diagram with specific load conditions is helpful in the process
- GOAL: Replace moment, M, in bending stress equation ($\sigma = \frac{Mc}{I}$) with force (F or P)
    - Draw a shear-moment (V-M) diagram with a particular load condition
    - Use *maximum moment* from the moment diagram in the bending stress equation
    - Simplify result in terms of applied load

SHEAR AND MOMENT DIAGRAMS

There are two approaches to creating a V-M diagram: (1) the more comprehensive method of developing the shear and moment equations and plotting the equations as functions or (2) a quick method of assessing key points and recognizing the contributions of integration or differentiation.

- Creating a *V-M* diagram (two approaches)
  1. Develop general shear and moment equations
     - Draw a free body diagram (FBD) of a specific beam load condition
     - Resolve reaction forces
     - Create a *virtual cut* in the beam (previously discussed in continuous systems vibrations)
     - Develop generalized shear and moment equations from the virtual cut for the entire beam
     - Plot generalized shear and moment curves using equations
  2. Quickly assess values at key points (approach presented here)
     - Draw FBD of beam with specific beam load condition
     - Resolve reaction forces
     - Solve values only at key points
     - Draw curve shapes based on derivative relationships from continuous systems chapter

Example 11.2

Consider a cantilever beam with an applied point load as shown in Figure 11.4. Quickly create a shear-moment (*V-M*) diagram to identify maximum loads and moments at key positions.

- Draw beam load configuration, FBD and *V-M* axes per Figure 11.4
- Include the deflection axis, if desired
- Recall derivative and integral relationships as indicated in Figure 11.4

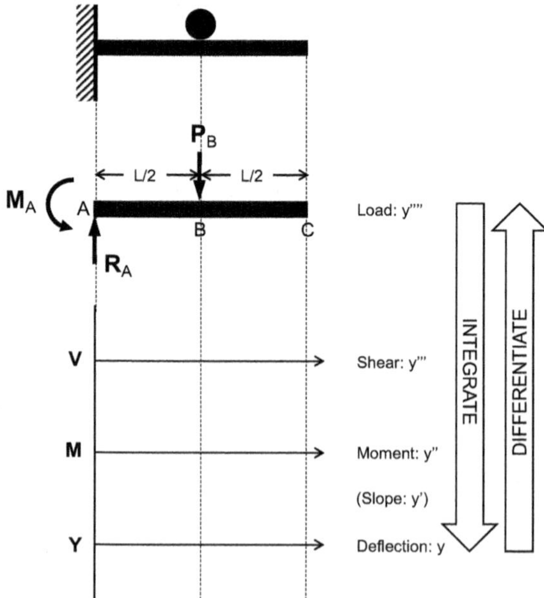

FIGURE 11.4 Shear-moment-deflection graph.

There are TWO *separate* sign conventions to consider when building a *V-M* diagram: (1) the FBD sign convention, and (2) the *V-M* sign convention. The FBD and *V-M* sign conventions should be considered separately. If the FBD sign convention dictates forces are summed positively in the "down" direction, this does not necessarily mean internal shear forces acting downward are also positive. For the *V-M* sign convention, it is usually best to work from a reaction force, either left or right, and consider how the positive or negative convention *contributes* to the overall values *accumulating* in the beam, as shown in Figure 11.5.

- Two separate sign conventions to consider
  1. FBD direction sign convention
     - Designate positive directions for summing both forces and moments
     - The specified direction does not matter as long as it is consistent
     - If the result is negative, it means the force or moment was *drawn in the wrong direction*
     - This sign convention does not dictate positive or negative values on the *V-M* diagram
  2. *V-M* value sign convention referenced in Figure 11.5
     - Shear and moments can have changing positive or negative values along the beam
     - If a shear force contributes to clockwise beam rotation, the contribution is *positive*
     - If a shear force contributes to CCW beam rotation, the contribution is *negative*
     - If a moment contributes to a *"smiling beam,"* contribution is *positive*
     - If a moment contributes to a *"frowning beam,"* contribution is *negative*
     - This sign convention dictates positive and negative values on the *V-M* diagram

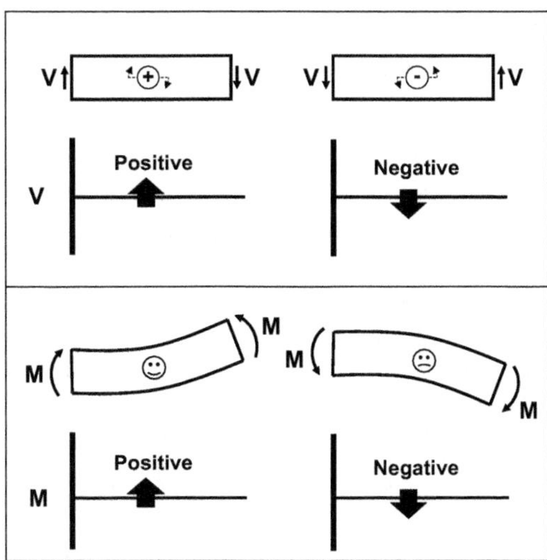

FIGURE 11.5 Shear-moment sign conventions.

- The two different sign conventions are *independent* parts of the solution process
  - Do not confuse and intermix the two separate sign conventions
- To continue building the *V-M* diagram for Figure 11.4, determine reactions for FBD
  - Sum forces in the positive *y*-direction (up): $(+\uparrow)\Sigma F_y = 0$
    - $+R_A - P_B = 0$
    - $R_A = P_B \leftarrow$ Reaction at Pt. A is just the applied load
  - Sum moments in the positive CW-direction: $(+\circlearrowright)\Sigma M_A = 0$
    - $-M_A + (P_B)\left(\frac{L}{2}\right) = 0$
    - $M_A = P_B L/2 \leftarrow$ Moment at Pt. A is applied load at the applied distance
  - Draw a shear (*V*) diagram as shown in Figure 11.6, following the *V-M* sign conventions for positive and negative in Figure 11.5
    - Pt. A: R_A is *correctly drawn up* based on FBD sign conventions
    - R_A is contributing to *positive clockwise rotation* based on the *V-M* sign convention
    - Draw shear upward (positive) by a magnitude of R_A on the shear plot
    - R_A has a magnitude of P_B based on static equilibrium; use the applied <u>load</u> value (P_B)
    - Draw a shear straight across until the P_B location since there is no change from Pt. A to B
    - Nothing occurs past Pt. B, so shear is back down to zero
  - Draw moment (*M*) diagram as shown in Figure 11.6, following *V-M* sign conventions for positive and negative in Figure 11.5
    - Pt. A: M_A is *correctly drawn* based on FBD sign conventions
    - M_A is contributing to a *negative frown* based on the *V-M* sign convention
      - Note that M_A is *not* pushing the beam up into a smile (just the opposite)
      - Moments require counteracting moments for static equilibrium
      - M_A will have a counteracting moment; together they create the frown
    - Draw the moment straight down (negative) by a magnitude of M_A on the moment plot
    - M_A has a value of $P_B L/2$ based on static equilibrium (use this load value on the diagram)
    - Note that nothing happens past Pt. B, so the moment must decrease to zero at B
    - Draw *some kind of line or curve* from M_A to zero at Pt. B per Figure 11.6
      - Based on the derivative relationships, moment is the integral of shear
      - Shear was drawn <u>straight</u> across with a <u>constant</u> value, say $V = 5$
      - The integral of a constant picks up an independent variable, $\int V = M = 5x$

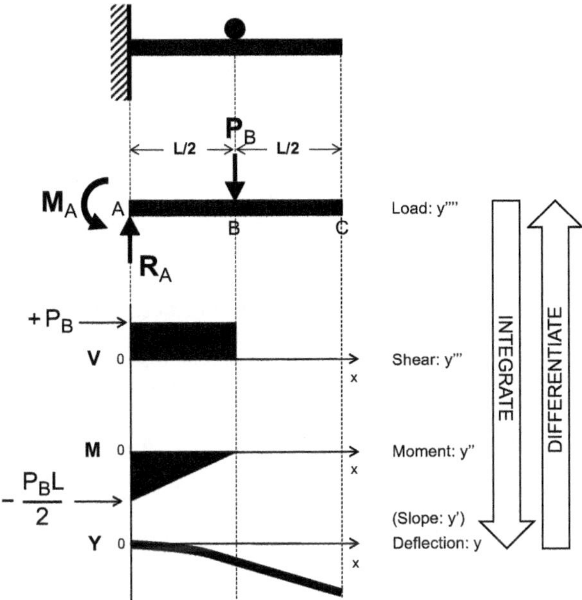

FIGURE 11.6 Completed shear-moment diagram.

  - *The line from negative M_A has a* <u>*positive linear*</u> *slope (+5) to the zero point (B)*
- Results of the *V-M* diagram are shown in Figure 11.6
  - *Maximum values* of moment and shear and their *locations* are now clear
  - Deflection (with slope) can also be drawn as desired

Example 11.3

Consider a non-destructive cantilever bending test to determine elastic modulus on a rectangular aluminum test specimen as shown in Figure 11.6. A test cell is used with the ability to plot applied load versus displacement. What is the maximum allowable load to ensure a non-destructive test?

- EXPERIMENTAL OBJECTIVE: Create stress-strain plot to determine elastic modulus, *E*, without exceeding yield stress
- GIVEN: A 7075T6 aluminum bar, 20 cm in length, 0.5 cm thick, 3 cm wide
- FIND: Calculate maximum applied load to remain within elastic region
  - Look up yield stress for 7075T6 aluminum: 503 MPa
  - Calculate the maximum load cell *force* based on material yield stress
    - Beam is in bending; use the bending stress equation: $\sigma = \frac{Mc}{I}$
    - Since the load cell applies force, rewrite the stress equation in terms of load (not moment)

- Maximum moment based on Figure 11.6 *V-M* diagram: $P_B L/2$ (magnitude)
- Moment of inertia can be calculated directly or simplified using variables
- Make substitutions in the bending stress equation and simplify
  - $\sigma = \frac{Mc}{I}$, where $I = bh^3/12$ for rectangular moment of inertia cross-section
  - $\sigma = \frac{\left(P_B L/2\right)\left(h/2\right)}{bh^3/12}$, where h is thickness (height) and b is width (base)
    - Maximum beam bending stress occurs at "c" which is h/2
  - $\sigma = \frac{3P_B L}{bh^2}$ ← Bending stress in terms of applied *load*, not moment
- Note that each load condition must be uniquely considered and derived
- Enter the given values to solve for the limiting load to remain in the elastic region
  - $503e6 = \frac{3P_B(0.2)}{(0.03)(0.005)^2} \rightarrow P_B = 629\text{ N}$

- To complete the experimental objective:
  - Record load cell applied force and displacement at desired intervals *below* 629 N
  - Convert load cell applied force to stress using the modified bending stress equation
  - Use a strain gauge to measure strain since a load cell cannot provide strain
- LOAD CELL ISSUE #2: Load cells do not offer a direct bending strain measure
  - Refer back to Figure 11.3 for typical load cell configurations
    - Load cells can directly measure tensile strain as displacement (ε formula required)
    - For bending, load cell *displacement* is not test specimen *strain*
    - A strain gauge offers a measure of material strain due to bending
  - Strain gauge(s) should be placed at the maximum stress location
    - A *V-M* diagram shows location of max moment (max stress and max strain)
    - For the cantilever point load in Figure 11.6, this occurs at Pt. A on the beam
      - For Figure 11.6, the strain gauge should be place on beam top at Pt. A
    - For 3-point bending in Figure 11.3, maximum moment occurs in middle of beam

Example 11.4

Consider a non-destructive 3-point bending test to determine the elastic modulus of a rectangular steel test specimen, as shown in Figure 11.7. A load cell is used with the ability to plot applied load versus displacement. What is the maximum allowable load to ensure a non-destructive test?

- EXPERIMENTAL OBJECTIVE: Create stress-strain plot to determine elastic modulus, *E*, without exceeding yield stress
- GIVEN: A 4130 steel rectangle, 5 cm in length, 6 mm thick (height), 2 cm wide
- FIND: Calculate the maximum applied load to remain within elastic region
  - Look up yield stress for 4130 steel: 435 MPa
  - Calculate the maximum load cell *force* based on material yield stress
    - Beam half-thickness: $c = 0.006/2 = 0.003\ m$
    - Moment of inertia (rectangle):

$$I = \frac{bh^3}{12} = \frac{(0.02)(0.006)^3}{12} = 3.60 \times 10^{-10}\ m^4$$

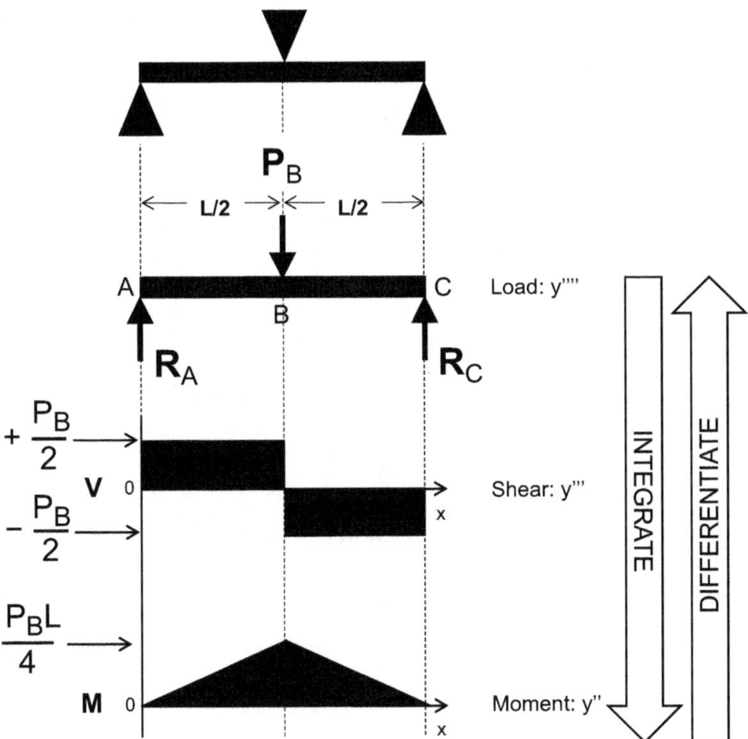

FIGURE 11.7 Shear-moment diagram for 3-point bending.

- Bending stress equation:

$$\sigma = \frac{Mc}{I} \rightarrow 435 \times 10^6 \; Pa = \frac{M(0.003)}{3.60 \times 10^{-10} \; m^4} \rightarrow M = 52.2 \; Nm$$

- Maximum load cell *moment* should be less than 52.2 Nm to remain in elastic region
- However, load cells typically apply *force* and not *moments*
  - Maximum moment of 52.2 Nm must be rewritten in terms of applied force
- Create a quick *V-M* diagram, as shown in Figure 11.7
  - Configure an empty *V-M* diagram with axes and key points, as show in Figure 11.4
  - Draw the load condition at top of *V-M* diagram
  - Draw the FBD below the load condition
  - Solve the reactions by inspection
  - Both R_A and R_C equally share the applied P_B load (one-half each)
    - Maximum shear within the beam is $\pm P_B/2$
  - Moment must begin and end at zero since there are no reaction moments
  - Maximum moment is <u>not</u> ~~$(P_B)(L/2)$~~ from the FBD
    - Moment is the integration of the <u>shear curve</u>
    - Moment accumulates as shear is integrated from left-to-right
    - The full $P_B/2$ *shear* is accumulated once integration reaches $L/2$
    - Maximum moment is therefore $(P_B/2)(L/2) = P_B L/4$
  - Moment is positive because the applied load causes the beam to *smile*
    - Moment is also positive because integral to positive shear is positive slope, and moment increases from zero into positive territory
  - The maximum allowable moment was worked out at $M = 52.2$ Nm
  - Equate the maximum moment from the *V-M* diagram to this value and solve for load
    - Maximum *V-M* moment set equal to max allowable moment: $P_B L/4 = 52.2 \; Nm$
    - Enter given values: $P_B = (4)(52.2 \; Nm)/L$
    - Solve for applied load: $P_B = (4)(52.2 \; Nm)/(0.05 \; m)$
    - Maximum allowable load to remain in elastic region: $P_B = 4176 \; N$

STRAIN GAUGES

To complete a stress-strain curve for a bending sample, a strain gauge can provide a measure of strain from increased resistance as the material stretches.

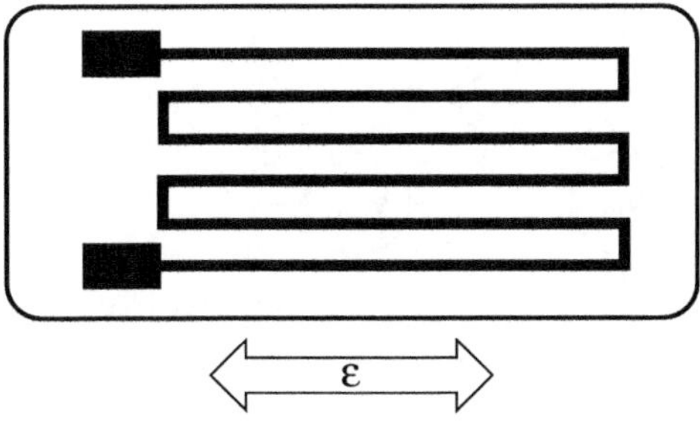

FIGURE 11.8 Common strain gauge configuration.

- Strain gauges are similar to variable resistors that increase resistance as they stretch due to strain
- Aligned lengthwise with strain direction as shown in Figure 11.8
- Typically have base values of 120 Ω or 350 Ω
- Bond to the surface of a part to stretch (strain) along with the part itself
- Usually bonded at locations of highest stress (highest strain)
- There are two common use cases by measuring strain via resistance with Hooke's law, $\sigma = E\varepsilon$
  - Elastic modulus is known: Gauge resistance → strain → determine applied stress
  - Applied stress is known: Gauge resistance → strain → determine elastic modulus
- Strain gauges require an electrical circuit such as a Wheatstone bridge with a voltmeter

WHEATSTONE BRIDGE

The change in resistance due to strain measured by a strain gauge is a very low order of magnitude. The Wheatstone bridge is an electrical circuit that improves the challenge of measuring very low electrical values.

- An electrical circuit is configured as two voltage dividers, as shown in Figure 11.9
- The output of the bridge is voltage, based on changing resistance values
- Two common use cases for a Wheatstone bridge
  - Determine an unknown resistance (not addressed here)
  - Measure very small changes in resistance, such as from a strain gauge
- A *balanced bridge* outputs zero volts regardless of the input voltage
- A balanced bridge in Figure 11.9 ($V_{OUT}=0$) occurs when either

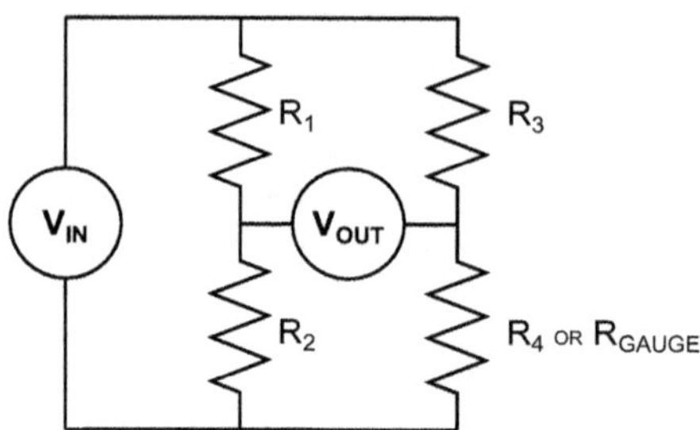

FIGURE 11.9 Wheatstone bridge circuit.

    - All resistances are equal ($R_1 = R_3 = R_2 = R_4$)
    - Top resistance equal and bottom resistances are equal ($R_1 = R_3$ and $R_2 = R_4$)
- Reason why a Wheatstone bridge facilitates reading small strain gauge changes
    - Consider a multimeter with five-digit precision measuring *resistance* directly
    - For a 350 Ω strain gauge, three digits of *overhead* are used just for baseline value
    - Only two digits remain to measure small changes in strain gauge resistance
    - Balancing a bridge to zero permits all five digits to read small strain gauge changes
- A Wheatstone bridge requires *voltage* measurements rather than resistance measurements
    - Voltage → resistance → strain → solve for either E or ε (via Hooke's law)
- A typical Wheatstone bridge circuit is given in Figure 11.9
- Formula to convert voltages (V_{IN} and V_{OUT}) into strain gauge resistance in Equation 11.4

$$\Delta R_{GAUGE} = \left[\frac{R_3}{\left(\frac{V_{OUT}}{V_{IN}}\right) + \left(\frac{R_2}{R_1 + R_3}\right)} \right] - R_3 \tag{11.4}$$

- Formula to convert strain gauge resistance (R_{GAUGE}) into strain in Equation 11.5

$$\varepsilon = \frac{\Delta R_{GAUGE} / R_{GAUGE}}{GF} \tag{11.5}$$

where: GF is the *gauge factor*

- The gauge factor is a manufacturer supplied value, typically around a value of 2.0
- Important notes for Equations 11.4 and 11.5 based on the Wheatstone bridge configuration
  - Equation 11.4 correctly results in increasingly <u>*negative voltages*</u> for increasing resistances
  - Voltage values are in the millivolt range

Example 11.5

Consider a strain gauge attached to the bottom of a rectangular steel test specimen in 3-point bending to measure maximum strain. The strain gauge is configured in a Wheatstone bridge per Figure 11.9. The bridge is initially balanced to zero volts with an input of 5 V. The strain gauge has a base value of 120 Ω and GF = 2.1. A load is applied, and the voltmeter indicates –3.018 mV. What is the strain?

- EXPERIMENTAL OBJECTIVE: Determine elastic modulus, *E*, with stress-strain plot
- GIVEN: A 4130 steel rectangle, 5 cm in length, 6 mm thick (height), 2 cm wide
- FIND: Determine strain from a voltmeter connected to a strain gauge in a Wheatstone bridge
  - Use Equation 11.4 for the Wheatstone bridge:

$$\Delta R_{GAUGE} = \left[\frac{R_3}{\left(\frac{V_{OUT}}{V_{IN}}\right)+\left(\frac{R_2}{R_1+R_3}\right)}\right] - R_3$$

  – Enter given values: $\Delta R_{GAUGE} = \left[\frac{120}{\left(\frac{-3.018e-3}{5.0}\right)+\left(\frac{120}{120+120}\right)}\right] - 120$

  – Solve for change in strain gauge resistance: $\Delta R_{GAUGE} = 120.29 - 120.00$

  – Strain gauge resistance change: $\Delta R_{GAUGE} = 0.29\ \Omega$

  - Use Equation 11.5 to convert resistance into strain: $\varepsilon = \frac{\Delta R_{GAUGE}/R_{GAUGE}}{GF}$

  – The change in resistance is *relative to* the change in strain

    – For this example, the bridge was initially balanced to zero with zero strain

    – ΔR_{GAUGE} = new resistance – previous resistance = 0.29 – 0.00 = 0.29 Ω

  – Enter given values: $\varepsilon = \frac{0.29/120}{2.1}$

  – Strain: $\varepsilon = 1.2e{-3}$ (unitless or mm/mm)

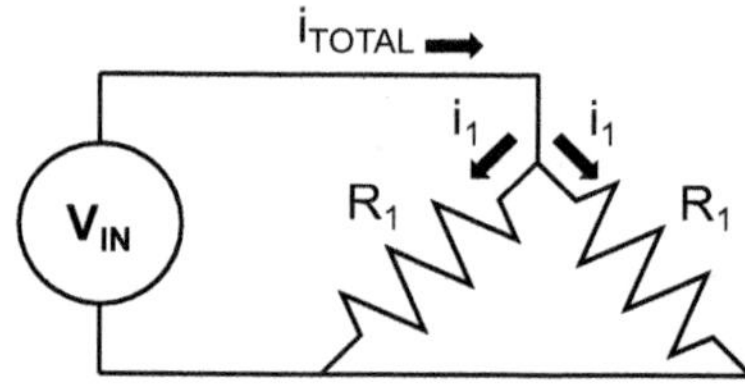

FIGURE 11.10 Current down parallel paths with equal resistance.

DERIVATION. The Wheatstone bridge circuit is best understood in terms of voltage drops or voltage potentials. If voltage is the same between two points, regardless of the actual voltages, there is no voltage drop and a voltmeter will read zero. The algebra required to arrive at Equation 11.4 can get tedious. Setting the strain gauge at location R_4 shown in Figure 11.9 eases some of the algebra compared to the R_3 location. Particular attention should be paid to negative signs, formulas and configurations with polarity presented in this chapter and the companion lab. With this formulation, as strain increases, the voltage drop becomes increasingly negative. This results in increasing strain gauge resistance with increasing strain.

- To understand the bridge circuit, begin with the simpler circuit in Figure 11.10
    - If both resistors are equal, $R_1 = R_1$, the total current *must* be split equally down both paths
        - Kirchhoff's current law accounts for all currents into and out of a node: $i_{TOTAL} = i_1 + i_2$
    - Slightly modify the circuit by adding two more resistors with equal values in Figure 11.11
        - Same current runs through both sides (paths)
        - Same resistances through both sides
        - Same voltage drop through both sides
    - Regardless of actual voltage drop value, it is the exact same down both resistor series
    - Since voltage is the same along both resistor series, voltage *across* series is zero
    - The final configuration is Figure 11.12 where V_{OUT} is zero for a balanced bridge
- Equation 11.4 is derived as follows using the nodes and notation in Figure 11.12
    - Ohm's law: $V = IR$ (voltage = current*resistance)
    - R_1 and R_2 are connected in series → same current flows *through both* R_1 and R_2
        - Current through both i_1 and i_2: $i_{1\to 2} = i_1 = i_2$
    - Use Ohm's law for current through both i_1 and i_2 ($i_{1\to 2}$)
        - $i_{1\to 2} = V_{IN} /(R_1 + R_2)$
    - R_3 and R_G are connected in series → same current $i_{3\to G}$ flows through both R_3 and R_G
    - Use Ohm's law for current through both i_3 and i_G ($i_{3\to G}$)
        - $i_{3\to G} = V_{IN} /(R_3 + R_G)$
    - The voltage drop (potential) across R_1 (from *A* to *B*)
        - The voltage at B: $V_{A\to B} = (i_{1\to 2})(R_1)$

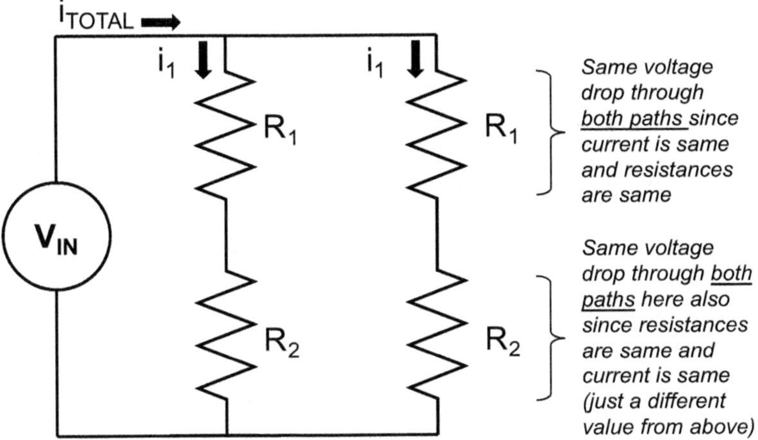

FIGURE 11.11 Wheatstone bridge circuit has equal potentials across both paths with series resistors of same values. Measuring across an equal potential results in a zero value.

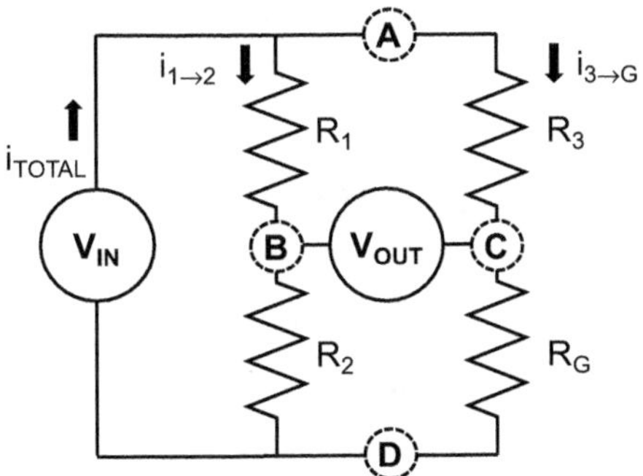

FIGURE 11.12 Wheatstone bridge circuit with nodes.

- Substitute in the previous relationship for $i_{1\to2} = V_{IN}/(R_1 + R_2)$ from above
- $V_{A\to B} = (V_{IN}\ R_1)/(R_1 + R_2)$

- The voltage drop (potential) across R_3 (from A to C)
  - The voltage at C: $V_{A\to C} = (i_{3\to G})(R_3)$
  - Substitute in the previous relationship for $i_{3\to G} = V_{IN}/(R_3 + R_G)$ from above
  - $V_{A\to C} = (V_{IN}\ R_3)/(R_3 + R_G)$
- The voltage potential *between* B and C is the difference *between* the voltage at B and C
  - V_{OUT} *is* $V_{B\to C}$: $V_{B\to C} = V_{A\to B} - V_{A\to C}$
  - Use the substitutions for $V_{A\to B} - V_{A\to C}$ from above
  - $V_{OUT} = [V_{IN}\ R_1/(R_1 + R_2)] - [V_{IN}\ R_3/(R_3 + R_G)]$
  - $V_{OUT}/V_{IN} = [R_1/(R_1 + R_2)] - [R_3/(R_3 + R_G)]$
- Solve for R_G to arrive at Equation 11.4

EXERCISES

A square tube of unknown material is planned for a project. The tube is measured at 1 cm on each side with a wall thickness of 1 mm. A short section of the tube is cut and placed into a load cell for a 3-point bending test, 6 cm long, to ascertain the material. A strain gauge with GF = 2.1 and R_{GAUGE} = 120 Ω is used in a Wheatstone bridge of 120 Ω resistors with 5 V applied to the circuit. Determine the following.

1. What is the maximum moment the tube section will encounter as a function of load?
2. Where should the strain gauge be located to measure strain at its maximum moment?
3. What is the cross-sectional moment of inertia formula and value?
4. The Wheatstone bridge is perfectly balanced. What is the output voltage?
5. The first voltage measurement is −0.1729 mV.
   a. What is the strain gauge resistance?
   b. What is the change in strain gauge resistance?
6. Three load versus strain gauge voltages are recorded per Table 11.1. Complete the table.
7. If the full stress-strain curve is plotted similar to the upper-left curve in Figure 11.2, what materials are possibilities and what materials can be ruled out?
8. What is the material modulus?
9. The material begins to yield at 915 N. What is the material?

TABLE 11.1
Load versus Strain Data

Load (N)	Stress (Pa)	ΔR (Ω)	Strain
150		0.0497	
300		0.0993	
500		0.1660	

12 Photoelasticity

Analytical stress methods usually identify only key stress points. Photoelasticity offers a visual depiction of an entire stress field. Labs #4 and #5 utilize photoelastic techniques to visually analyze stresses relative to theoretical and computational methods.

Photoelasticity is an optical technique to both qualitatively and quantitatively assess stresses within a part. A transparent replica of the actual part is created from a suitable material. The transparent material must exhibit two different indices of refraction under stress, as is typical of polymers. Basic photoelastic techniques have been expanded to surface coatings of non-transparent parts and digital image analysis techniques. The emphasis in this discussion is on the physics underlying the technique to clarify how light indicates lines of constant stress. Although quantitative analysis is possible and briefly introduced, this discussion is primarily limited to qualitative considerations.

- Photoelasticity is an optical technique to study the entire stress field in a model
- Advantages of photoelasticity
  - Simple, inexpensive, non-destructive analysis
  - Provides an accurate indication of the complete stress field
  - Works with irregular shapes that are difficult to model computationally
  - Offers quantitative and qualitative results
- Disadvantages of photoelasticity
  - Requires a plastic (transparent) version of the part
  - Quantitative calculations are tedious
  - Generally limited to planar 2D parts (3D possible but cumbersome)
- Light propagates as a transverse wave
  - Although electric and magnetic fields are not "*particles*," their vectors are perpendicular to the propagation direction of the wave, which meets the definition of a transverse wave
    - The electric field vibrates perpendicular to wave direction as shown in Figure 12.1
    - The magnetic field vibrates perpendicular to wave direction as shown in Figure 12.1
  - Electric and magnetic fields are orthogonal to each other
  - Electric and magnetic fields radiate outward in all directions, as shown in Figure 12.1
    - The sine wave actually *balloons out* (expands and collapses) in all directions

DOI: 10.1201/9781003329237-12

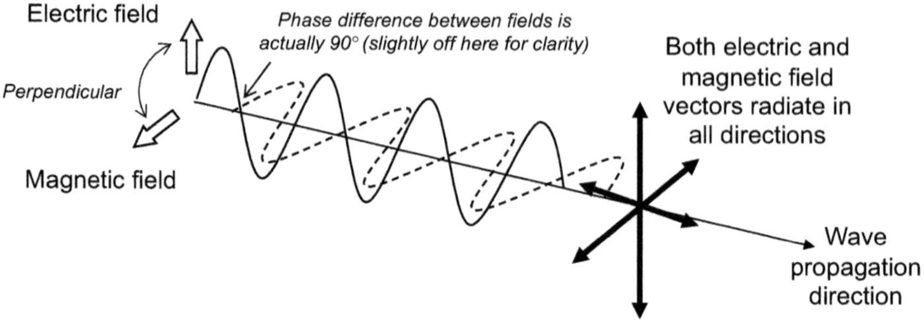

FIGURE 12.1 Propagating light wave with electric and magnetic fields in all directions.

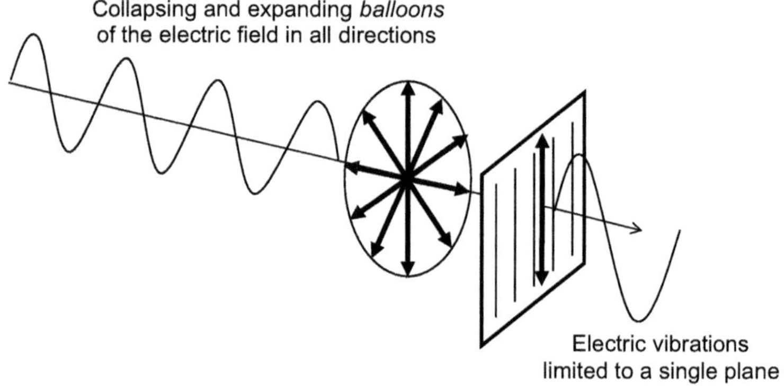

FIGURE 12.2 Electric vibrations polarized to only a single plane.

POLARIZATION AND BIREFRINGENCE

All discussions about light propagation are now limited only to the electric field as the means of propagation. The magnetic field is energy storage for the electric field and is not considered further. Materials with two indices of refraction are called birefringent. Birefringence can occur naturally within a material or be induced by stressing an otherwise non-birefringent material. Polarized light provides the means to visualize birefringence in a material.

- Polarization limits the electric vibration vector to a single plane by absorbing all other directions as shown in Figure 12.2
  - A polarizing sheet is similar to a slotted fence
- Light transmission through a medium slows down
  - Light is fastest in a vacuum (3×10^8 m/s)
  - Light slows down when it transits any other medium
  - Light, as a wave, is *re-transmitted* through material interaction

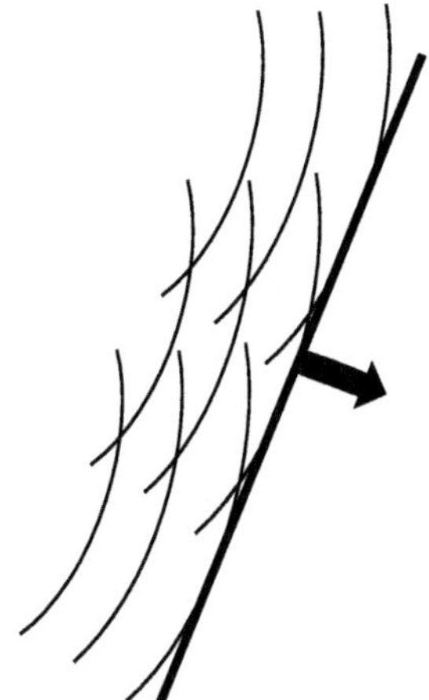

FIGURE 12.3 Wavefront of traveling light.

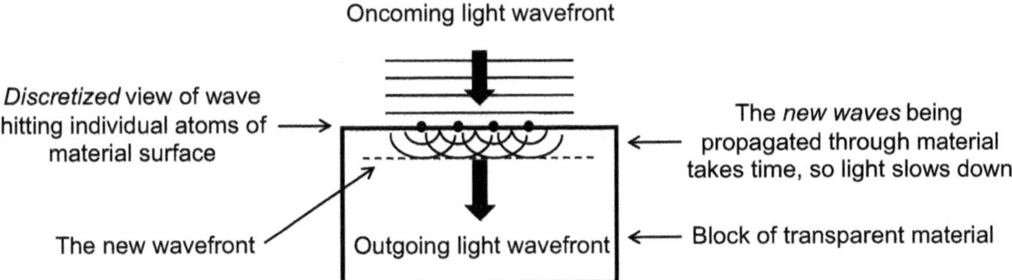

FIGURE 12.4 Redevelopment of wavefront at interface of a different medium.

- Huygens's principle[1] describes how the wavefront of light is transmitted through materials
  - Oncoming light waves arrive as a wavefront, as depicted in Figure 12.3
  - A wavefront is *discretized* at a surface and *re-transmitted* to reform a wavefront
  - A new wavefront propagates through material as shown in Figure 12.4
  - The rate at which light slows down through a medium is the index of refraction
    - Speed of light in a vacuum: c
    - Speed of light in a substance: v
    - Index of refraction: $n = c/v$
  - A light wavefront arriving at an *angle* to a material changes direction, as in Figure 12.5
    - The change in speed changes the wavefront based on the index of refraction

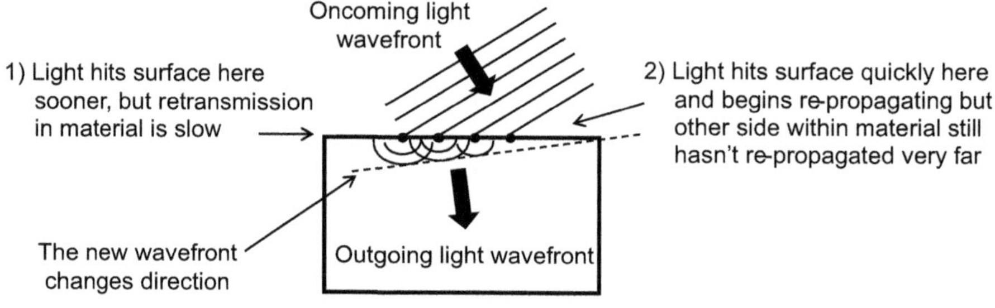

FIGURE 12.5 Wavefront changes direction at medium interface due to speed differential.

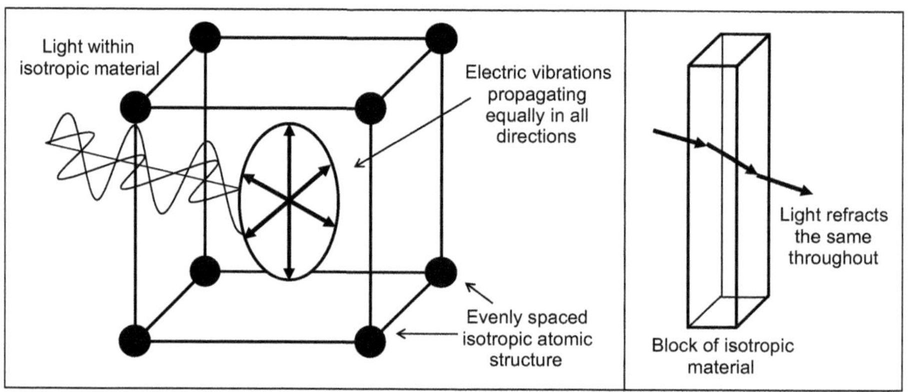

FIGURE 12.6 Light travels consistently throughout an isotropic material.

- Light propagates equally within an isotropic material since atomic structure is symmetric
  - Isotropic materials have a uniform atomic structure, as shown in Figure 12.6
  - Electrical vibrations of light transmission are equally affected in all directions
  - Light rays (direction of the wavefront) travel through medium with one index of refraction
- Light propagates *unequally* within an anisotropic material
  - Anisotropic atomic structure is non-symmetric as shown in Figure 12.7
  - Atomic structure is denser in one direction compared to other directions
  - Electric vibrations are slower in one direction compared to other directions
  - Light rays (direction of wavefront) travel through medium with two indices of refraction
  - Anisotropic materials have a *slow axis* and a *fast axis* of propagation
  - Icelandic spar is a naturally occurring anisotropic material with two indices of refraction
  - Materials with two indices of refraction are *birefringent*

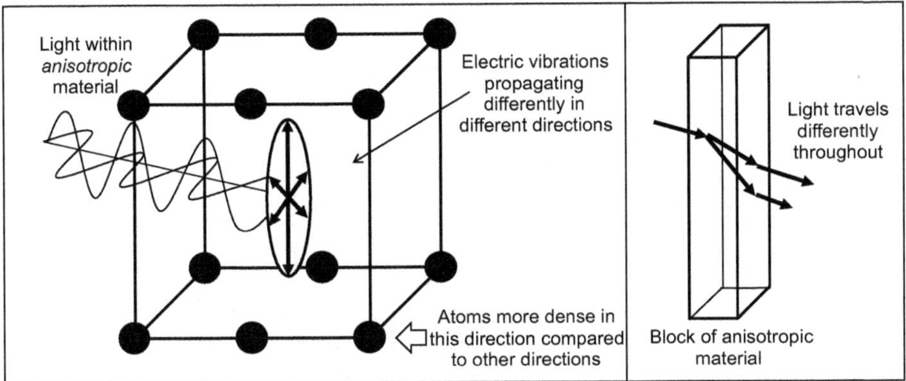

FIGURE 12.7 Light travels faster in one direction compared to the other in anisotropic materials.

- Unstressed plastic materials have long polymer chains but are generally isotropic through
  - Unstressed isotropic plastics have a single index of refraction
- Stressed plastics distort the polymer chains and become slightly anisotropic
  - Stressed plastics with distorted polymers have varying indices of refraction throughout
  - The change in refractive index is a *direct function* of the stress causing the change
  - Light vectors change direction according to changing indices of refraction from stress
  - Polarization can be used to remove a component direction of varying light vectors
  - Light and dark patterns are the result of the remaining light vector components per Figure 12.8
- The light and dark patterns from the photoelastic effect are isoclinics and isochromatics
  - Isoclinics and isochromatics provide information about stress difference and direction
  - Isoclinics are dark lines of constant principal stress direction
    - A 2D object has stress in both x- and y-directions
    - Dark lines appear only where vectors are completely cut by both polarizers
    - Other stress vectors, and therefore light vectors, have varying intensities of lightness
    - Therefore, isoclinics indicate everywhere the principal stresses have the same orientation
  - Isochromatics are lines or areas of constant color visible with a white light source
    - Isochromatics indicate where principal stress differences are the same

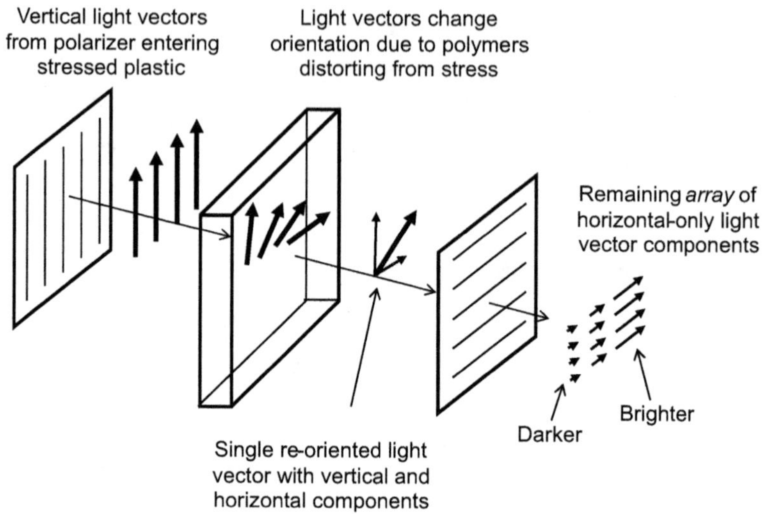

FIGURE 12.8 Polarization removes light vector components to see stress distributions.

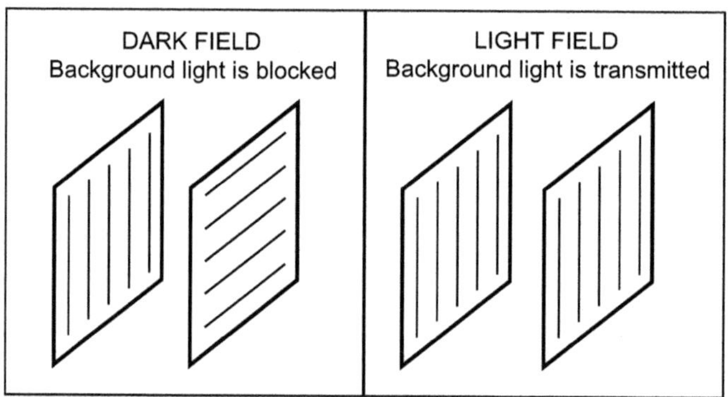

FIGURE 12.9 Orientation of polarizers establish either a light field or dark field.

  - Isoclinics give direction information; isochromatics provide magnitude information
  - Isochromatics only indicate where the *difference* between principal stresses is the same
- Photoelastic fringe patterns can be set up as a light field or dark field per Figure 12.9
  - The orientation of polarizer sheets with respect to each other determines a light or dark field
  - When polarizers are aligned, all background light passes
  - When polarizers are oriented at 90° to each other, background light is blocked
  - Light is transmitted through parts based on stress vector re-orientation

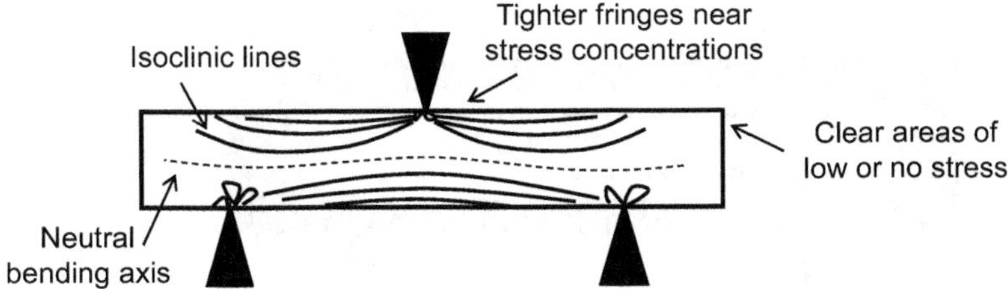

FIGURE 12.10 Photoelastic fringe pattern showing isoclinics on a 3-point bending beam.

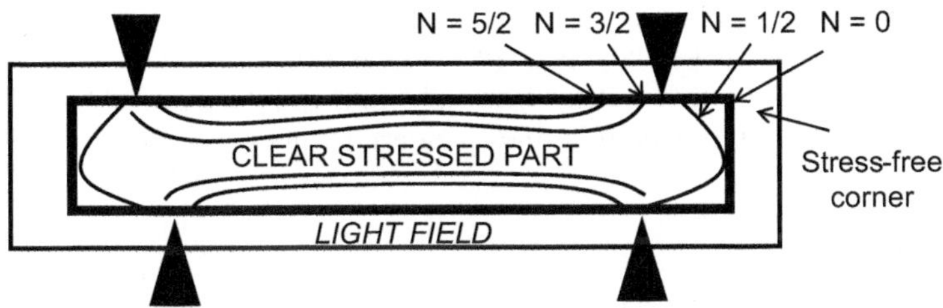

FIGURE 12.11 Fringe numbers in a light field.

- Figure 12.10 depicts an example of a photoelastic fringe pattern from a beam in 3-point bending
- The *stress optic law*[2] in Equation 12.1 permits quantitative analysis of the photoelastic stress field

$$\sigma_{xx} - \sigma_{yy} = \frac{f_\sigma N}{h} \tag{12.1}$$

where: σ_{xx} is principal stress in x-direction
σ_{yy} is the principal stress in y-direction
f_σ is the material fringe constant
N is the fringe number
h is the thickness of the plastic model

  - The fringe constant, f_σ, is experimentally determined for a material with a known load
  - The fringe number, N, is a particular isoclinic counted up from a zero-stress location
- How to determine the fringe number from a photoelastic light field in Figure 12.11
  - Identify a corner or edge free of stress: $N = 0$
  - Find the first dark isoclinic line and count up by half an integer: $N = 1/2$
  - Subsequent dark isoclinic lines increase by integers thereafter: $N = 3/2$, $N = 5/2$, …

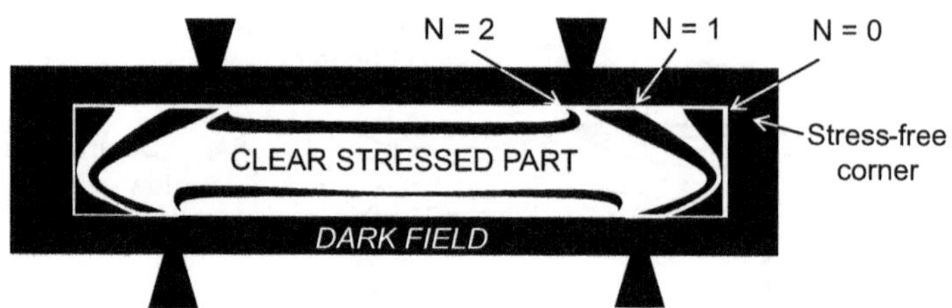

FIGURE 12.12 Fringe numbers in a dark field.

- How to determine the fringe number from a photoelastic dark field in Figure 12.12
  - Identify a corner or edge free of stress: $N=0$
  - Find the first dark isoclinic line and count up by integers: $N=1$
  - Subsequent dark isoclinic lines increase by integers: $N=2$, $N=3$, …

EXERCISES

1. Consider a rectangular part under perfect tensile stress. How many isoclinics should be visible through photoelastic techniques?
2. TRUE or FALSE. The fringe number for isoclinics begins at $N=0$ for any corner free from stress?
3. TRUE or FALSE. The fringe number for isoclinics begins at $N=0$ for the neutral axis in pure bending?
4. TRUE or FALSE. Any point along a single isoclinic will have the same stress?
5. What do tight groups of isoclinics mean?

NOTES

1 Hewitt, P. (1981). *Conceptual physics: a new introduction to your environment.* Little, Brown and Company. p.406.

2 Dally, J. & Riley, W. (1978). *Experimental stress analysis.* McGraw-Hill Book Company. p.411.

13 Fractures

Failure due to fracture is sometimes overlooked in the basic design process. A part designed within yield stress limits may still fail due to fracture from stress concentrations. Lab #5 evaluates the fracture behavior of ductile and brittle materials relative to their ultimate stress.

Structural components can fail in a variety of ways. Designing a part within a material's yield stress is the most fundamental structural design constraint. However, fracture failure modes must also be considered, especially during the service life of a part. To reduce the risk of fracture, stress concentrations should be reduced in the initial design. Small flaws or cracks due to manufacturing processes or service life damage should also be factored into the design. Fracture should always be evaluated as a possible failure mode in addition to exceeding yield stress.

- Materials can fail in a variety of modes
- Two typical failure modes of interest in design
  - Yield stress: material fails when yield stress is exceeded
  - Fracture: material fails due to fracture *prior* to yield stress
- Consider normal tensile loading on two similar rectangular test specimens in Figure 13.1
  - Specimen *A* is flawless while specimen *B* has a small edge cut or crack
  - Depending on the material, two scenarios are possible
    - Both specimens A and B might fail at yield stress, σ_{YS}
    - The edge flaw might cause specimen B to fail much *lower* than σ_{YS}
- The fracture potential of a material is referred to as *fracture toughness*
  - Brittle materials have low fracture toughness
  - Ductile materials have high fracture toughness
  - Fracture toughness indicates the stress required to propagate a flaw
- Flaws can be cracks, voids, material inclusions, weld defects or scratches
  - Assume flaws are not completely avoidable in the fabrication or service of part
  - A common design practice is to *assume* a flaw of some given size exists
- Fracture failure has three modes, depicted as examples in Figure 13.2
  - Mode I: tensile or opening – the most common mode
  - Mode II: shear or sliding – typical for fasteners in shear
  - Mode III: tearing – same as tearing open a bag of chips
    - It is always easier to tear open a bag via Mode III with an initial cut

Fracture failures are catastrophic failures. One moment a part is maintaining a load, and the next moment it breaks and the structure fails. Fracture failures are a brittle failure mode where a crack rapidly propagates until the part is completely

DOI: 10.1201/9781003329237-13

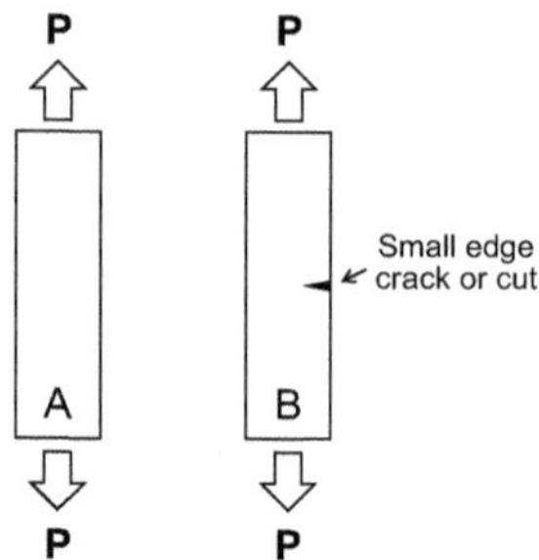

FIGURE 13.1 Comparing effect of an edge crack on otherwise identical parts.

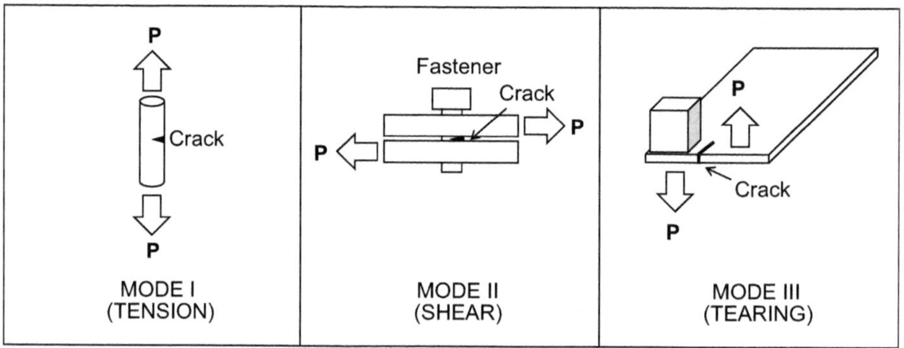

FIGURE 13.2 Three fracture modes.

compromised. The catastrophic result is similar to exceeding *ultimate* tensile stress, where a part is typically torn into two pieces.

FAILURE MODES

Engineers design based on yield strength, not ultimate strength. Since both fracture failure and exceeding ultimate stress result in physically *broken* parts, discussions will mostly consider ultimate stress to compare with fracture stress. Consider the following two different failure stresses for a part:

- σ_{ULT} is the ultimate failure stress due to tension
- σ_C is the critical fracture failure stress due to one of three fracture modes

If stress within the part exceeds σ_{ULT}, the part breaks. If stress within the part exceeds σ_{C}, the part also breaks, but it may have already broken due to first exceeding σ_{ULT}. Both stresses are based on material properties. Ultimate stress is a straightforward published look-up value. However, critical stress requires computation since the look-up value is fracture toughness, K_{IC} (for Mode I). Consider the two material properties for AL6061T6 aluminum[1]:

- Ultimate failure stress: $\sigma_{ULT} = 310$ MPa
- Fracture toughness: $K_{IC} = 29 \text{ MPa}\sqrt{\text{m}}$

If the aluminum part is in tension, it will fail at an ultimate stress of 310 MPa. The same part will fail due to fracture based on the type of crack, size of the crack and dimensions of the material. To determine the critical fracture stress, these parameters are entered into the *critical stress formula* in Equation 13.1.

$$\sigma_{IC} = \frac{K_{IC}}{Y\sqrt{\pi a_c}} \tag{13.1}$$

where: σ_{IC} is the critical (fracture) stress for Mode I
K_{IC} is the material Mode I fracture toughness
Y is a look-up value based on crack geometry
a_c is the "half-crack" length (crack dimension)

Example 13.1

Consider a small aluminum bar in tensile loading with a small edge crack 0.5 mm long, similar to specimen B in Figure 13.1. Based on the part dimensions given below, at what *ultimate stress* will the part fail? Based on the part dimensions given, at what *critical stress* will the part fail?

- GIVEN: An AL6061T6 rectangle, 20 cm in length, 5 cm wide, 1 mm thick, 0.5 mm crack
- FIND: 1) Ultimate failure stress
  - Look up ultimate stress for 6061T6 aluminum: 310 MPa
  - ANSWER: The part fails whenever stress reaches 310 MPa
- FIND: 2) Critical failure stress
  - Look up fracture toughness for 6061T6 aluminum: $K_{IC} = 29 \text{ MPa}\sqrt{\text{m}}$
  - Look up edge crack properties: $Y \approx 1.12$
  - Enter all values into Equation 13.1
  - $\sigma_{IC} = \dfrac{29e6 \text{ Pa}\sqrt{\text{m}}}{1.12\sqrt{\pi(0.5e-3 \text{ m})}}$
  - $\sigma_{IC} = 653$ MPa
  - ANSWER: The part fails whenever stress reaches 653 MPa

With this example, the part fails due to ultimate stress at 310 MPa and never reaches the critical fracture stress of 635 MPa. This is due to a relatively small crack in a reasonably ductile material. If the crack is larger or the material is more brittle, such as AL7075T6, the critical fracture stress is lower. If the critical fracture stress calculates to a value lower than the ultimate stress, then fracture is the governing failure mode.

It is worth noting that Equation 13.1 can be implemented in a few different ways. Equation 13.1 can be used as is, where applied stress is checked against critical stress as in Example 13.1. Equation 13.1 can also be solved for fracture toughness, K_{IC}, and the result can be checked directly against the material property.

- Applied stress can be checked against the critical stress (is $\sigma > \sigma_C$?)
- Stress intensity, K_I, can be checked against fracture toughness, K_{IC} (is $K_I > K_{IC}$?)
- The maximum acceptable crack length can be checked against the actual crack length (is $a > a_C$?)

DERIVATION. A complete derivation of the critical fracture stress formula is lengthy and beyond the scope of this discussion. The derivation set up in Figure 13.3 highlights some important details about solving for fracture stress.

- Theoretical Mode I (tensile) fracture stress assumes the following per Figure 13.3
  - An infinite flat plate
  - An elliptical hole (crack) of total width "2a," where "a" is the *half-crack* length
    - For central cracks, the crack length is a total of "2a" per Figure 13.3
    - For edge cracks, the crack length is "a"
    - All formula variants follow this convention
- The generalized stress intensity formula based on geometry in Figure 13.3 is Equation 13.2

$$\sigma_{yy} = \frac{\sigma_0 \sqrt{\pi a}}{\sqrt{2\pi r}} f(\theta) \tag{13.2}$$

where: σ_{yy} is the stress at any point in the *y*-direction
σ_0 is the far-field applied stress
a is the "half-crack" length (crack dimension)
r is the distance from center of crack to point of interest
$f(\theta)$ is a trigonometric function of θ in Figure 13.3

- Equation 13.2 provides the total tensile (*y*-direction) stress field throughout the plate
  - Equation 13.2 is generally not used directly for engineering calculations
  - Equation 13.2 is ultimately refined into Equation 13.1

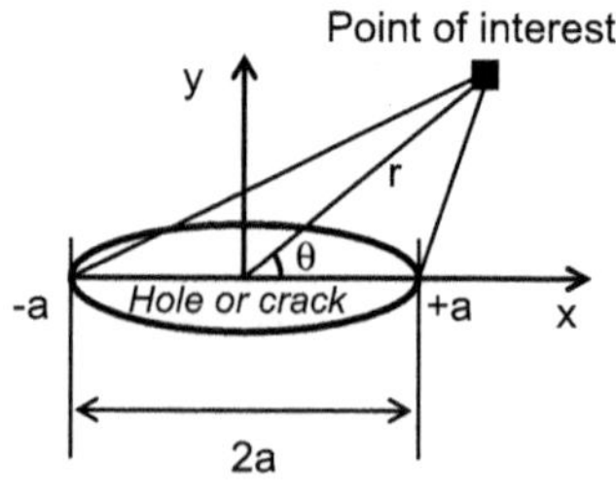

FIGURE 13.3 Geometry for theoretical fracture stress development.

- A solution similar to Equation 13.2 is also available for stress in the *x*-direction, σ_{xx}
  - The *x*-direction stress is parallel to the crack
  - Stress parallel to a crack is not significantly impacted by the crack
  - The *x*-direction stress is usually ignored as inconsequential
- The term $f(\theta)$ in Equation 13.2 is a trigonometric function based on angle to crack
  - Where $f(\theta) = \cos\left(\frac{\theta}{2}\right)\left[1 + \sin\left(\frac{\theta}{2}\right)\sin\left(\frac{3\theta}{2}\right)\right]$

- Engineering design is less concerned with the total stress field function in Equation 13.2
  - Stress intensity increases drastically close to a crack tip
  - Engineering design is concerned with high stress governed by the crack tip
  - A *stress intensity factor* (SIF) simplifies Equation 13.2 for crack-tip stress calculations
- Parameters in Equation 13.2 can be conveniently grouped as a SIF
  - The parameters $\sigma\sqrt{\pi a}$ tend to recur in formulations for both *x*- and *y*-directions
  - Stresses tend to be proportional to $\sigma\sqrt{\pi a}$
  - Define "*K*" as a SIF and set equal to $\sigma\sqrt{\pi a}$ for Equation 13.3

$$K = \sigma_0\sqrt{\pi a} \tag{13.3}$$

where: K is the SIF
σ_0 is applied far-field stress
a is the "half-crack" length (crack dimension)

- Equation 13.3 is only valid as the theoretical SIF for a central crack in an infinite plate
- SIFs are localized crack-tip stresses only and not the entire stress field from Equation 13.1
- Equation 13.3 must be corrected for other than a central crack in an infinite plate
  - Stress intensity deviates from Equation 13.3 as a function of width and crack geometry
  - Modify Equation 13.3 with the correction, $Y\left(\frac{a}{w}\right)$, for Equation 13.4

$$K = Y\left(\frac{a}{w}\right)\sigma_0\sqrt{\pi a} \tag{13.4}$$

- The term $Y\left(\frac{a}{w}\right)$ in Equation 13.4 is a geometry and crack-specific function based on
  - Plate width for actual finite width plates
  - Crack length and geometry
- The term $Y\left(\frac{a}{w}\right)$ is also just written as "*Y*" per Equation 13.1

  - The (*a/w*) indicates *Y* is a function of crack length, geometry and plate width
  - The parentheses are merely *function notation* and do not call for a calculation
- The term $Y\left(\frac{a}{w}\right)$ is a published look-up value for each unique crack configuration
- Equation 13.4 is easily rewritten for stress in Equation 13.5

$$\sigma = \frac{K}{Y\sqrt{\pi a}} \tag{13.5}$$

  - Equation 13.5 is a general form of Equation 13.1
- Equation 13.5 should include a SIF subscript based on mode
  - Mode I: use K_I instead of K in Equation 13.5
  - Mode II: use K_{II} instead of K in Equation 13.5
  - Mode II: use K_{III} instead of K in Equation 13.5
  - SIF units: $\text{Pa}\sqrt{\text{m}}$

CRITICAL FRACTURE STRESS

The *critical* SIF is the stress intensity associated with fracture failure. The critical stress is the design limit stress for fracture, notwithstanding any appropriate safety factors.

- A small "*c*" subscript is added to each SIF mode to indicate the critical SIF
  - Critical SIFs for Modes I, II and II: K_{IC}, K_{IIC} and K_{IIIC}
- A small "*c*" subscript is likewise added to sigma and crack length to indicate critical stress
  - Critical stress associated with a critical SIF: σ_c
  - Critical crack length associated with a critical SIF: a_c
- The critical stress a part can withstand is a *material property* called *fracture toughness*
  - A material's fracture toughness for tension is its "K_{IC}"
  - Fracture toughness is a look-up property published for various materials
  - Some typical fracture toughness values:
    - Aluminum: 13–28 $\text{MPa}\sqrt{\text{m}}$
    - Steel (4340): 50 $\text{MPa}\sqrt{\text{m}}$ ← Steel's high fracture toughness not prone to fracture
    - Glass (soda-lime): 0.7-0.8 $\text{MPa}\sqrt{\text{m}}$ ← Low fracture toughness prone to fracture
    - Concrete: 0.2–1.4 $\text{MPa}\sqrt{\text{m}}$
- When evaluating critical stress with Equation 13.5, include the "*c*" subscripts (σ_c and K_{IC})

Example 13.2

Consider a rectangular section of steel under tensile loading with an edge crack 0.5 cm long. The applied stress is 250 MPa. Will the part fail?

- GIVEN: Steel, $\sigma_0 = 250$ MPa, edge crack length: $a = 0.5$ cm, other dimensions not specified
- FIND: Check ultimate stress, then critical stress and compare to steel fracture toughness
  1. Failure due to tensile load:
     - Look up yield stress (design stress) for steel: $\sigma_{YS} = 300$ MPa
     - Applied stress < yield stress ($\sigma_0 < \sigma_{YS}$)
     - **No failure**
  2. Failure due to fracture:
     - Mode: K_I
     - Look up fracture toughness for steel: $K_{IC} = 120$ MPa$\sqrt{m}$
     - Edge crack
       - Look up the appropriate published SIF formula
       - $K_I = 1.12\ \sigma_0 \sqrt{\pi a}$
       - For an edge crack in a semi-infinite plate, $Y(a/w) \approx 1.12$
     - Calculate SIF
       - $K_I = 1.12\ (250e6)\sqrt{\pi(0.5e-2)}$
       - $K_I = 35$ MPa$\sqrt{m}$
     - SIF, $K_I < K_{IC}$ (35 MPa$\sqrt{m}$ < 120 MPa$\sqrt{m}$)
     - **No failure**

Example 13.3

Consider a 6-cm wide rectangular section of cast iron in tensile loading with a central crack of 0.5 cm long. At what stress will the part fail?

- GIVEN: Cast iron, central crack length: $2a = 0.5$ cm, width = 6.0 cm
- FIND: Check ultimate stress then critical stress and compare to steel fracture toughness
  1. Failure due to tensile load
     - Look up yield stress for cast iron: $\sigma_{YS} = 275$ MPa
  2. Failure due to fracture
     - Mode: K_I
     - Look up fracture toughness for cast iron: $K_{IC} = 10$ MPa$\sqrt{m}$
     - Central crack, finite plate:
       - Look up the published SIF formula
       - $K_I = \sigma_0 \sqrt{w * \tan\left(\pi a/w\right)}$
     - Calculate the critical stress based on fracture toughness
       - $K_{IC} = \sigma_C \sqrt{w * \tan\left(\pi a/w\right)}$
       - Formula requires "a," but center cracks are given as "2a"; divide by "2"

- $10e6 = \sigma_C \sqrt{0.06 * \tan\left(\pi(0.0025) / 0.06\right)}$
  - Solve for $\sigma_C = 112$ MPa
- Critical stress, $\sigma_C < \sigma_{YS}$ (112 MPa < 275 MPa)
- Part fails due to fracture at 112 MPa

- Things to watch for
  - "2a" is the measured crack length for central cracks → check if the formula requires "a"
  - "a" is the measured crack length for edge cracks
  - If fracture stress is higher than yield stress → part fails due to normal yield or ultimate
  - Ensure correct units for calculator trig functions
    - From the above equation:
    - (0.0025/0.06) = 0.13 ← These are radians
    - Calculator expects radians (OK): tan(0.13) = 0.13
    - Calculator expects degrees (NOT OK): tan(0.13) ≠ 0.0022
    - Convert to degrees (OK) tan(0.13rad) = tan(7.5°) = 0.13

EXERCISES

An overhead shelf has rectangular brackets of AL6061T6 aluminum, as shown in Figure 13.4. Unfortunately, one of the brackets had a hole drilled through it and now has cracks from previous loads. You planned to place a heavy box on the shelf, but you are unsure if the shelf will fail due to fracture. The Mode I SIF is approximated with the following formula: $K_I = Y\left(\frac{a}{R}\right)\sigma\sqrt{\pi a}$. The hole radius, R, is 3 mm, and the half-crack length, a, is 3.6 mm. The look-up function for $Y\left(\frac{3.6}{3}\right)$ is 0.9851.

Determine the following:

1. The bracket has dimensions of 3 cm wide by 1 mm thick. What is the applied stress on the bracket if the box weighs 100 lbf?

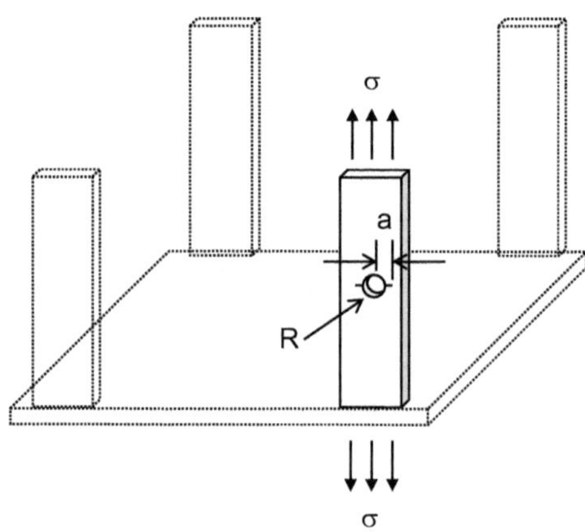

FIGURE 13.4 Center hole with cracks.

2. Will the shelf fail due to a fracture with the box?
3. How much load can the shelf handle prior to failure?

NOTE

1 Materials data repository. (2014). *Additional properties of aluminum alloy 6061.* National Institute of Standards and Technology. https://materialsdata.nist.gov/handle/11115/242.

14 Non-Destructive Inspection

Even properly designed parts can fail during service due to unforeseen flaws and degradation. Non-destructive inspection techniques help ensure structural components remain within their design parameters. Lab #6 utilizes ultrasound as a common NDI technique to detect flaws within a material.

Once a structural part or component is designed and put into service, it is subjected to the stress, wear and tear of normal use and abuse. To help ensure these parts do not lose structural integrity during their service lives, routine inspection is required. This is particularly true of the aerospace industry, where weight limitations demand higher performance and in-service failure can have catastrophic consequences.

The only true way of knowing when a structural part will fail is to stress it to failure. The part is then no longer serviceable, so this is not much of an option. Non-destructive techniques are industry-standard approaches to inspecting for defects or cracks before they compromise the structure.

- Destructive testing is testing a material or part to failure to determine specific properties
  - This type of testing is important during the design process to verify design parameters
  - Performance of composite lay-ups tends to vary based on process, so destructive sample testing is particularly important to ensure design parameters are accurate
  - Parts are no longer serviceable after destructive testing
- Non-destructive inspection checks for flaws or defects without damaging the usability of part
- Several mostly interchangeable terms are used for non-destructive inspection
  - NDI: non-destructive inspection
  - NDT: non-destructive testing
  - NDE: non-destructive evaluation
- NDI is extensively used throughout industry, especially aerospace

TYPES OF NDI

Several forms of NDI are available; some of the more common types are included as follows.

- X-ray: radiograph of a part to detect defects
  - Advantages: check internal flaws, density changes and misalignment

DOI: 10.1201/9781003329237-14

  - Disadvantages: poses a radiation hazard, expensive, requires access to both sides of components for illuminating and imaging, and does not indicate the depth of a defect
- Dye penetrant: chemical colorant to visually indicate the presence of a surface crack
  - Advantages: low cost, portable and relatively simple
  - Disadvantages: defects must be open to the surface, and the part must be thoroughly cleaned
- Eddy current: measures changes in the electric field due to flaws in conductive materials
  - How it works[1]
    - Coil in probe induces current flow in conductive material
    - *Eddy currents* are created within the inspected part and measured within the probe coil
    - Energy is required to create eddy currents, which are measured by the device as
      - Resistance and inductance
    - A crack is a discontinuity that inhibits eddy current formation
    - Fewer eddy currents are created due to cracks that appear as
      - Decreased resistance since less energy is required for fewer eddies
      - Increased inductance across crack
      - Total energy is conserved – just a phase change in the complex plane
      - Everything is measured *within* the probe, not the inspection material
  - Advantages: fast, inexpensive and non-contact (requires close proximity)
  - Disadvantages: conductive materials only and shallow depths only
- Vibration: checks resonant frequency changes due to changes in mass, stiffness and damping
  - Advantages: quick pass/fail, no consumables and tests the entire part at once
  - Disadvantages: does not indicate location of flaw and requires baseline data
- Laser tomography: detects distortion between stressed and unstressed states via laser light
  - Advantages: checks large areas at once and detects sub-surface flaws
  - Disadvantages: expensive equipment
- Ultrasound: high-frequency sound waves detect hidden cracks, voids and delaminations
  - Advantages: fast and checks the entire thickness of a part
  - Disadvantages: sensitive to surface conditions and complex parts are difficult

ULTRASOUND

Ultrasonic inspection techniques are particularly useful, well-developed and broadly applied throughout various fields. Ultrasonic sound waves are used for medical imaging as well as detecting material flaws. Just as with ripples from a rock thrown into calm water, transmitted high-frequency sound waves travel cleanly outward from the source through a homogeneous medium. It is not until the wave encounters an obstruction, material boundary or change in density that the wave bounces off the inhomogeneity and returns an echo of some magnitude. These waves can travel as normal (longitudinal) or shear (transverse) waves.

An imaging ultrasound transducer does not send out a continuous ultrasonic wave. Instead, it sends out a short ultrasonic pulse and then listens for a return echo. No one stands over a canyon or near a cave and yells out a long, continuous "Heeyyyyyyyy". The return echo would be hard to hear over the continuously transmitted source. Even if the echo was heard, the only useful information would be the time between the first utterance and its return. Everything else would be a muddy mix of transmitted and returning sound waves. It is much more natural to yell a sharp "Hello", and then listen for the echo. The scheme is to send a pulse of sound, listen and time the return, then send another pulse and repeat the process. For ultrasonic techniques, the timing is usually expressed in terms of micro- and milliseconds.

There are two frequencies for ultrasonic imaging. Sound travels as waves, so the first frequency is the high-frequency pulse of sound. This high frequency is the quick "hello" yelled into the canyon (*stand too close to the canyon edge and the frequency might get higher yet*). The next frequency is the lower frequency rate at which these pulses are sent out. This rate specifies the listening interval. This lower frequency listening interval provides distance information based on elapsed time and the speed of sound through the medium. Details of the higher frequency pulse are often ignored and characterized only as a pulse. The pulse repetition frequency (PRF) is the lower frequency rate at which these pulses are sent, which provides the information of interest. All velocities and computations are based on the PRF rather than the high-frequency emission. Since time measurements are based on reflected signals, time-of-flight (TOF) is always *twice* the actual distance, which must be factored into calculations.

- Ultrasonic NDI transducers transmit a high-frequency sound wave through material
  - Wave propagation occurs as molecules are displaced and carry the wave
    - Normal (longitudinal): molecules push-pull on atoms, front-to-back, to propagate a wave
    - Shear (transverse): molecules tug up-down on atoms, side-by-side, to propagate a wave
  - An ultrasonic pulse returns distance from the timing between transmission and return pulse
    - Distance is a measure of material thickness or depth of a defect reflecting a sound wave
    - Distance requires knowing the *velocity* of wave propagation and the pulse time-of-flight
    - General distance relationship: $Distance(m) = velocity\left(\frac{m}{s}\right) * time(s)$
    - Time-of-flight distance relationship: *Distance* (m) $= velocity\left(\frac{m}{s}\right) * \left(\frac{1}{2}\right) TOF(s)$

- The velocity of sound propagation within a material is called *celerity*
- Typical velocities (celerity) for longitudinal waves (shear waves are slower)
  - Air: ~340 m/s
  - Water: ~1500 m/s
  - Metal: ~ 6000 m/s
- Velocity is a function of
  - Material density, ρ
    - Dense, tightly packed atomic structures propagate waves faster
  - Bulk modulus, B, a measure of elastic compressibility
    - *Squishy*, compressible substances propagate sound slower
- Velocity equations all take the same basic form, regardless of medium
  - String celerity: $v = \sqrt{T/\mu} \leftarrow v = \sqrt{Tension\ or\ stiffness / Density}$
  - Air celerity:
    $v = \sqrt{RT/M} \leftarrow v = \sqrt{Gas\ Law\ 'stiffness' / Molecular\ density}$
  - Solids celerity: $v = \sqrt{B/\rho} \leftarrow v = \sqrt{Bulk\ (elastic)\ modulus / Density}$
- Bulk modulus for solids can be converted to elastic modulus
  - Compared to elastic modulus, bulk modulus is not published for all materials
  - Equation 14.2 can be used instead of Equation 14.1
- The celerity for sound waves in a solid in terms of bulk modulus is given in Equation 14.1

$$v = \sqrt{B/\rho} \tag{14.1}$$

where: v is the speed of sound within a material
B is the bulk modulus of material
ρ is the material density

- The celerity for sound waves in a solid in terms of elastic modulus is given in Equation 14.2

$$v = \sqrt{\frac{E(1-\upsilon)}{\rho(1+\upsilon)(1-2\upsilon)}} \tag{14.2}$$

where: v is the speed of sound within a material
E is the Young's (elastic) modulus of material
ρ is the material density
υ (nu) is Poisson's ratio

- The derivation of the velocity equation (not presented here) is lengthy and involves
  - Stress-strain relationship, equations of motion and wave PDE

Example 14.1

Consider an experiment to verify the speed of sound within a block of 6061T6 aluminum. What is the theoretical velocity?

- GIVEN: A 6061T6 aluminum block
- FIND: Speed of sound in aluminum
  - Published speed of sound[2]: $v = 6420\,m/s$
  - Finding velocity with Equation 14.1 using properties of aluminum[3]
    - Bulk modulus not readily published for AL6061T6
    - Bulk modulus for aluminum in general: $B = 62e9$ Pa to 106e9 Pa
      - Aluminum alloys range from soft (compliant) to stiff
      - AL6061T6 is more stiff than compliant, but not the stiffest aluminum alloy
    - Density: $\rho = 2700\,kg/m^3$
    - Using the high-end published value: $v = \sqrt{106e9\ Pa \Big/ 2700\ kg/m^3}$
    - Velocity for this higher-end modulus value of aluminum: $v = 6266\,m/s$
      - This value is likely high since AL6061T6 is not the stiffest available alloy
  - Finding velocity with Equation 14.1 using elastic modulus
    - Published elastic modulus for AL6061T6: $E = 69e9$ Pa
    - Published Poisson's ratio for AL6061T6: $\nu = 0.33$
    - Velocity: $v = \sqrt{\dfrac{69e9(1-0.33)}{2700\ (1+0.33)(1-2*0.33)}}$
    - Velocity within AL6061T6: $v = 6153\,m/s$
      - This result appears reasonable relative to the high-end calculation

Example 14.2

Consider an experiment using ultrasound to determine the thickness of a block of 6061T6 aluminum compared to the value measured with calipers.

- GIVEN: A 6061T6 aluminum, measured at 2.00 cm thick with verified velocity, $v = 6153\,m/s$
- FIND: Thickness of 6061T6 aluminum using ultrasound
  - Experimentally measured TOF of ultrasonic pulse through AL6061T6: $t = 6.46\ \mu s$
  - Thickness: $Distance(m) = velocity\left(\frac{m}{s}\right) * \left(\frac{1}{2}\right) TOF(s)$
  - $Thickness(m) = 6153\ m/s * \left(\frac{1}{2}\right)(6.46e-6\ s)$
  - Ultrasonic measured thickness: $t = 1.99\,cm$
  - % Error relative to caliper measurement: $\%E = \dfrac{2.00-1.99}{2.00} = 0.5\%$

EXERCISES

1. List any three forms of NDI and the associated advantages of technique.
2. An ultrasonic pulse is sent through the air to a wall and is measured when it returns 2.4 ms later. How far away is the wall?
3. A block of unknown homogeneous metal is measured at 5.0 cm thick. Ultrasound is used to measure time-of-flight through the material, where TOF = 49.7 μs. What is the metal?
4. Assume the block of metal in Question #3 is a 5.0 cm cube with a mass of 1.4 kg.
   a. What is the density?
   b. Does this confirm the type of metal from Question #3?
   c. What is the bulk modulus?

NOTES

1 National bureau of standards. (1977). *Eddy current nondestructive testing.* U.S. Department of Commerce. https://www.govinfo.gov/content/pkg/GOVPUB-C13-a71a3e527f40884fc6963cb511f71196/pdf/GOVPUB-C13-a71a3e527f40884fc6963cb511f71196.pdf.

2 Engineering ToolBox, (2004). *Solids and metals - Speed of sound.* https://www.engineeringtoolbox.com/sound-speed-solids-d_713.html.

3 Engineering ToolBox, (2008). *Aluminum alloys - Mechanical properties.* https://www.engineeringtoolbox.com/properties-aluminum-pipe-d_1340.html.

15 Instruments and Sensors

Instrumentation is essential to recording and measuring the results of an experiment. Engineers should be familiar with typical laboratory equipment. Lab #8 explores configuring thermocouples as sensors which is covered in this chapter.

Most modern test equipment are digital electronic devices, with high precision and accuracy, little to no warm-up time and very low loading. However, every device has an appropriate application along with limitations, so it is helpful to understand the equipment available for common applications. A Device Under Test is commonly abbreviated as DUT, and a *measurand* is the quantity being measured.

TEST EQUIPMENT

Electrical test equipment generally measures only voltage, and if a different unit such as ohms is required, a conversion is performed within the instrument for the appropriate output. Bench equipment is more accurate and sensitive compared to handheld devices.

The following devices do not constitute a comprehensive list. Rather, it provides an introduction to basic equipment and a starting point for further study.

- Voltmeter (units: volts, V)
  - What it is: measures voltage potential, the fundamental electrical motive force
    - Has high impendence to prevent loading from current flowing through the instrument
  - How to operate: two probes measure a voltage potential or voltage drop across a device
    - Must be set to measure either direct or alternating current in the correct voltage range
- Ammeter (units: amps, A)
  - What it is: measures current flow
    - Has low impendence to prevent loading as current flows through the instrument
  - How to operate: two probes must *break into* a circuit to sense current flow
    - Inductance versions are available that do not require breaking into the circuit to measure
    - Must be set to the correct measurement range
    - Ammeters are fuse-protected, and fuses are easily blown by a mistaken setting

DOI: 10.1201/9781003329237-15

- Ohmmeter (units: ohms, Ω)
  - What it is: measures resistance
    - Has high impendence to prevent loading from current flowing through the instrument
  - How to operate: two probes measure resistance across a device
    - Ohmmeters have internal power sources to apply a reference voltage
    - *Voltage is measured* with reference current; *resistance is calculated* from Ohm's law
    - Cannot be used on an active circuit with a power source since ohmmeters supply power
- Kelvin ohmmeter or 4-wire ohmmeter (units: ohms, Ω)
  - What it is: measures very low resistance and measures all resistances accurately
    - Has high impendence to prevent loading from current flowing through the instrument
    - Has four probes: two outside probes carry current, and two inside probes measure voltage
    - Probe wires do not contribute to the overall measured resistance
  - How to operate:
    - Connect two current source probes across a device
    - Connect two voltage measurement probes across the device but inside the current probes
    - Separating the current and voltage removes probe wire resistance for greater accuracy
- Oscilloscope (units: volts, V, and time, s)
  - What it is: measures both voltage and time in a time-based signal
    - Displays voltage along the vertical axis and time along the horizontal axis
    - Has selectable impedance settings, typically high impedance and 50 Ω standard
  - How to operate: two probes measure a voltage potential across a device
    - Can be set to measure a continuous signal waveform or a triggered event for timing
    - Has a high frequency response to measure down to nanosecond timing events
    - Can measure and display periodic waveforms
- Frequency generator or function generator (units: volts, V, and Hertz, s^{-1})
  - What it is: generates a time-based signal at a given voltage and amplitude
    - Used in conjunction with an oscilloscope to determine the frequency response of a device
    - Has a 50 Ω output impedance
  - How to operate: two probes supply a time-varying signal across a device
    - Multiple waveforms are usually selectable
    - Used to determine device behavior due to different responses at different frequencies
- Vector network analyzer (units: volts, V, time, s, Hertz, s^{-1}, phase, °, signal strength, dB)

- What it is: measures changes in a time-based signal through a device
  - Multiple display options are available to include the magnitude and phase of a signal
  - Similar to light, alternating current signals are electromagnetic waveform
  - Similar to light hitting a window, signals can either reflect or transmit a signal
  - Scattering parameters, S1, S2, S3 and S4, are measures of reflection and transmission
- How to operate: two or more probes to measure a signal across a device
  - Multiple signal displays are possible
  - Two primary displays include signal magnitude and phase shift through a device

- Lock in amplifier (units: volts, V, time, s, Hertz, s^{-1}, phase, °)
  - What it is: precisely measures a frequency signal in a high noise environment
    - Displays the magnitude and phase of a signal
    - Provides a reference frequency to precisely measure low signal-to-noise responses
    - Less flexible than a VNA but more precise for specific SNR applications
  - How to operate: two probes provide reference and measurement signals across a device
    - Displays magnitude and phase of the signal
- Data acquisition device, DAQ (units: volts, V, time, s)
  - What it is: measures and records interval samples of one or more signal channels
    - Measures analog or digital signals for desired durations and sample rates
  - How to operate: connect sensor device leads to DAQ terminals
    - DAQs typically measure voltage potentials
    - Conversions to appropriate units are based on the sensor datasheet or DAQ configuration

Data acquisition devices are versatile in that they can record multiple signal sources at specified sampling rates for any desired duration. This capability is useful for recording changing heat distributions along a heat transfer device or pressure differences due to fluid flow through ductwork.

DAQs to measure thermocouple inputs usually require hardware specifically designed for thermocouples to properly isolate the signal and minimize noise. DAQ software configuration varies depending on the manufacturer. Thermocouple configuration must be selected based on thermocouple type. Multi-channel DAQs sometimes require specifying a total *throughput* sample rate for samples-per-second-per-channel. In this case, a 1-Hz sample rate for a single channel equates to a one-sample-per-four-second sample rate with four channels (0.25 Hz).

SENSORS

Sensors vary widely in type and technology used. Sensors must be selected based on compatibility with the operating environment, sensor operating limits and frequency response. Expected values within the operating range must be worked out ahead of time to ensure a sensor is selected with sufficient sensitivity without exceeding the device limits. Only a few examples are offered here with DAQ applications in mind.

- Load sensor or force sensor
  - What it is: a family of sensors outputting voltage as a function of applied force
    - Used in conjunction with a voltmeter or DAQ
    - The datasheet provides a transfer function for converting the output voltage into force
  - How to operate: affix sensor into an appropriate load cell configuration
    - External power is required
    - Flexure of the load cell structure can be a source of error
- Strain gauge
  - What it is: a thin wire swatch that changes resistance according to stretch
    - Measures material strain
  - How to operate: securely adhere to the object and measure resistance change based on strain
    - Resistance can be read directly, but changes are too small for most direct measurements
    - A Wheatstone bridge facilitates very low voltage measurements instead of resistance
- Pressure sensor
  - What it is: a family of sensors that output voltage as a function of pressure
    - Measures either absolute, gauge or differential pressure, depending on the sensor
    - Can measure static, total or stagnation pressures directly, depending on conditions
    - May have barbs for connecting tubes to the remote pressure sensing location
  - How to operate: securely position the sensor or tube in the pressure measurement location
    - The datasheet provides a transfer function for converting output voltage to pressure
    - Ensure the sensor selected is within the limits of the anticipated pressure to avoid device rupture

POWER SOURCE

Most sensors require a stable voltage input, and they output a varied voltage based on the sensor's transfer function. Since many sensors operate within a 3.3 V to 5.0 V range, a computer's USB port can be sourced as a convenient and stable 5 V power supply for sensor measurements outside of a lab setting. If 3.3 V are required, a simple

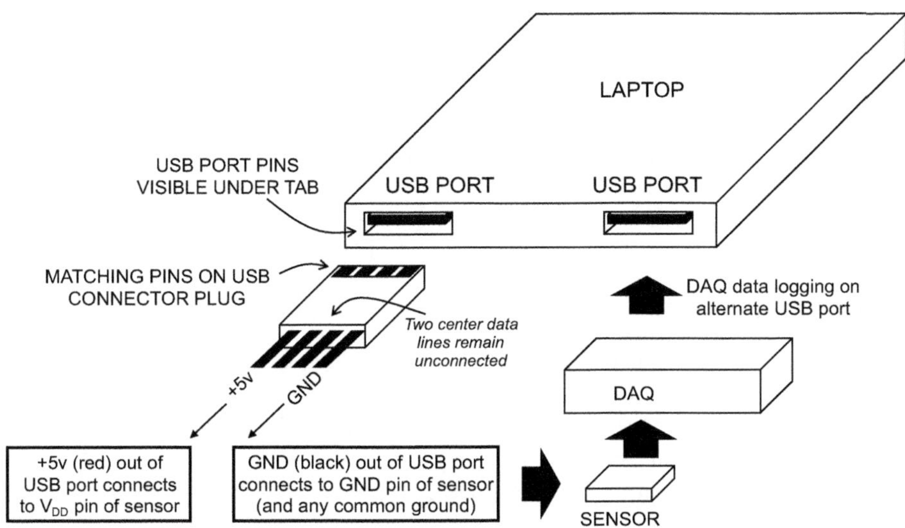

FIGURE 15.1 Utilizing a USB port for a low-power, stable 5 V sensor power source.

two-resistor voltage divider can step down the USB voltage accordingly. Batteries are also a stable source of power for sensors. Vehicle electrical systems are not stable nor quiet enough for sensor applications without a dedicated discrete component regulator and filtering. When recording data outside of the lab environment, Figure 15.1 offers a suitable solution for USB power.

Example 15.1

An aircraft is instrumented for temperature and pressure at key locations to improve operating performance from sea level to 10,000 feet. All sensors are connected to a DAQ with a laptop to run the software and record data at one sample per second. A reference ambient pressure is required on one of the DAQ channels.

- GIVEN: Sensor requirement for absolute pressure from sea level to 10,000 feet elevation
- FIND: Suitable pressure sensor and power source

Solution

- Convert altitudes to pressure range: 100 kPa (sea level) to 70 kPa (10K′)
- Assess desired resolution and frequency: ±1.5 kPa at 1 Hz (one sample per second)
- Research available pressure sensors to meet requirements: KP21F1701[1]
  - Input voltage: $V_{DD} = 5\,\text{V} \rightarrow$ configure power according to Figure 15.1
  - Transfer function: $P_{ABS} = \dfrac{V_{OUT}}{a * V_{DD}} - \dfrac{b}{a}$, where $a = 0.0081$ and $b = -0.00095$

- Device-specific transfer functions are provided in device datasheets
- Program transfer function into DAQ or spreadsheet

- At 3 V output: $P_{ABS} = \frac{V_{OUT}}{a * V_{DD}} - \frac{b}{a} \rightarrow P_{ABS}$

$$= \frac{3.0}{0.0081 * 5.0} - \frac{-0.00095}{0.0081} \rightarrow P_{ABS} = 74.2\ kPa$$

THERMOCOUPLES

Thermocouples are a particularly useful means of temperature measurement for a wide variety of applications. As a sensing device, thermocouples do not require a power source since they generate a voltage potential based on temperature. The voltage is quite low, within the millivolt range, so a powered device is required to read the signal. Either a quality voltmeter or DAQ with dedicated thermocouple measurements and isolation is required.

- Thermocouples
  - What it is: two different metal wires joined to measure temperature as a function of voltage
    - A temperature sensing junction is formed where wires are joined together
    - A junction can be formed by twisting or welding wires together
    - A dissimilar metal junction creates a small millivolt potential as a function of heat
    - Thermocouple metal types have different color coding, with red always negative
    - Thermocouples are simple, reliable, inexpensive and rugged temperature sensors
    - Thermocouples have a fast response, wide temperature range and linear response
  - How to operate: use a dedicated instrument or create a fundamental circuit
    1. Using a dedicated thermocouple measurement instrument
       - Connect thermocouple wires to measurement instrument; the red wire is negative
       - Select the appropriate thermocouple type; K-type is most common
       - All thermocouple measurement devices have built-in reference temperatures
    2. Create a fundamental circuit as shown in Figure 15.3 (also, reference Lab #9)
       - Create a temperature sensing junction with a thermocouple wire pair
       - Create a reference temperature junction with a same-type wire pair
       - Complete the circuit as shown in Figure 15.3 with a millivolt meter
       - Use a published thermocouple table with Equation 15.1 to determine temperature

- Three fundamental laws of thermocouples[2]
  1. Law of homogeneous materials: at least two dissimilar metals are required
     - Both a dissimilar metal contact *and* temperature differential are required
  2. Law of intermediate materials: voltage difference is zero at uniform temperatures
     - Permits other metals and use of measuring equipment without changing results
     - Isothermal portions of the circuit have no impact on results, regardless of metals
  3. Law of intermediate temperatures: voltages must add up with other junctions
     - Permits adding a reference sensing junction for standardized results
     - Junction voltages must still sum to the published totals according to Equation 15.1
     - Equations 15.1(a–c) provide alternative views of the same equation

$$V_{TOTAL} = V_{REF} + V_{SENSE} \tag{15.1a}$$

$$V_{TABLE\ VALUE} = V_{TABLE\ VALUE\ AT\ REF\ TEMP} + V_{VOLTMETER\ VALUE} \tag{15.1b}$$

$$V_{0 \to T2} = V_{0 \to T1} + V_{T1 \to T2} \tag{15.1c}$$

  - The three laws of thermocouples can be summarized as follows: it takes *both* a dissimilar metal contact *and* heat (more precisely, a temperature differential within the circuit) to generate a voltage, and this voltage is *limited* by the particular metals, so additional temperature junctions must still add up to the limiting published voltages.
- Basic thermocouple circuit is shown in Figure 15.2

If the measurement instrument in Figure 15.2 is a dedicated thermocouple device, the circuit would be complete since the instrument provides an internal temperature voltage reference. However, if the measurement device is a simple *voltmeter*, it would not provide a usable temperature reference. For example, if a simple voltmeter is used in Figure 15.2 to measure ambient temperature, the conversion of volts to degrees is zero, regardless of ambient. The first law of thermocouples requires a temperature differential to drive the circuit, so the result is always zero. If the sensing junction in Figure 15.2 was instead subject to a temperature different from ambient, then a voltage potential would be created. However, the result is still not practical since another temperature sensing device is also required to measure the ambient reference.

The solution to the preceding issue is to standardize thermocouple voltages to the ice point of water at 0°C. The ice point of water is a repeatable reference, and all thermocouple voltage reference tables, such as Table 15.1, are based on the 0°C reference.

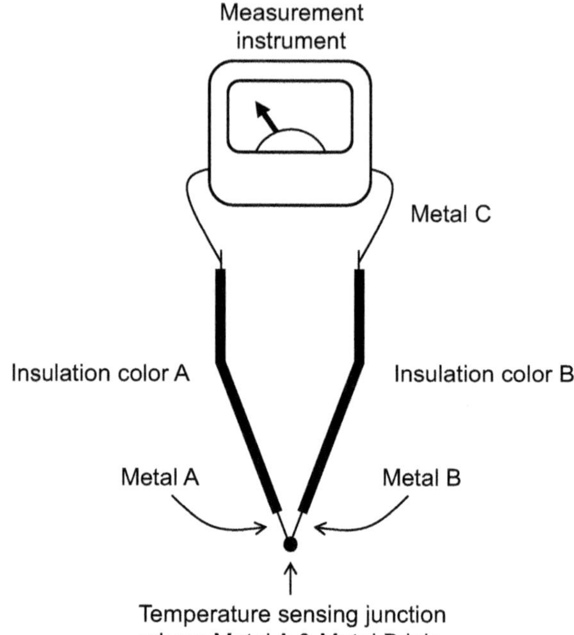

FIGURE 15.2 Basic thermocouple circuit with two dissimilar metals joined together.

TABLE 15.1
Type J Thermocouple Voltages at 0°C Reference, Generated at Fifth-Order Polynomial Resolution (Coefficient Equation by Recktenwald)

Voltages for Type J Thermocouple at 0°C Reference

°C	0	1	2	3	4	5	6	7	8	9
0	0.000	0.050	0.101	0.151	0.202	0.253	0.303	0.354	0.405	0.456
10	0.507	0.558	0.609	0.660	0.711	0.762	0.814	0.865	0.916	0.968
20	1.019	1.071	1.122	1.174	1.226	1.277	1.329	1.381	1.433	1.485
30	1.537	1.589	1.641	1.693	1.745	1.797	1.849	1.902	1.954	2.006
40	2.059	2.111	2.164	2.216	2.269	2.322	2.374	2.427	2.480	2.532
50	2.585	2.638	2.691	2.744	2.797	2.850	2.903	2.956	3.009	3.062
60	3.116	3.169	3.222	3.275	3.329	3.382	3.436	3.489	3.542	3.596
70	3.650	3.703	3.757	3.810	3.864	3.918	3.971	4.025	4.079	4.133
80	4.187	4.240	4.294	4.348	4.402	4.456	4.510	4.564	4.618	4.672
90	4.726	4.780	4.835	4.889	4.943	4.997	5.051	5.106	5.160	5.214
100	5.269	5.323	5.377	5.432	5.486	5.541	5.595	5.650	5.704	5.759

Source: Recktenwald, G. (2020, February 27). *Conversion of Thermocouple Voltage to Temperature.* https://web.cecs.pdx.edu/~gerry/epub/pdf/thermocouple.pdf).

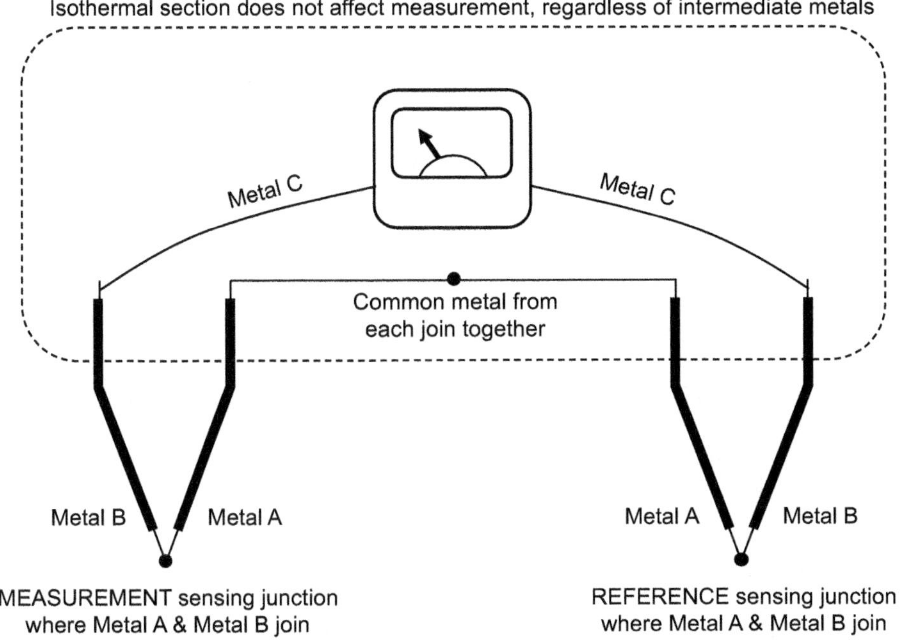

FIGURE 15.3 Temperature compensated thermocouple circuit.

A standardized temperature-referenced thermocouple circuit is shown in Figure 15.3. This circuit joins two thermocouple junctions, one of which can be kept at the ice point reference. The resulting voltage potential is driven by the temperature differentials between the two junctions, and the remaining circuit of connectors and instruments does not affect the results since those portions are usually isothermal in the ambient environment. The reference temperature junction in Figure 15.3 can be maintained at 0°C in an ice bath, or it can be done electronically with most instruments using a cold junction reference circuit.

If the reference temperature in Figure 15.3 is not 0°C, then Equation 15.1(a, b or c) must be used in conjunction with the thermocouple voltage tables. Equation 15.1 provides an offset based on the non-standard reference temperature.

Example 15.2

A Type J thermocouple circuit is created similar to Figure 15.3.

- GIVEN: Reference temperature T_0 at 0°C and measurement temperature T_1 at 58°C
- FIND: Voltage expected using the standard Type J voltage table with Table 15.1

Solution

This is a simple look up value since the reference is 0°C

- First, read all table information to confirm thermocouple type
- Read 10's digits down (to 50) and 1's digits across (to 8)
- At 58°C, voltage = 3.009v

Example 15.3

A Type J thermocouple circuit is created similar to Figure 15.3.

- GIVEN: Reference temperature T_0 at 10°C and measurement temperature T_1 at 58°C
- FIND: Voltage expected using the standard Type J voltage table with Table 15.1

Solution

This requires Equation 15.1 since the reference is *not* 0°C

- First, read all table information to confirm thermocouple type
- Read the table value for a reference temperature of 10°C: $V_{REF} = 0.507\,V$
- Read the table value for a maximum possible value of 58°C: $V_{TOTAL} = 3.009\,V$
- Use Equation 15.1 to solve for the difference measured by the voltmeter
    - $V_{TOTAL} = V_{REF} + V_{SENSE}$
    - $3.009\ V = 0.507\ V + V_{SENSE}$
- Voltmeter reading: $V_{SENSE} = 2.502$

The above result makes sense since a thermocouple circuit can only produce a certain voltage based on temperature and cannot somehow create more. Since the circuit reference temperature of 10°C is higher than the published data referenced at 0°C, there is less temperature difference to drive the circuit, so the voltmeter reads a lower value according to Equation 15.1. If the circuit reference temperature was increased up to the measurement temperature of 58°C, there would be no temperature differential to drive the circuit, and the voltmeter would read zero. Equation 15.1 therefore also gives this result of zero.

EXERCISES

For thermocouple questions, refer to Figure 15.3 for configuration and Table 15.1 for Type J thermocouple data.

1. The reference junction is maintained at 0°C. The voltmeter output, V_3, is 3.062 millivolts. What temperature is sensed at the measurement junction?

2. The reference junction is maintained at 15°C. The measurement junction is 40°C. What voltage (millivolts) is measured in the thermocouple circuit?
3. TRUE or FALSE. A thermocouple circuit with a reference temperature junction higher than the measurement junction results in a negative voltage?
4. What instrument is best for very low signal-to-noise ratio measurements?
5. As signal frequency increases, components can respond with different capacitance, inductance and resistance values. What instrument is best suited for measuring the signal response of a device subjected to high frequencies?
6. What simple electrical circuit can be used to step down a 5-V USB power source to 3.3 V?

NOTES

1 Infineon. (n.d.). *KP215F1701 Datasheet.* Infineon technologies. https://www.infineon.com/cms/en/product/sensor/pressure-sensors/barometric-bap-manifold-map-sensors/kp215f1701/.

2 Figliola, R. & Beasley, D. (2015). *Theory and design for mechanical measurements.* Jon Wiley & Sons. p.345.

16 Analyzing Data Sets

Multiple values from experimental data sets result in a single mean value. Basic statistical tools should be used to assess the data and evaluate any outliers to have confidence in the mean.

A single experiment with a single data point is usually not convincing since experiments are inherently subject to errors and inexact measurements. Just like re-running a set of calculations to be sure of a result, conducting multiple trials or multiple replications of an experiment provides more certainty in the outcome. Statistical tools provide well-established means to organize and evaluate the resultant data sets to help increase credibility. While not every experiment may directly support a hypothesis, the burden of a convincing data set is much greater for those experiments. This discussion is limited only to data sets resolving a single truth value based on normal distributions. Correlating data sets with curve fits or other tools is beyond the scope of this discussion.

- Theory and hypothetical relationships require data for validation
  - Theory can exist without data, but a suitable theory should be predictive of data
  - Data support a theory when experimental results correlate with theoretically predictive results
- Data are obtained in two basic ways
  - Passive observation by *watching* natural phenomenon and recording parameters
    - Observation leads to correlation and theoretical relationships but not causality
  - Active experimentation by setting up an experiment to determine causality
    - Experimental data elevates correlation to causality
    - Causality is supported by comparing the test parameter data set to the control data set

DISTRIBUTION OF DATA

Assuming systematic bias errors are well controlled within an experiment, random error will always remain due to the inexact nature of measurements involving physical parameters. The variation from one result to the next from an experiment is the distribution. An experimental data set influenced only by random error results in a normal distribution with a central tendency. The central tendency is the mean value and the most credible result of a data set.

DOI: 10.1201/9781003329237-16

- Data sets
  - Most experiments involve more than one data point to be convincing
  - Multiple experimental trials increase insight into actual truth value
  - Multiple experimental trials result in data set variances
    - Measurements are always inexact numbers
    - Random error is inherent in physical measurement and processes
  - Assuming experimental bias is eliminated, repeated trials of the same experiment influenced by uncontrolled, independent random variables, usually result in a *hit-or-miss* near the truth value
    - Results with random errors tend to cluster around a central mean (the truth value)
- Central limit theorem (CLT)[1]
  - *If a random variable is the sum of independent random variables, then for a sufficiently large sample size, the results are normally distributed*
- Histograms
  - A histogram is a distribution of results
  - A data set of multiple trials is often best presented in the form of a histogram
  - If only random errors influence results, the histogram is a *normal* distribution
  - A histogram is a graphical collection of separate trial results from the same experiment
    - The vertical axis is the number of occurrences, the frequency of events or experimental results
    - The horizontal axis is the value of interest, divided into *bins*
  - Histogram bins are discretized ranges of values on the horizontal axis
    - Individual results are *stacked* in the appropriate bins
    - The range of values in bins can be set as desired
    - Small bin ranges provide greater resolution but may be unsuited for small data sets
    - Larger bin ranges have less resolution but may be more suitable for small data sets
  - The number of occurrences (frequency) indicates values accumulated *per bin*

Example 16.1

An experiment is designed to verify the boiling temperature of water on a standard day at sea-level pressure. The expected truth value is 100°C and a sample size of 20 is selected for reliable results of a *well-behaved* data set (refer to Chapter 5: Sample Size). Measurements are taken, and results are *accumulated* into a histogram chart as shown in Figure 16.1.

- Distributions displayed by a histogram can take several forms, shown in Figure 16.2
  - *Continuous*: a smooth curve described by a mathematical function
  - *Discrete*: a collection of individual, discrete events that form a pattern

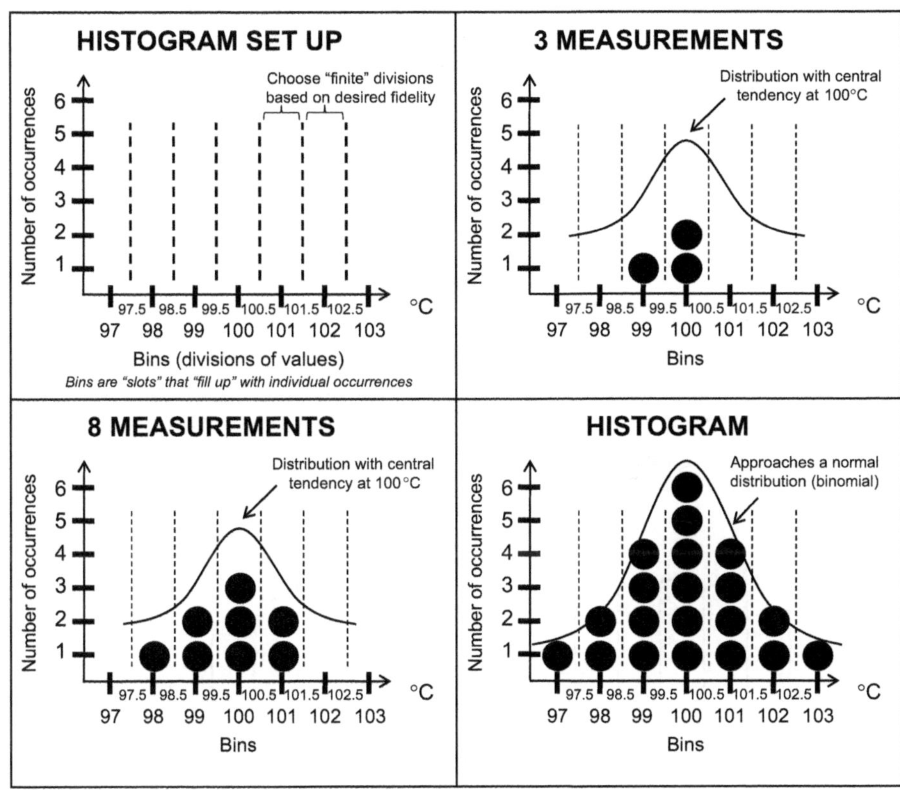

FIGURE 16.1 Creating a histogram chart and how data is organized in the chart.

- *Normal* (also, Gaussian): a bell-shaped continuous function with a central mean
- *Binomial*: a bell-shaped collection of discrete events, similar to a normal distribution
- *Uniform* (also, rectangular): a rectangular-shape where any event is as likely as another
- *Exponential*: an exponential function typical of time occurring between events

THE NORMAL DISTRIBUTION

Since the normal distribution is a fundamental aspect of most data sets, many statistical tools are available for analysis. Proper analysis of experimental data sets requires determining the mean and variance (standard deviation squared) and assessing outlying data points as possible mistakes or valid results.

- The normal distribution is a symmetric and continuous mathematical function

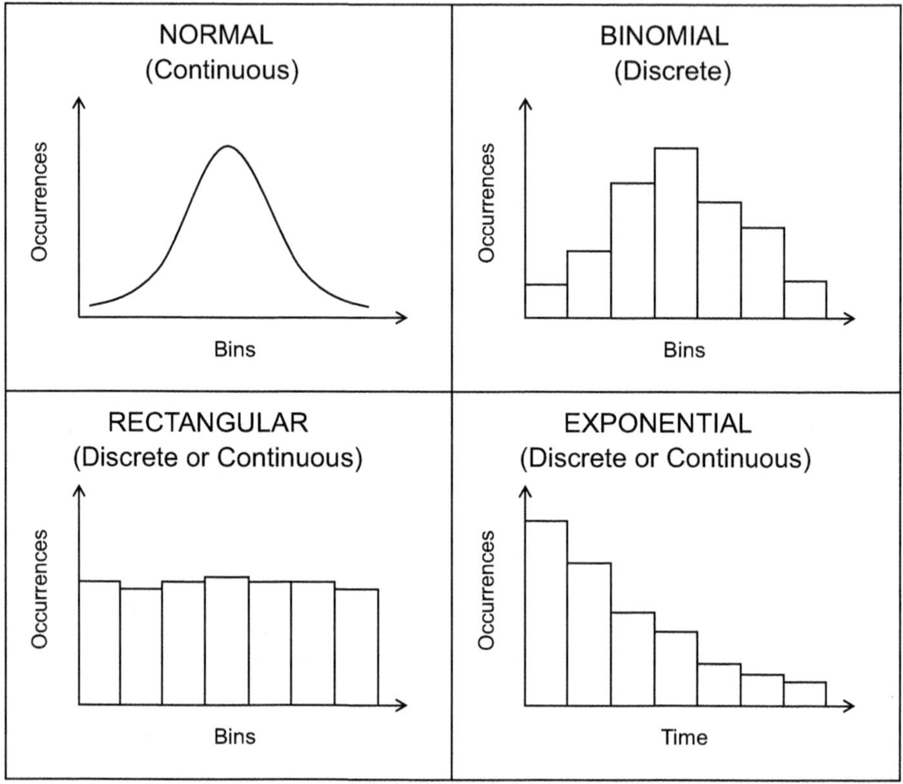

FIGURE 16.2 Common event distributions.

- The *standard* normal distribution is a normal distribution with a mean of zero
- Standard normal distribution function: $y(x) = \frac{1}{\sqrt{2\pi}} e^{-\frac{1}{2}x^2}$
- Plotting the function in graphing programs results in the normal *bell curve* shape
- The area under curve is a *probability distribution* of the data
    - Statistical analysis usually involves determining the area under the normal bell curve
    - Standardized tables provide area under the curve; integration is not required
- Normal distributions have several characteristic parameters, as shown in Figure 16.3
    - *Mean*, μ, is the average of all values occurring at the central tendency value
    - *Standard deviation*, σ, is distance from the mean on the horizontal axis
    - *Variance*, σ^2, is a measure of variability and *flatness* of the curve

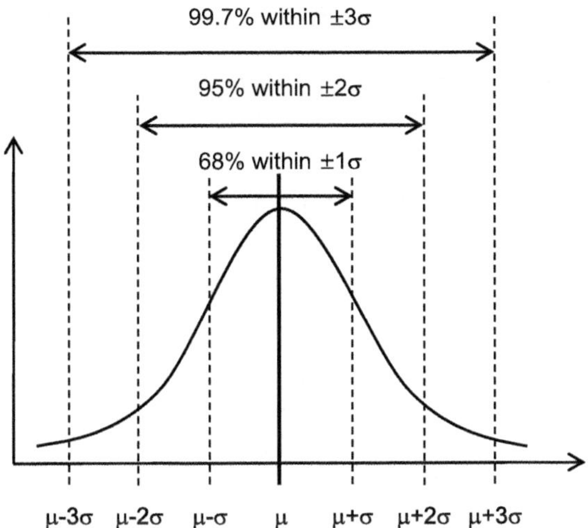

FIGURE 16.3 Parameters of the normal distribution.

- A standard deviation, σ, can be any value, but 1σ, 2σ and 3σ are common references
  - 68% of all normally distributed data should occur within 1σ (both plus *and* minus 1σ)
    - Given 100 samples, 68 are within 1σ, 32 are outside of 1σ
    - Also stated accurately, there is a 68% probability of any data point falling within 1σ
    - Also correct, $\pm 1\sigma$ contains 68% of all area under the normal distribution curve
    - 68% is typically cited for 1σ, but 68.26% is more accurate from tabular data
  - 95% of all data should occur within 2σ
    - Given 100 samples, 95 are within 2σ, 5 are outside of 2σ
    - Also stated accurately, there is a 95% probability of any data point falling within 2σ
    - Also correct, $\pm 2\sigma$ contains 95% of all area under the normal distribution curve
    - 95% is typically cited for 2σ, but 95.44% is more accurate from tabular data
  - 99.7% all data should occur within 3σ
    - Given 1000 samples, 997 samples are within 3σ, 3 are outside of 3σ
    - 99.7% of all area under the normal distribution curve is contained within 3σ

- Normal distribution tables provide the above values (integrated area under the normal curve)
- Normal distribution tables are *normalized* to a maximum value of *one* (1.0000)
  - 100% of all data under the curve equals 1.00
  - Tables provide a full range of values other than the easily memorized 1σ, 2σ and 3σ

Most statistical discussions revolve around the concept of *population*, which is not always applicable to engineering data sets. Classical statistics often consider a single, large pool of data, the population, where a subset is extracted *from* the population for analysis. Hypotheses are tested on the subset, relative to the population. While large quality control experiments may still test subsets from an overall population production run, most engineering experiments have little to do with extracting subsets from a population. Designing and running engineering experiments is about generating one or two small but complete data sets with no reference population. Since much of the statistical terminology references *population versus sample* for the choice of a statistical tool, an engineering experimenter should instead focus on the size of the data set size to choose the correct statistical tool.

- **Mean** of a normally distributed (binomial) data set is calculated with Equation 16.1
  - Mean is the most representative of the truth value from a data set of n values
  - Both μ and x_m are common notations for arithmetic mean (x_m is typical for summations)

$$x_m = \frac{1}{n}\sum_{i=1}^{n} x_i \qquad (16.1)$$

where:

x_m = *arithmetic mean (also, μ)*
n = *number of samples*
x_i = *individual sample values*

- **Standard deviation** is calculated with either Equation 16.2 or 16.3
  - Standard deviation is a measure of how close or far data is relative to the arithmetic mean
  - Equation 16.1 is appropriate for *small data sets* or *an unknown population*
    - Use for small engineering experiment data sets
    - Also suitable for analyzing *sample sets* extracted from an *unknown population*
    - Provides an *unbiased* result for small data sets

$$\sigma = \sqrt{\frac{1}{n-1}\sum_{i=1}^{n}(x_i - x_m)^2} \qquad (16.2)$$

where:

σ = *standard deviation*
Remaining definitions per Equation 16.1

- Equation 16.2 is appropriate for everything else
  - Use for a *known large population* or *large* data sets
  - This is the *biased* equation since it tends to underestimate value for small data sets

$$\sigma = \sqrt{\frac{1}{n}\sum_{i=1}^{n}(x_i - x_m)^2} \quad (16.3)$$

where:

σ = standard deviation

Remaining definitions per Equation 16.1

  - "*Large*" is not precisely defined when choosing between Equation 16.1 or 16.2
  - A sample size of 30 is the cutoff between the t-test sample size and the z-test normal distribution
  - Large data sets should be greater than 30 and closely match a normal distribution
  - Most discrete engineering data sets will use Equation 16.1
- Equations 16.1 and 16.2 are computed as follows
  - Inner parentheses: subtract the mean from the individual value and then square the result.
  - Repeat for $i = 2$: subtract the mean from the next value, then square the result and sum with the previous result
  - Repeat to $i = n$: subtract the mean from the next value, then square the result and sum with the previous result
  - Divide summation by n (or $n - 1$), take the square root

OUTLIERS AND CHAUVENET'S CRITERION

The normal distribution is an ideal mathematical function, and discrete data sets may or may not quite conform to the function. While reasonably behaved large data sets generally follow the ideal distribution, data points can appear somewhat removed from the rest of the distribution. Such points are outliers and require special consideration. From an experimental perspective, an experimenter should have a reasonable understanding of how well errors were controlled and whether a mistake was made. Data should not be discarded from a well-controlled experiment. *Data that can be traced to a mistake should be discarded.* The gray area with no clear answer is open for debate, and opinions vary on whether data should ever be discarded. The decision should never compromise the ethical reporting of results, but judgment and knowledge of the experiment should weigh the possibilities of a potentially bad data point. Chauvenet's criterion[2] is one of a few statistical techniques to help assess whether an outlying data point should be expected.

- Occasionally, experimental results yield data points well outside of expected values

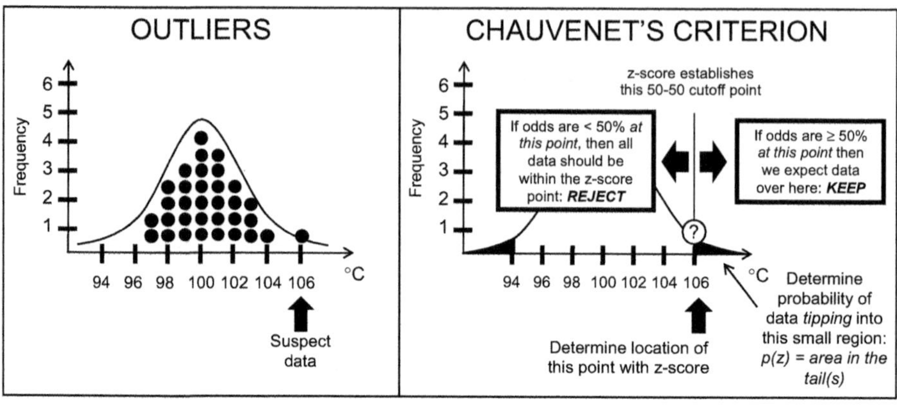

FIGURE 16.4 Chauvenet's criterion.

- Validly obtained data cannot be arbitrarily discarded
- Data resulting directly from a known or discovered error should be discarded and reacquired
- An anomalous data point could be the result of an unknown error, which can be assessed for discard
- Options to assess whether to discard an uncertain outlier
  - Chauvenet's criterion
  - Grubbs' test (*not discussed*)
  - Dixon *Q*-test (*not discussed*)
  - Pierce's criterion (*not discussed*)
- Chauvenet's *(show-ven-āyz)* criterion, as visually depicted in Figure 16.4
  - Assumes a normal distribution
  - Answers the question, "What are the odds we expect to see this data point?"
    - If odds are **less than 50%** of expecting the data point →**reject**, because odds are against
    - If odds are **50% or better** of expecting the data point →**keep**, because it is likely
  - Probability for normal distribution needs to be determined <u>at the outlier point of interest</u>
    - A *z*-score determines *where* the outlier point is located
    - *z*-score tables are then used to determine *probability* at that location
  - Not much data are expected past *z*-score location in the distribution tails, so the data point can be rejected if it does not meet the 50% threshold indicating it should be at that point
  - A *t*-score (vs *z*-score) is typically more suitable for discrete experimental data sets; however, a *z*-score has a tighter criterion than a *t*-score, and therefore, a *z*-score is more appropriate as an outlier test
  - The criterion can only be applied once per data set

- Steps in applying Chauvenet's criterion
  1. Assess all likely reasons for an outlying data point and proceed if data capture is suspect
     - Validly obtained data should not be discarded without due consideration
  2. Calculate the arithmetic mean using all data points with Equation 16.1
  3. Calculate the unbiased standard deviation using all data points with Equation 16.2
  4. Calculate how far the suspect data point is from the mean via the z-score in Equation 16.4

$$z_i = \frac{(x_i - x_m)}{\sigma} \tag{16.4}$$

where:

z_i = z-score for individual data point (point of interest)
z-score is the number of standard deviations a value is from the mean
z = z-score = (deviation ÷ standard deviation) ← All equivalent

  5. With the z-score → refer to an appropriate z-score table to get probabilities
  6. Adjust the probability value from the table to get appropriate tail data
     - If the data can be wrong, *either* as too high or as too low, it is a two-tailed test
     - If the data can be wrong, *only* as too high or *only* as too low, it is a one-tailed test
     - The test requires getting the *area under the curve* for either one or both tails, accordingly
     - The area under the curve in tails is the probability of data in the tails
  7. Multiply the probability by the number of samples in the data set
     - All table data are normalized to 1.0 and must be multiplied by the number of data samples
  8. Apply the criterion
     - If result of calculations ≥ 0.5 KEEP (greater than 50% odds data point is likely)
     - If result of calculations < 0.5 REJECT (less than 50% odds data point is likely)
  9. Recalculate the new mean and standard deviation if a data point is discarded
     - It may be possible that more than one outlier is discarded during the one allowable pass
     - The test cannot be applied again once a new mean and standard deviation are computed
- Understanding data from standard statistical tables, as visually depicted in Figure 16.5
  - Area under the normal distribution curve is a probability distribution

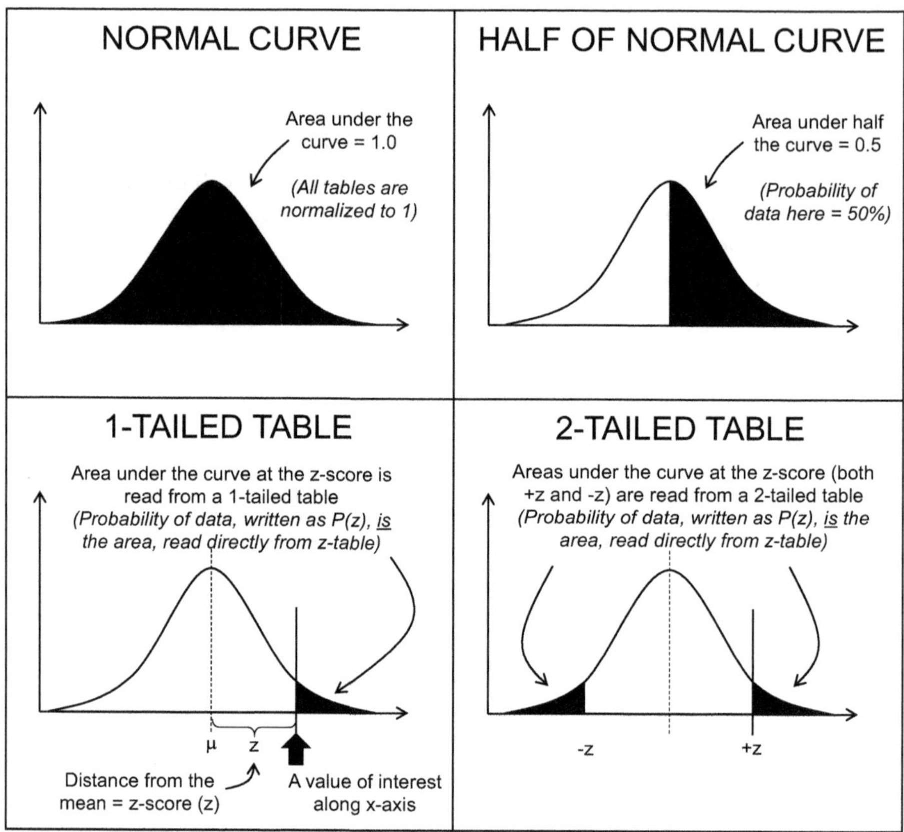

FIGURE 16.5 Relating areas under the curve to probability returned from a z-table.

- All tables are normalized to a maximum value of 1.0, where 100% of the "area under the curve" = 1
  - There is a 100% chance that all normally distributed data falls under the normal curve
  - There is a 50% chance that normally distributed data falls either left or right of the mean
  - Alternatively, 50% of all normally distributed data is either left or right of the mean
- A z-score identifies a point of interest along the horizontal axis
  - A z-score is a standard deviation
  - 1σ (one standard deviation) is equivalent to a z-score of $z = 1.0$
- A z-table provides the area under the curve at the z-score value
  - A one-tailed z-table returns area under one tail (left or right, but values are symmetric)
  - A two-tailed z-table returns area under both tails (twice a one-tailed value)

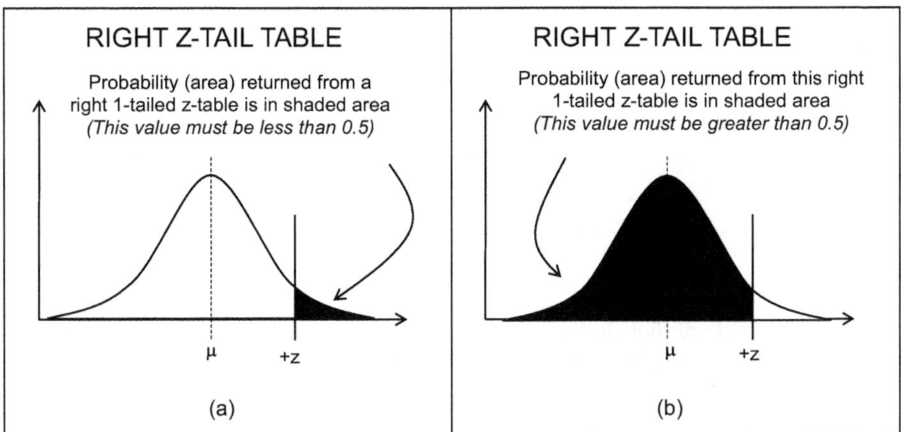

FIGURE 16.6 (a and b) Two options for presenting the same data from a *z*-table.

- The area returned from a *z*-table *is* the probability of data in that area under the curve
- Area under curve = *z*-score table value = probability → written as $P(z)$, called a *p*-value
- *z*-tables are configured differently, so *read* the table graphic to ensure proper values
  - *Data under the curve ALWAYS equals one and data under half the curve equals 0.5*
  - Figure 16.6 shows two approaches to presenting one-tailed data
- Graphic (a) must return a value less than 0.5 since it is less than half the curve
- Graphic (b) must return a value greater than 0.5 since it is more than half the curve
  - For a given *z*-score, both curves in Figure 16.6 are equal by: $P(z)_a = 1 - P(z)_b$

Example 16.2

Use Chauvenet's criterion to assess an outlying data point. The heat pipe lab (Lab #12) uses Schrader valves to hold vacuum instead of pressure, as designed. An experiment was designed to test how much vacuum force high-pressure Schrader valves can hold from a random sample of 12 valves. The experiment was crudely constructed with visual observations rather than an instrumented test. It bears considering that there was error in data measurements.

- GIVEN: The force holding data set in Table 16.1.
- FIND: Determine the extreme data points and assess the *potential* to discard them with Chauvenet's criterion. Even if the criterion is met, all factors surrounding the nature of the experiment need to be weighed

TABLE 16.1
Vacuum Force Holding Limit for 12 Randomly Selected High-Pressure Schrader Valves

Valve	1	2	3	4	5	6	7	8	9	10	11	12
Force (g)	376	364	412	409	306	382	391	352	358	418	382	389

before discarding. In this case, the experiment is acknowledged as being crudely configured, with the possibility of poor measurements. Apply the steps for Chauvenet's criterion.

1. Data point #5 appears well removed from the next nearest value
2. Using Equation 16.1, the arithmetic mean is: $x_m = 378$ gram-force
3. Using Equation 16.2 for a small sample set, the standard deviation is: $\sigma = 31$
4. Using Equation 16.4, the *z*-score for point #5, 306 g, is: $z_i = -2.33$
   - The negative value places the data point left of the mean
   - Normal distribution data tables are symmetric, so the negative sign is inconsequential
   - With $z = 2.3$, the probability, $P(z)$, from the *z*-table *must* be between 95% and 99.7% since 2.3 is the number of standard deviations, which falls between the easily memorized 2σ (95%) and 3σ (99.7%) values from Figure 16.3
5. All normal distribution data tables return the *same* probabilities (area under the curve)
   - Table values differ only in terms of the area tabulated in each table
   - Choose a table based on the values needed; in this case, a two-tailed table
   - If the necessary table is not available, transform the available table values as necessary
   - Table 16.2 is available but does not provide tail data directly and must be transformed
6. Figure 16.8 depicts the probability required compared to the values available in Table 16.2
   - The probability (area) returned from Table 16.2 at $z = 2.33$ is: $P = 0.4901$
   - Since *half* the area under the normal curve is *always* 0.5
   - One-tail probability is: $P_1(z = 2.33) = 0.5 - 0.4901 = 0.0099$
   - Two-tail probability is twice the one-tail value: $P_2(z = 2.33) = 2(0.0099) = 0.0198$
   - The probability of data in the tails for data point #5, $z = 2.33$, is $P = 1.98\%$
7. There is less than 2% chance the data should be in the tails *at* data point #5 (306 gram-force)
   - Data point #5 is teetering on the 1.98% probability, and there are 12 total samples

TABLE 16.2
***Z*-Score Values (Probability or Area under Normal Curve) Associated with Figure 16.7**

Z	0.00	0.01	0.02	0.03	0.04	0.05	0.06	0.07	0.08	0.09
0.0	0.0000	0.0040	0.0080	0.0120	0.0160	0.0199	0.0239	0.0279	0.0319	0.0359
0.1	0.0398	0.0438	0.0478	0.0517	0.0557	0.0596	0.0636	0.0675	0.0714	0.0753
0.2	0.0793	0.0832	0.0871	0.0910	0.0948	0.0987	0.1026	0.1064	0.1103	0.1141
0.3	0.1179	0.1217	0.1255	0.1293	0.1331	0.1368	0.1406	0.1443	0.1480	0.1517
0.4	0.1554	0.1591	0.1628	0.1664	0.1700	0.1736	0.1772	0.1808	0.1844	0.1879
0.5	0.1915	0.1950	0.1985	0.2019	0.2054	0.2088	0.2123	0.2157	0.2190	0.2224
0.6	0.2257	0.2291	0.2324	0.2357	0.2389	0.2422	0.2454	0.2486	0.2517	0.2549
0.7	0.2580	0.2611	0.2642	0.2673	0.2704	0.2734	0.2764	0.2794	0.2823	0.2852
0.8	0.2881	0.2910	0.2939	0.2967	0.2995	0.3023	0.3051	0.3078	0.3106	0.3133
0.9	0.3159	0.3186	0.3212	0.3238	0.3264	0.3289	0.3315	0.3340	0.3365	0.3389
1.0	0.3413	0.3438	0.3461	0.3485	0.3508	0.3531	0.3554	0.3577	0.3599	0.3621
1.1	0.3643	0.3665	0.3686	0.3708	0.3729	0.3749	0.3770	0.3790	0.3810	0.3830
1.2	0.3849	0.3869	0.3888	0.3907	0.3925	0.3944	0.3962	0.3980	0.3997	0.4015
1.3	0.4032	0.4049	0.4066	0.4082	0.4099	0.4115	0.4131	0.4147	0.4162	0.4177
1.4	0.4192	0.4207	0.4222	0.4236	0.4251	0.4265	0.4279	0.4292	0.4306	0.4319
1.5	0.4332	0.4345	0.4357	0.4370	0.4382	0.4394	0.4406	0.4418	0.4429	0.4441
1.6	0.4452	0.4463	0.4474	0.4484	0.4495	0.4505	0.4515	0.4525	0.4535	0.4545
1.7	0.4554	0.4564	0.4573	0.4582	0.4591	0.4599	0.4608	0.4616	0.4625	0.4633
1.8	0.4641	0.4649	0.4656	0.4664	0.4671	0.4678	0.4686	0.4693	0.4699	0.4706
1.9	0.4713	0.4719	0.4726	0.4732	0.4738	0.4744	0.4750	0.4756	0.4761	0.4767
2.0	0.4772	0.4778	0.4783	0.4788	0.4793	0.4798	0.4803	0.4808	0.4812	0.4817
2.1	0.4821	0.4826	0.4830	0.4834	0.4838	0.4842	0.4846	0.4850	0.4854	0.4857
2.2	0.4861	0.4864	0.4868	0.4871	0.4875	0.4878	0.4881	0.4884	0.4887	0.4890
2.3	0.4893	0.4896	0.4898	0.4901	0.4904	0.4906	0.4909	0.4911	0.4913	0.4916
2.4	0.4918	0.4920	0.4922	0.4925	0.4927	0.4929	0.4931	0.4932	0.4934	0.4936
2.5	0.4938	0.4940	0.4941	0.4943	0.4945	0.4946	0.4948	0.4949	0.4951	0.4952
2.6	0.4953	0.4955	0.4956	0.4957	0.4959	0.4960	0.4961	0.4962	0.4963	0.4964

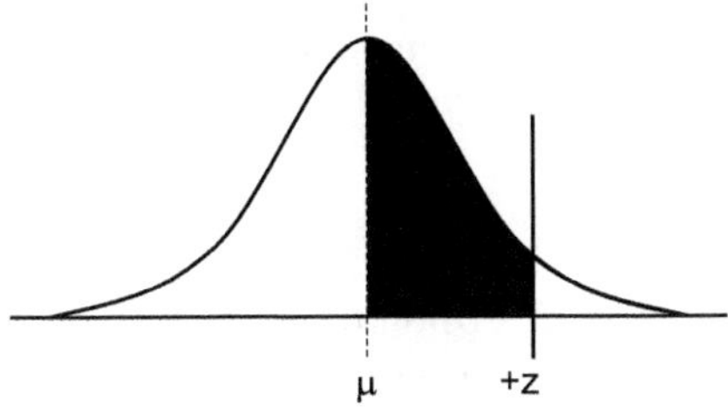

FIGURE 16.7 Graphic for Table 16.1 depicting data provided in Table 16.2.

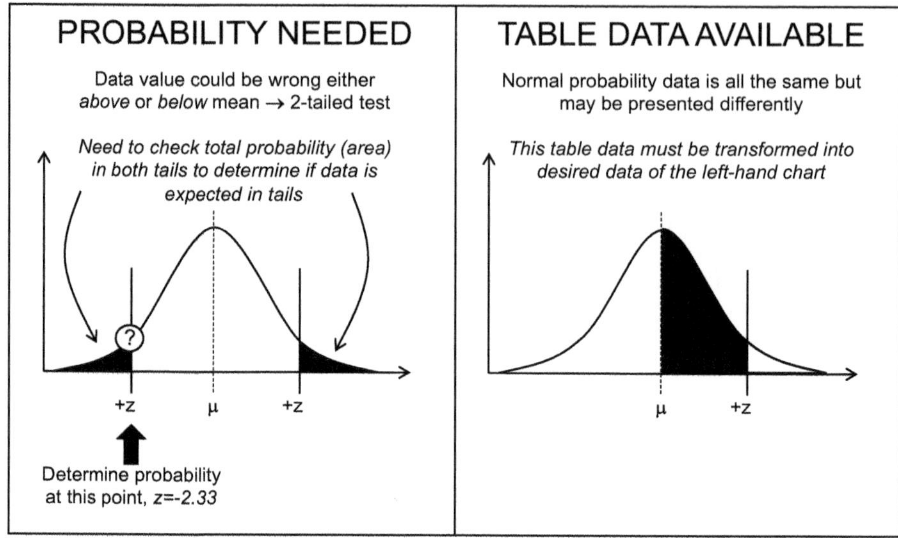

FIGURE 16.8 How to manipulate probability values (area) based on available tables.

- The number of samples that should be in the tails: $n_{tails} = 12(0.0198) = 0.24$
- Only about a quarter of <u>one</u> data point should be in tails

8. Apply the 50-50 criterion
   - If n_{tails} was 0.50, we would be obligated to give benefit of the doubt and concede this one data point might have tipped into the tails since it has a 50% (0.5) chance
   - However, the odds are only 24% that this point would end up in the tails
   - REJECT the data point: $0.24 < 0.5$ (0.5 is the criterion)
9. Recalculate the new mean and standard deviation once the outlier has been discarded
   - Revised data set: $n = 11$, $x_m = 385$, $\sigma = 22$
   - As expected, the standard deviation has now tightened and improved from $\sigma = 31$ to $\sigma = 22$

- For this example, if the total number of samples were $n = 26$ instead of only 12, with all other values the same, data point #5 should be retained since likelihood increases with more data
  - With 26 samples: $n_{tails} = 26(0.0198) = 0.51 \rightarrow 0.51 > 0.50 \rightarrow$ *KEEP*

EXERCISES

A factory produces rivets to fit in 2.0mm holes. Any rivet larger than 2.05mm will not fit in the 2.0mm holes, whereas smaller diameter rivets will fit and expand just fine. Rivets are produced with normally distributed tolerances. Only rivets larger than 2.05mm must be discarded.

1. Using a statistical analysis for discarded rivets, should a one-tailed or two-tailed test be used?
2. What percentage of rivets should be within one standard deviation?
3. Refer to Table 16.2. What percentage of rivets should be within 1.5 standard deviations?

An experiment is designed to measure the speed of sound through a solid metallic cylinder. Four measurements are taken, as shown in Table 16.3. Answer questions 4 through 7.

TABLE 16.3
Ultrasonic Time-of-Flight Measurements for a Metallic Solid

Sample #	1	2	3	4
Time (μs)	21.9	21.8	21.7	21.9

4. Determine the arithmetic mean of the sample set.
5. Determine the standard deviation of the sample set.
6. Data point #3 (21.7 μs) is lower than the other values. If the experiment was conducted during a single lab session under controlled conditions, should data point #3 be considered for outlier analysis?
7. Assume the experiment was not conducted during a single day, and the data set might have been subjected to errors in an uncontrolled environment. Apply Chauvenet's criterion to data point #3 and determine whether it can be discarded as an outlier.

NOTES

1 Hines, W. & Montgomery, D. (1980). *Probability and statistics in engineering and management science.* John Wiley & Sons. p.181.

2 Holman, J. (1984). *Experimental methods for engineers.* McGraw-Hill Book Company. p.72.

17 Evaluating the Hypothesis

Experiments supporting a hypothesis must provide a convincing data set. A statistical test can be applied to the data set to assess if results are sufficiently clear compared to the reference value.

There is a difference between correlation and causality. Just because two events or phenomena are correlated does not mean one caused the other or vice versa. Correlation is good for observations which lead to hypothesizing models, but it is also a basis for superstitions. Establishing causality is the foundation of modern society through predictable and repeatable relationships. A hypothesis contends a causal relationship, and an experiment is meant to demonstrate the hypothesized relationship. *Proof* of a causal relationship through experimentation is too strong a word since experiments cannot test every scenario, and a hypothesis survives only until a single event undermines it. A well-designed experiment includes accepted statistical approaches to properly support a hypothesis.

QUANTIFYING THE HYPOTHESIS

From a data analysis point of view, a hypothesis is one thing only: the mean value of a data set; a single number. An experiment designed to support a hypothesis generates a data set, which in turn provides a single mean value. The fundamental concept of an experiment is to change a parameter (or parameters) to determine if there is a causal relationship. This concept of testing for change requires having a baseline for comparison. This baseline can be a control data set, an established value or a preexisting data set. Therefore, when it comes to hypothesis testing, there are always *two* numbers: a *control* value (or control set mean value, a single number or a generally accepted value) and an *experimentally* generated mean value. The control value is the *null hypothesis*. The experimentally generated data set mean is the *alternative* hypothesis. The experimental hypothesis is always an alternative to the established null *stake-in-the-ground* reference. *We never have the option to accept or reject our own alternative hypothesis since we put it forward in the first place.* Rather, we're either stuck with the null because our data isn't convincing, or we can reject the null because our data is clear in relation to the null. When population data sets are involved, it's safe to say, "*it's all about the null*," since we need discernment *from* the null, and levels of significance are <u>set on the population (null) data set</u>, *not* the samples. These concepts are illustrated in Figure 17.1, where the hypotheses are the resultant mean values from a data set. An unconvincing study that fails a hypothesis test has a data set that is too close and mostly indistinguishable from the null data set. A convincing study shows a clearly acceptable distinction between the null and alternative data sets.

DOI: 10.1201/9781003329237-17

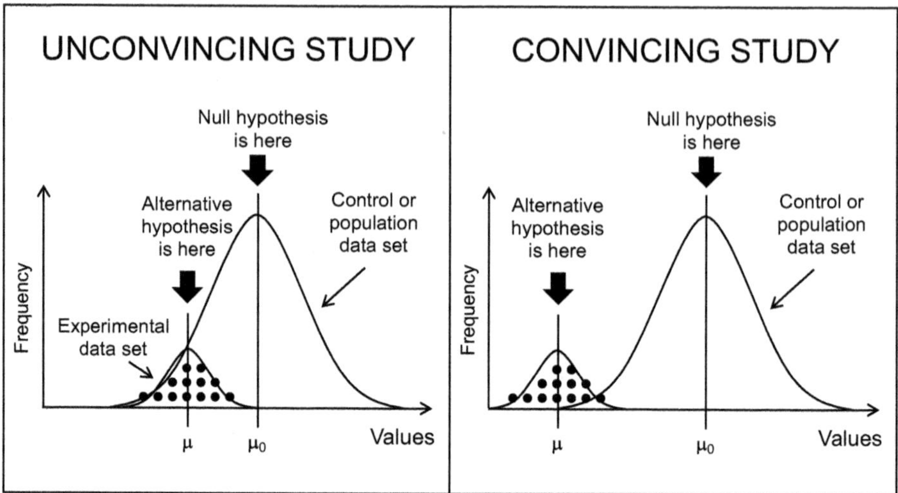

FIGURE 17.1 An unconvincing study fails a hypothesis test when there is little distinction between the control and the experimental data sets. A convincing study shows discernment.

- A hypothesis is a single number, the mean of a distribution
  - Null hypothesis: a pre-existing value, control value, or mean of a control data set
  - Alternative hypothesis: the mean from an experimental data set
- Null and alternative hypotheses live in a cloud of uncertain values (the distribution)
  - A convincing study has sufficient separation between alternative and null hypotheses
- Hypothesis testing is in relation to null since the null exists prior to the experimental data set
  - The alternative hypothesis cannot be rejected or accepted
  - Only the null hypothesis can be rejected or not rejected
  - "*Is our experimental data good enough to reject the null?*"
    - "*If we can reject the null, it leaves us with our alternative.*"
    - "*If we cannot reject the null, we are stuck with the null.*"
  - "*It's all about the null.*"

STATISTICAL VERSUS EXPERIMENTAL STUDIES

For testing a hypothesis, consider the three separate scenarios in Figure 17.2. The first scenario is a statistician's hypothesis test about a sample set taken from a population. The second scenario is the experimenter's data set compared to a single value, and the third scenario is the experimenter's data set compared to a control or reference data set.

The first statistical study scenario in Figure 17.2 refers to and uses the classic terminology of "*population*." In this scenario, a hypothesis is conjectured about a subset of the population, and a statistical test is performed to see if the subset is in fact

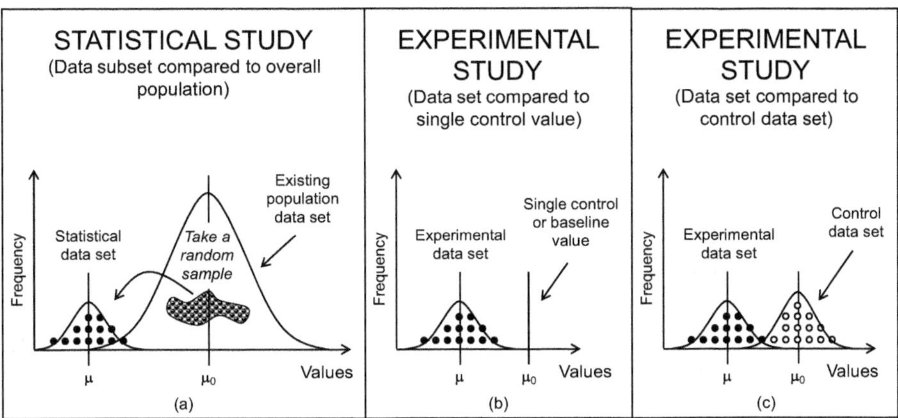

FIGURE 17.2 (a–c) Three hypothesis test scenarios.

clearly distinguished from the overall population. However, since most experiments generate their own data sets relative to a control, the term *population* is unsuitable and is mostly dismissed for experimenters. This leaves the two experimental study scenarios in Figure 17.2 to consider for hypothesis testing: comparing a single data set to a single value or comparing two data sets.

- Two typical hypothesis-testing scenarios for the experimenter
  - *Comparing* an experimental data set mean to a single value
  - Comparing an experimental data set mean to a control set mean

LEVEL OF SIGNIFICANCE

The statistical level of significance establishes the discernment between an experimental data set and the null data set. As in Figure 17.1, the results of an experiment should provide a data set sufficiently clear from the control to be convincing. For statistical studies, the level of significance is set on the population or null data set since that is the fixed reference for discernment. While most any level of significance can be selected, the generally accepted level is 5%, or 1% for a particularly stringent requirement. As indicated in Figure 17.3, a 5% level of significance sets bounds on the null data set, where 95% of all population or null values occur within the 5% bounds and only 5% of data falls into the tails outside of these bounds.

For a data set to pass a hypothesis test, the mean must meet the established level of significance. The alternative hypothesis, which is the mean of the experimental data set, must fall outside the level of significance markers as shown in Figure 17.4.

Figures 17.3 and 17.4 establish the statistical definitions for the level of significance necessary for hypothesis testing. However, these tests are typically based on statistical studies relative to large population null data sets. For the experimenter in scenario (b) of Figure 17.2, there is no existing population data set to set the level of significance. Therefore, a statistical *one-sample t-test* can be used for the hypothesis test. In scenario (c) of Figure 17.2, a *two-sample test* is most suitable.

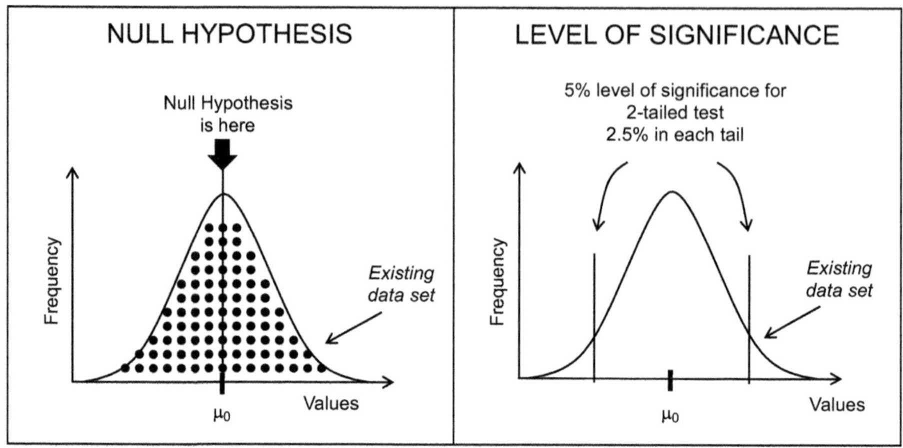

FIGURE 17.3 Null hypothesis and level of significance.

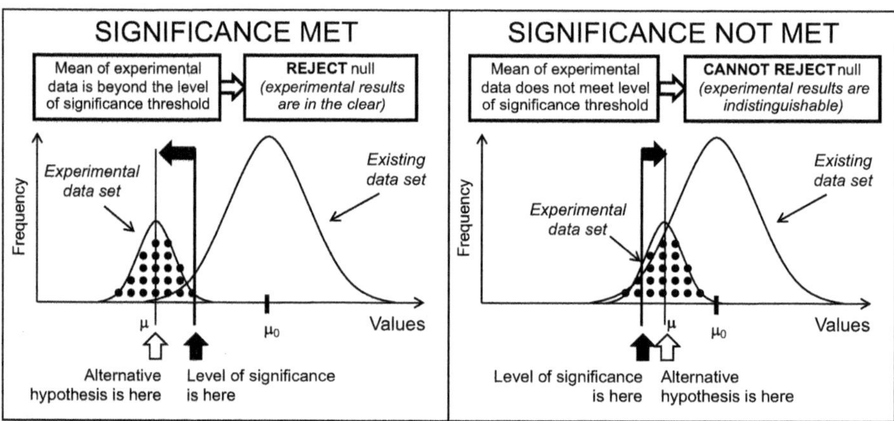

FIGURE 17.4 Meeting or failing to meet the level of significance.

HYPOTHESIS TESTING RELATIVE TO A SINGLE VALUE

Testing a data set mean, relative to a single value, refers to the first experimental study scenario, (b), in Figure 17.2. In some cases, only a single control value is available or an experiment suggests a new or improved relationship relative to an existing single value. A famous example of this hypothesis test scenario was astronomer Arthur Eddington's test of Einstein's general theory of relativity in 1919.[1] The established single value based on Newtonian physics called for starlight deflection in the vicinity of the sun during an eclipse of 0.87 arc-seconds. The Einstein prediction called for a 1.75 arc-second deflection. Three different data sets were captured during the experiment to ultimately demonstrate the validity of Einstein's theory. As a hypothesis test, the data set should have provided a mean of 1.75 arc-seconds relative to the single reference value of 0.87 arc-seconds. This hypothesis test passes when the data set distribution yields a mean of 1.75 arc-seconds that has sufficient discernment from the single existing value of 0.87 arc-seconds.

Two options are available for this test: a *t*-test for small data sets or a *z*-test for large data sets and a tighter criterion. The same discussion from Chapter 5 for sample size hypothesis testing is valid here.

- Option #1: the *t*-test for small sample sizes
  - Establish the level of significance first, usually 5% based on standard statistical practices
  - Determine if the test is a one- or two-tailed test
    - One-tail: the experimental mean can only be less than or only be greater than the reference
    - Two-tail: the experimental mean can be either less than or greater than the reference value
  - Determine the degrees of freedom from the sample size to enter a standard *t*-table
  - Establish the *marker to exceed* by looking up the *t*-test value from a *t*-table
  - Use Equation 17.1 to compute the data set *t*-score for comparison to the *t*-table marker

$$t = \frac{x_m - \mu_0}{S/\sqrt{n}} \tag{17.1}$$

where: t = *t*-value to compare with a *t*-table value
x_m = experimental data set mean
μ_0 = reference or control value
S = standard deviation of small data set
n = number of samples

  - Check the *t*-score from Equation 17.1 against the *t*-test from a *t*-table
    - If the *t*-score from Equation 17.1 is greater than the table *t*-test, the hypothesis passes
    - If the t-score from Equation 17.1 is less than the table *t*-test, the hypothesis fails

Example 17.1

Consider a notional re-creation of Eddington's experimental test.

- GIVEN: Control reference value, $\mu_0 = 0.87''$ and the data set in Table 17.1.
- FIND: Determine whether the data set in Table 17.1 is sufficient to support the hypothesis that light is deflected 1.75″ due to the sun's gravity at the 95% confidence level (5% level of significance) compared to the reference value of 0.87″

TABLE 17.1
Arc-Second Measurements from Star Photos during an Eclipse

Measurement No.	1	2	3	4	5
Arc-secs (″)	1.01	1.77	1.21	1.84	1.68

- Set the level of significance at 5%, which means 0.05 of data is allowed in both tails
- This is a two-tailed test since experimental data could be wrong above or below the reference
- Determine degrees of freedom (refer to Chapter 5): $DOF = n - 1 = 4$
- Enter any standard *t*-table to determine the level of significance and the cutoff value
  - $DOF = 4$, and two-tailed significance = 0.05
  - *t*-test value from any *t*-table: $t_{Table} = 2.776$
  - $t = 2.776$ is the level of significance marker as described in Figure 17.4
  - If *t*-score from Equation 17.1 does not exceed $t = 2.776$, the data is not significant

At this point, it is worth nothing that the hypothesis test value is established at $t = 2.776$. *This value is established almost completely independent of the data set.* The only consideration for data is the number of samples where the degrees of freedom are identified as $DOF = 4$ ($n - 1$). The DOF establishes the variance in the distribution, but the distribution is standard and otherwise independent of the data. Referring to Figures 17.3 and 17.4 and the discussions from Chapter 5, the *t*-value merely marks a point on the horizontal axis of the distribution, which could be on the left tail and also the right tail of a two-tailed test. However, the *t*-test table returns a totalized value of all tail data as the cutoff. If the *t*-value calculated from the data with Equation 17.1 exceeds the *t*-score from the table, the mean of the experimental data set, with its messy distribution of values, has pushed far enough away from the reference value and the hypothesis passes the test.

- Calculate the mean of the data set (Equation 16.1): $x_m = 1.502$
- Calculate the sample standard deviation (Equation 16.2): $S = 0.3691$
- Calculate the *t*-score (Equation 17.1): $t = \dfrac{1.502 - 0.87}{0.3691/\sqrt{5}} \rightarrow t = 3.8288$
- $t_{score} = 3.8288 > t_{test} = 2.776 \rightarrow$ Hypothesis passes and the data is acceptable
- "Hypothesis passes" only because the null is rejected

All table data is standard-normal, so the distribution is a normal distribution (normalized to 100%) with an adjustable variance based on the number of samples. The table data mean is standardized to zero. Once the data set *t*-value is computed, there is no requirement to adjust for the mean since Equation 17.1 already standardizes and normalizes the data *t*-score, so it is comparable and ready to be compared with standard-normal table values.

- Option #2: the z-test for large sample sizes
  - As the number of test samples increases, the t-table values approach the z-table values
  - Data sets with $n \geq 30$ can use a z-table
  - The process is exactly the same as with the previous example
  - Use the z-score formulation in Equation 17.2 to compute the data z-score

$$z = \frac{x_m - \mu_0}{\sigma/\sqrt{n}} \tag{17.2}$$

where: z = z-value to compare with a z-table value
σ = standard deviation of large data set
All other parameters same as Equation 17.1

Example 17.2

Consider everything exactly the same as in Example 17.1, except there are 30 experimental data points. The goal is to find the z-score that returns 95% of all data, which is 0.95 area under the standard-normal curve. This task is best completed with a computational tool since retrieving the desired value directly from a z-table requires working the table *backwards*. However, understanding this process assures a greater understanding of z-tables.

Refer to Figure 16.7 and Table 16.2 for this example. At the desired 5% level of confidence, 0.95 of all data should appear under the normal curve. Table data is the integrated area under the curve. Figure 16.7 indicates that for a given z-score, Table 16.2 will return up to half the area under the curve. We need 0.95 of the area, but since the table is only working with 50% or less, we need to enter the table with 0.95/2, which is 0.4750. Searching the table for 0.4750, a z-score of 1.96 matches this value. In other words, in order to get 95% of our data (95% confidence, 5% significance), we need 0.4750 under the curve as shown for this particular table, multiplied by 2, and the z-score is 1.96. The z-score for 5% confidence is always $z = 1.96$ because 95% of the data is always between ± 1.96 standard deviations (z-values) for the standard-normal curve, regardless of how a z-table may choose to present this value.

Comparing hypothesis cutoff values from Example 17.1, $t = 2.776$, and Example 17.2, $z = 1.96$, 95% of all data should be within a tighter distribution of ± 1.96 standard deviations with more samples using the z-table. The higher t-value of ± 2.776 standard deviations away from the mean is necessary to capture the same 95% of data confidently with the wider variance of the smaller sample set.

- Calculate z-score (Equation 17.2): $z = \dfrac{1.502 - 0.87}{0.3691/\sqrt{30}} \rightarrow z = 23.5$
- $z_{score} = 23.5 > z_{test} = 1.96 \rightarrow$ Hypothesis clearly passes and the data is acceptable
- "Hypothesis passes" only because the null is rejected

With a large, normally distributed data set, the data set mean and control reference are well separated, and Einstein's theory is supported. (*Apologies to Einstein for suggesting an inaccurate mean in these examples since his value was exact!*)

HYPOTHESIS TESTING BETWEEN TWO SAMPLE DATA SETS

The second experimental study scenario, (c), in Figure 17.2 requires considering different variances between two small experimental data sets. The desire is to consider the variations in both data sets to ensure the hypothesis meets the significance level.

This hypothesis test begins as with the previous hypothesis tests by establishing the significance level and degrees of freedom and then entering a t-value table to get the threshold value. Since there are now two data sets, the degrees of freedom are the sum of both sample set DOFs. Once the t-value is pulled from a t-table as the hypothesis test threshold value, compute the two-sample t-test for comparison with Equations 17.3 through 17.5.[2]

$$t = \frac{x_{m1} - x_{m2}}{SE} \tag{17.3}$$

$$\text{Where}: SE = S_P\sqrt{\frac{1}{n_1} + \frac{1}{n_2}} \tag{17.4}$$

$$\text{Where}: S_P{}^2 = \frac{s_1{}^2(DOF_1) + s_2{}^2(DOF_2)}{DOF_1 + DOF_2} \tag{17.5}$$

where: t = t-value to compare with a t-table value

x_{ml} and x_{ml} = data set means from controls and test specimens

SE = standard error calculation

$S_P{}^2$ = pooled variance between sample sets

$S_1{}^2$ and $S_2{}^2$ = data set variances (square of standard deviations)

n = number of samples

$DOF = n - 1$ (degrees of freedom)

Example 17.3

Consider an experiment to measure very low resistances in a metal test specimen. The experiment tests a hypothesis about the change in resistance based on the influence of a crack. Two data sets are taken, one for the control and one for the change condition with the crack.

- GIVEN: Experimental data set parameters in Table 17.2
- FIND: Determine whether the data sets in Table 17.2 are sufficient to support the hypothesis
- SOLUTION: Follow basic steps with previous examples, using Equations 17.3 through 17.5
  - This is a one-tailed test since the change in values can *only be greater* than the control
    - Resistance can only increase due to a crack, not decrease

TABLE 17.2
Resistance Measurements before and after a Material Crack

	Control Set	Change Set
Number of samples	15	15
Mean	0.1400	0.1500
Standard deviation	0.020	0.023

- Set the level of significance at 5% which means 0.05 of data is allowed in tail
- Determine degrees of freedom (refer to Chapter 5): $DOF = n - 1 = 14$
- Enter any standard *t*-table to determine the test hypothesis significance cutoff value
  - $DOF = 14$ and one-tail significance = 0.05
  - *t*-test value from any *t*-table: $t_{Table} = 1.761$
  - $t = 1.761$ is the level of significance marker requirement to exceed for significance
- Pooled variance with Equation 17.5: $S_P{}^2 = \dfrac{0.14^2(14)+0.15^2(14)}{14+14}$ $\rightarrow S_P{}^2 = 4.65E-4$
- Standard error with Equation 17.4: $SE = 0.0216\sqrt{\dfrac{1}{15}+\dfrac{1}{15}} \rightarrow SE =$ $7.87E-3$
- *t*-score with Equation 17.3: $t = \dfrac{0.14-0.15}{7.87e-3} \rightarrow t = -1.27$

The *t*-score is negative, which is inconsequential since the distributions are symmetric. The (absolute value) *t*-score between the two data sets of $t = 1.27$ does not meet the threshold for significance from the *t*-table of $t = 1.761$. The change in resistance with the test parameter compared to the control data set is not significant, and the null hypothesis cannot be rejected (there is no change).

AN UNSUPPORTED HYPOTHESIS

In the previous example (Example 17.3), there was insufficient distinction between the control data set and the test parameter data set. When plotted, the two data sets mostly overlap, similar to Figure 17.4 where the significance level is not met. If the hypothesis is not supported by the data, it may be due to an incorrect hypothesis, an inadequate experimental setup or insufficient data. If data shows a slight but insufficient change, the experimental setup should be examined for errors and measurement device fidelity, both of which can lead to poor data and large variances. More samples can also improve the resolution of potential differences. Increasing the sample size also tends to improve the *power of the study*. For a given mean and level of significance, a higher power study will simply have more data beyond the level of significance for a more compelling case.

- Hypothesis tests can fail due to
  - The hypothesis itself
  - The experimental setup (errors and measurement fidelity)
  - Small or insufficient data sets
  - Type II (Beta) error

Data sets only provide *insight* into the truth. Selecting a 95% confidence level, or any other confidence level, still leaves room for error. If data is not normally distributed, there will be errors. Since data is discrete, it can only approach a normal distribution. If test parameter data intending to support the hypothesis does not meet the threshold significance, yet the hypothesis should have passed, this is a Type II or Beta error. In more correct terminology, the null hypothesis was not rejected when it was actually wrong, so the alternative hypothesis should have remained. Conversely, if the data appears to support the hypothesis, yet incorrectly so, this is a Type I or Alpha error. In other words, incorrectly rejecting the null in favor of our alternative hypothesis is a Type I error.

- Errors in rejecting or not rejecting the null
  - "*It's all about the null*," only the null can be rejected or not rejected
  - The alternative hypothesis cannot be directly accepted or rejected
  - Data is only insight into the truth
  - Data is a *probability* distribution, not a *certainty* distribution
  - Given the ever-present uncertainty in the data, it is possible to make an error in the decision
    - Rejecting the null when it was actually correct → Type I (Alpha) error
    - Failing to reject the null when it was actually wrong → Type II (Beta) error

EXERCISES

1. TRUE or FALSE. If two events are correlated, there is a causal relationship.
2. TRUE or FALSE. The mean of an experimental test parameter data set is the hypothesis.
3. TRUE or FALSE. If supported by data, an alternative hypothesis can be accepted.
4. At 5% significance, how much data is in the tails of a two-tailed test?
5. Answer the following about t-scores and t-tables for a test parameter data set, where $t = 2.12$
   a. If the t-table value is $t = 2.093$, is the null rejected?
   b. For 20 samples, how many degrees of freedom are in the test?
   c. Is this a one- or two-tailed test?

6. Consider an experiment an experiment to test whether a crack has a change on natural frequency as predicted by theory. For a given setup, theory predicts a change of 1.2 rad/s. Assess the following data set and make a decision about the hypothesis if the level of significance is set to 1%.

TEST No.	1	2	3	4	5	6	7
$\Delta\omega$	1.07	0.80	0.74	1.10	0.62	0.71	0.90

NOTES

1 Kennefick, D. (2019). *No shadow of a doubt: the 1919 eclipse that confirmed Einstein's theory of relativity*. Princeton University Press.
2 Montgomery, D. (2013). *Design and analysis of experiments*. John Wiley & Sons. p.38.

18 Virtual Experiments

Physical experiments require significant resources. The Monte-Carlo simulation offers a computational method to determine a statistically likely result from a series of uncertain events as an alternative to physical experiments. Lab #7 is a computer-simulated experiment using the Monte-Carlo algorithm to assess material failure stress.

The purpose of an experiment is to determine the outcome of changing one or more parameters from a set of conditions. The purpose of running an experiment more than once is to obtain a data set with a mean. Well-designed experiments eliminate systematic errors but are still left with random errors due to inexact measurements or other uncontrolled but quantifiable influences. However, physical experiments require resources, time and effort to configure and run to learn the likely outcome of changing parameters. In some cases, setting up and running a physical experiment may be impractical, or limitations may prevent obtaining more than one or two data points. Virtual simulations performed with computational algorithms provide a means to run an experiment where a physical experiment may be impractical or unnecessary.

MONTE-CARLO SIMULATIONS

The Monte-Carlo simulation begins with known relationships between parameters. For example, mechanical stress is load force divided by cross-sectional area. If force and area are precisely known, stress can be easily calculated. However, if a load or area is subject to variation, stress cannot be precisely known. If a new alloy is considered for design, the yield stress must be well defined and tested. If the machine applying the load has a range of uncertainty and the area measurement is subject to uncertainty, then the stress is also subject to uncertainty as a function of both load and area variations. The situation becomes even less certain when heat-treatment processes for the alloy are considered along with other variations in the manufacturing process. Several physical experimental tests can be performed with several samples for an acceptable mean value. Alternatively, all of the relationships can be entered into a computer algorithm with random variables for each of the sources of randomness, and the simulation can return the most likely result after 10^4, 10^5 or more iterations. The Monte-Carlo simulation utilizes the power of a computer to generate random inputs while creating a sample set far greater than that of a physical experiment to return a well-defined result.

- A stochastic algorithm is a randomly determined process
- A Monte-Carlo simulation is an example of a stochastic algorithm

DOI: 10.1201/9781003329237-18

- Randomized variable inputs in a stochastic algorithm can be the result of different distributions
  - Uniform (rectangular): any event is as likely as the other, such as rolls of a single die
  - Normal (Gaussian): central tendency events, such as measurements of boiling water
  - Poisson: probability of events occurring during a time interval
  - Exponential: intervals of time between events
  - Other less defined distributions
- Separate events with different distributions combine and propagate in less certain ways
  - Results may or may not be normally distributed
- Computers can simulate the most likely outcome from a series of different distributions
  - The algorithm randomizes the variable parameters and iterates a great number of times
- Monte-Carlo simulation
  - An algorithm (computer program) using random sampling to obtain a statistically likely result
  - Formally developed in the 1940s to model neutron propagation during Manhattan Project
  - Monte-Carlo was the code name used to maintain secrecy with the classified project
  - Monte-Carlo was chosen as a gambling location since the developer's uncle enjoyed gambling

Example 18.1

Refer back to the design of experiments (DOE) full factorial experiment from Table 4.5. In this experiment, the full design space captured the greatest impact of parameters in Run #3. However, there is no clear indication of the most likely damped natural frequency if these parameters vary between the limits of 3.5 and 6.5 as configured in the experiment.

- FIND: Use a spreadsheet program to determine the most likely damped natural frequency as independent parameters mass, stiffness and damping vary uniformly between 3.5 and 6.5
- SOLUTION: For the spreadsheet option, create columns (or rows) for independent variables similar to Table 18.1. Establish the relationship between dependent and independent variables. Create as many rows (or columns) for the number of experimental runs. Return the mean of all results.

Each of the independent variables, mass, stiffness and damping, have been chosen to vary with uniform distributions such that any value between 3.5 and 6.5 is just as likely at the other. A normal variation would require choosing a normally distributed random number versus a uniformly distributed number. For damped natural frequency, these uniformly random influences result in a mostly normal distribution

TABLE 18.1
Spreadsheet Configuration for Monte-Carlo Simulation

Run No.	m	k	c	c_{CR}	ζ	ω_n	ω_d
	RAND	RAND	RAND	$2\sqrt{mk}$	C_{CR}/c	$2\sqrt{k/m}$	$\omega_n\sqrt{1-\zeta^2}$
1	5.77	3.67	3.85	9.21	0.42	0.8	**0.72**
2	4.60	5.47	4.92	10.03	0.49	1.09	**0.95**
↓	↓	↓	↓	↓	↓	↓	↓
Mean							**0.83**

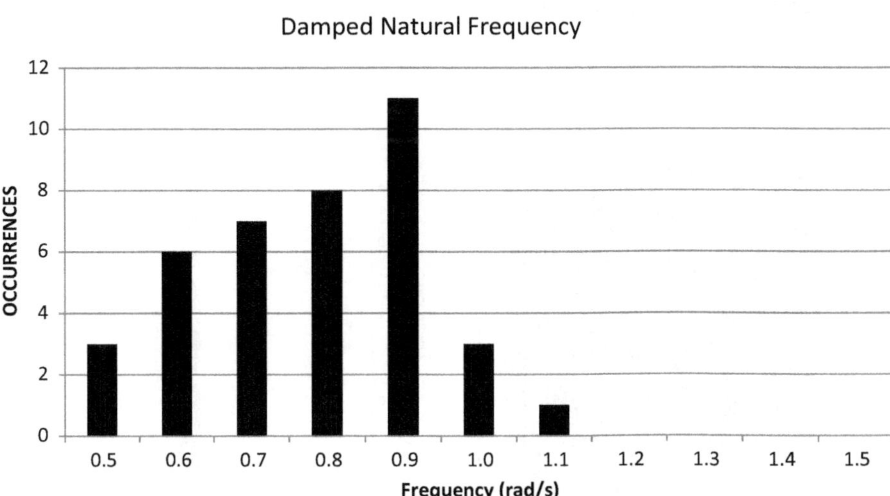

FIGURE 18.1 Result of Monte-Carlo simulation for variations in mass, stiffness and damping.

with a mean of approximately $\omega_d = 0.83$ rad/s. For a small sample set of $n = 39$ (39 run rows in the spreadsheet), the distribution curve is shown in Figure 18.1. If the distributions of mass, stiffness and damping change from uniform distributions to normal distributions, the result in Figure 18.1 will likely be different.

Example 18.2

What is the most likely mean value of successive rolls of a single die?

- FIND: Use a Monte-Carlo computer algorithm to determine the mean of the rolls of a single die
  - Distributions expected
    - Rectangular: the roll of a single die where any number is as likely as the other

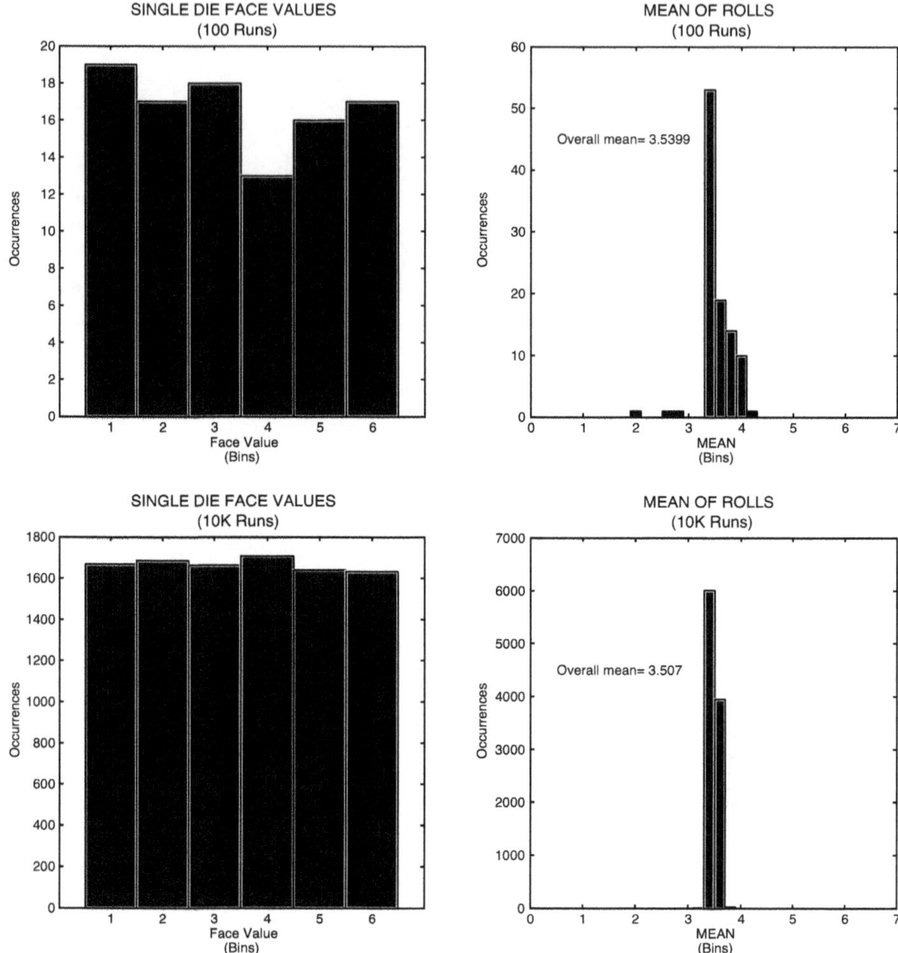

FIGURE 18.2 Monte-Carlo simulations for rolls of a die with mean values.

  - Normal: the average of equally likely numbers 1–6 should tend toward 3.5
- SOLUTION:
  - Algorithm
    - Define the variables subject to change: in this case, a single value called FaceValue
    - Specify the range (limits) within which the variables are subject to change: Low = 1; High = 6
    - Specify the probability function of each variable: uniform random integer for FaceValue
    - Allow each variable to randomly vary based on its probability function
    - Repeat many times
    - Output the most likely result (mean or median, as desired)
  - Computer algorithm

```
For i=1 to "a lot" 10^4, 10^5                        ←Run many iterations
    %OBTAIN THE RANDOM VARIABLE(S)                   ←Randomize each iteration
    FaceValue(i)=RandomInteger(Low to High)          ←Obtain dice roll value
    %COMPUTE DESIRED FUNCTION                        ←Run all necessary calculations
    SUMofROLLS=sum(FaceValue)                        ←Sum each die roll
   MEANofROLLS(i)=SUMofROLLS/i                       ←Calculate mean and enter into array
Next i                                               ←A single pass is now complete
Histogram(FaceValue)                                 ←Plot histogram of all FaceValues
Histogram(MEANofROLLS)                               ←Plot histogram of all means
```

- Results of the algorithm are shown in Figure 18.2 for both 100 and 10,000 iterations
- With more iterations, the results become more refined

MARKOV CHAINS

The Monte-Carlo simulation runs an entire experiment on each iteration, randomizing the inputs each time. A Markov chain is a state transition process having designated probabilities for moving from one condition or state to another. Movement from one state condition to another depends only on the probabilities of the current state options, independent of previous states. A Markov chain can simulate propagation through states by randomizing values at each state.

EXERCISES

1. What is a stochastic process?
2. What are the advantages of a Monte-Carlo simulation versus a physical experiment?
3. Will the results of a Monte-Carlo simulation always be normally distributed?
4. Name a few ways a Monte-Carlo simulation can be used other than just determining a mean.
5. For manufacturing simulations, should inputs be uniformly or normally distributed?

19 Unexpected Results

Researchers and experimenters should expect more failures than successes. Failed experiments are likely frustrating, but implementing a systematic approach to troubleshooting can help identify the problem. Experiments with negative results are still valuable for publication.

Despite careful planning, well-controlled errors and skill at conducting activities, experiments do not always go as planned. A hierarchy of issues begins with the broadest categories of *recognized* problems and *unrecognized* problems. Recognized problems provide an opportunity for further investigation and correction, while unrecognized problems result in a concluded experiment with no further consideration.

- Experiments can fail or return unexpected results
  - Recognized: correctable
  - Unrecognized: unrealized and uncorrectable

Unrecognized experimental failures or problems are due to wrongly appearing to achieve the intended result or objective. These types of experimental failures include Type I errors, Type II errors and human errors. Since these experiments proceed forward as if they were not problematic, the scientific process of open examination is the mechanism to hopefully uncover the issue over time.

- Unrecognized issues
  - Type I error: false positive ("*an innocent person is convicted*")
    - Incorrectly rejecting the null hypothesis in favor of a wrong alternative
    - Experiment incorrectly supports the contended experimental hypothesis
    - *Increase the required confidence level (lower significance) to reduce Type I errors*
  - Type II error: false negative ("*a guilty person is not convicted*")
    - Incorrectly keeping the null hypothesis and giving up the alternative
    - Experiment fails to prove the reality
    - *Increase the sample size to improve the power of the study and reduce Type II errors*
  - Human error
    - Flawed experiments, uncontrolled errors or unrecognized problems in the setup
    - *Use peer reviews to assess the experimental plan and setup*

Recognized experimental failures or problems are generally divided into three categories: failure of the experimental setup itself, results that are not as expected or

DOI: 10.1201/9781003329237-19

results that are inconclusive. If an experiment fails during the conduct of activities, it is usually due to an unforeseen situation during planning, equipment failure during the experiment or insufficient resources to complete activities. With improved insight into the process and a clear problem to address, the experiment can be reconfigured and re-attempted if time and resources permit. If the results are not as expected, it still may be due to a failed experiment, but it may also be due to an incorrect hypothesis and a deeper investigation into the issue is required.

- Recognized issues
  - Unable to complete experiment
    - Easy to recognize as failure with no ability to assess results or hypothesis
    - Process failure, equipment failure, setup failure or lack of control over variables
    - *Redesign experiment and re-attempt*
  - Results were not as expected
    - Hypothesis is assessable but data does not support hypothesis (a "wrong" hypothesis)
    - Experiment may or may not be a failure
    - Unexpected results may be due to the experiment itself
    - Unexpected results may be due to an incorrect hypothesis
    - *Further investigation is required*
    - *Design a new experiment, re-examine the hypothesis and seek peer reviews*
  - Experiment results inconclusive
    - Data is uncertain with limited ability to assess results or hypothesis
    - Insufficient control over variables, lack of equipment sensitivity or too few samples
    - *Redesign the experiment with improved fidelity*

REAL-TIME TROUBLESHOOTING

If an experiment is not proceeding as planned and not returning the expected results, real-time troubleshooting is necessary to assess the most likely cause. The first step is to verify the expected results from theoretical calculations or published values. Double-check all assumptions, values, calculations and definitely confirm correct units. Always have others verify calculations *independently*. These actions should always occur prior to the lab session, but if the experiment appears to be in order, reconfirm values prior to physical troubleshooting. Once expected values are confirmed to be correct among all peers, the best approach to troubleshooting an experimental setup is to remove as many variables as possible and test against known values. Start at one end of the experimental setup or the other. For example, with a stress-strain test using strain gauges and a Wheatstone bridge, a common problem is reading voltages in the range of five volts instead of only a few millivolts. Having properly worked out expected values during planning, it is easy to recognize the incorrect values. Begin at the strain gauge and disconnect all connections. Measure the resistance of the strain gauge for the expected 120 Ω or 350 Ω. If this checks, isolate sections of the Wheatstone bridge or check the bridge as a separate device. Add in elements of the experiment only after verifying each component returns the expected value.

- Troubleshooting actions during an experiment
  - Start at one end of the setup and isolate elements to check for known values
  - Check setup configuration, proper equipment, proper settings and software settings
  - Revisit frequency responses, voltage levels, units, sensitivities and modes of operation
  - Check for poor connections, solder joints, broken wires and burned-out fuses

Wrong settings on handheld multimeters often lead to burning out the fuse for amperage readings. Equipment may have fuses that are not obvious and are not obvious when burned out. Both the device under test *and* the test equipment itself should be suspect until one or the other is verified functional in the appropriate modes.

"*Failure*" is not always the best term to categorize setbacks during research or experimentation. Issues that initially appear as a failure might actually prove a success in later examination. The 1887 Michelson-Morley experiment to test for the presence of ether as the medium for light wave propagation[1] is sometimes referred to as *the most famous "failed" experiment.* The experimental objective was to detect ether, yet it failed to achieve that objective. The published report contributed to the body of science leading to Einstein's theory of relativity. Despite shortcomings of the word, "failure" is simply defined as an experiment that did not meet the objective or support the hypothesis, justly or unjustly. Since the experiment did not go as planned, there is likely frustration and a need to invest additional time, resources and effort. It is also important to note that the subject of the experiment cannot fail. Only the experiment or the hypothesis contended by the experiment can fail.

- Failures can occur in
  - Experimental design: materials, samples, equipment, interfaces and environment
  - Experimental processes: human actions, improper use of equipment and measurements
- Dealing with failure in research and experimentation
  - Discouraging but normal and part of the scientific process
  - Success is always published and acknowledged, yet failure is far more common
  - Failure always offers opportunity for enlightenment since it was not expected
  - Step back and reconsider the experiment with better controls or a different approach
  - Consider a different direction with the hypothesis
  - Always question whether failure is due to the experiment or the hypothesis

EXPERIMENT REDESIGN

If an experiment fails with no clear indication of the problem, additional investigation is required. Techniques from quality management systems (QMS) can aid in the systematic investigation of experimental design problems.

- Most engineering designs are iterative, and an experiment design should be iterative
  - Design → Build → Test and Evaluate → Repeat until succeeding or abandoning
- When evaluating failure in a process, consider the 6Ms to investigate all categories[2]
  - *Method*: processes, standard operating procedures, steps taken and sequences
  - *Machine*: equipment, calibrations, frequency response, warm-up time and drift
  - *Material*: test sample specifications or contamination and other consumable influences
  - *Measurement*: precision, units, correct equipment, loading and disturbances
  - *Man (people)*: observational errors, mistakes, incorrect procedures or lack of training
  - *Mother Earth (environment)*: external influences, uncontrolled variables and interference
- QMS Option #1: The 5-Why Root Cause Analysis[3]
  - Root cause: true cause of failure, not an intermediate cause or result of the failure
  - "5-Why" rule implies 3 to 5 "*why*" questions are often necessary to discover root cause
  - Ask "*why*" enough times to get through the symptoms to the root cause
  - Ask "*why*" problem occurred, then ask if failure could still occur, if yes, ask "why" again
  - Example: heat pipe Lab #12
    - Why did the heat pipe not work? Because it did not get hot compared to the control.
    - Why did it not get hot? Because the thermodynamic cycle did not work.
    - Why did the thermodynamic cycle not work? Because there was no vacuum.
    - Why was there no vacuum? Because the valve leaked.
    - Why did the valve leak? Because the valve spring force was inadequate.
    - Solution: use premium valves with higher spring force constants.
- QMS Option #2: Failure Mode and Effects Analysis (FMEA)[4]

- FMEA is a comprehensive risk analysis to identify risks and ways a process can fail
- Processes are *functions*, identifiable as: 1 verb + 1 noun
- Steps
  - Build a table of process steps with FMEA headings similar to Table 19.1
  - Identify potential failure modes for each component and function
  - Identify potential failure effects
  - Identify potential root causes for each failure mode
  - Identify process controls for each failure mode and variable
  - Determine rankings for severity, likeliness and detection for each effect
    - Severity of impact → *1 = least severe to 10 = most severe*
    - Likelihood of issue occurring → *1 = least likely to 10 = most likely*
    - Detectability of issue → *1 = most likely to 10 = least likely*
  - Calculate a risk priority number (RPN) by multiplying each of the above
  - Develop an action plan to revise the experiment

Table 19.1 is not intended to be a complete FMEA. The table is only an example with a few key steps from the heat pipe lab. The table also does not include the recalculation columns typical of a complete FMEA table. Based on the FMEA of the heat pipe lab in Table 19.1, the highest RPNs are associated with a cold solder joint and a hot solder joint. These failures are most likely due to poor technique with the torch. To improve success with the experiment, a practice session can be implemented, and a control step implemented requiring students to verify solder joints with the teacher's assistant prior to moving on to the next step in the experiment build process.

PUBLISHING FAILED EXPERIMENTS

Most research and experimentation efforts are geared toward successfully demonstrating new and useful relationships. Failed experiments are rarely published. However, publishing negative results still contributes to the field of science and offers the potential for future researchers and experimenters to avoid wasting time following a similar path.

- Failure to demonstrate a hypothesis is not a failure and should be considered for publication
  - Results are still of value, even if not originally as intended
  - If results are inconclusive, "*cast*" them to the scientific community for follow-up opportunity
  - Unexpected results can serve as education or a starting point for others
- Online open publishing platforms provide opportunities for "null result" publishing

TABLE 19.1
Partial FMEA Example of Heat Pipe Lab (Lab #12)

Process Step	Failure Mode	Failure Effect	Severity	Potential Causes	Occurrence	Existing Controls	Detection	RPN	Actions
Clean tubes	Not clean	Poor solder joints	10	Improper technique	5	TA* check	5	250	Training and practice session
Crimp	Insufficient	Leaking pipe	10	Improper technique	3	TA check	5	150	Training and practice session
Add flux	Too much	Fouls working fluid	7	No measurement	3	None	5	105	Measure flux
	Too little	Poor solder joint	6	No measurement	2	None	5	60	Measure flux
Solder	Cold joint	Pipe leaks	10	Insufficient torch heat	8	None	8	640	Training and TA check
	Hot joint	Pipe leaks	10	Too much torch heat	8	None	7	560	Training and TA check
	Too much	Clogged pipe	6	Too much solder	5	None	9	270	Wire check tube
	Too little	Pipe leaks	7	Insufficient solder	2	None	6	84	Training and TA check

*TA: Teacher's Assistant

EXERCISES

1. Can the subject of an experiment fail? Explain.
2. Should a failed experiment still be published? If so, why?
3. What is a Type I error and how can it be minimized?
4. What is a Type II error and how can it be minimized?
5. Consider an experiment for a vibrating cantilever beam (Lab #2). The experiment uses a polycarbonate beam. The natural frequencies measured during the experiment appear significantly different from the expected results. What troubleshooting steps should be considered?
6. If the natural frequencies of a vibrating polycarbonate beam are much lower than expected, use the 5-Why root cause analysis to help uncover the issue.

NOTES

1 Michelson, A. & Morley, E. (1887, November). On the relative motion of the earth and the luminiferous ether. *American Journal of Science*. 34(203), 333–345.

2 Hessing, T. (n.d.). *6Ms in six sigma*. 6σStudyGuide. https://sixsigmastudyguide.com/six-ms-6ms-or-5ms-and-one-p-5m1p/.

3 George, M., Rowlands, D., Price, M. & Maxey, J. (2005). *The lean six sigma pocket toolbook*. McGraw-Hill. p.145.

4 George, M., Rowlands, D., Price, M. & Maxey, J. (2005). *The lean six sigma pocket toolbook*. McGraw-Hill. p.270.

20 Reporting Numbers

Engineers need to write and present results using numbers. Numbers require correctness, both in their mathematical evaluations as well as in their presentation. Rules exist to ensure numbers do not imply more precision than actually exists.

Numbers have a variety of classifications and rules. For experimentation, two classifications of numbers are important, along with four presentation concepts. The two classifications are *exact* and *inexact* numbers, and the four concepts are significant digits, precision, uncertainty and accuracy.

To bring these concepts together, consider the scenario of a lab technician measuring a small cylindrical test specimen with calipers. There is exactly one test specimen with no ambiguity. This countable item results in an *exact* number. Alternatively, *measurements* will always yield an *inexact* result, no matter how precise. A ruler may give a result with an *uncertainty* of about plus or minus half a millimeter. A set of calipers will increase precision to within the manufacturer's published uncertainty of the calipers, such as ±0.001 mm. Measurement precision can be improved again with a laser interferometer, but ultimately, any measurement is inherently inexact and limited by the measurement device, no matter how precise.

A practical distinction is now made between *decimal precision* and *significant digits*. Decimal precision applies to a series of consistent unit measurements while the rules of significant digits apply to mathematical operations and how values propagate through the operation.

DECIMAL PRECISION AND SIGNIFICANT FIGURES

Consider the cylinder to have a truth value length of 25.000 mm and a true diameter of 5.000 mm. If length is measured with a ruler at 24.5 mm and diameter of 5.5 mm, the results have one *decimal of precision*, and the last digit is always uncertain. Given that uncertainty, it is inappropriate to include any additional digits of precision. Confusion occurs where the rules of significant digits might imply forcing a consistency between the three significant digits of 24.5 mm length and the two significant digits of 5.5 mm. These values are correctly indicated with consistent precision but with different significant digits.

If cylindrical volume is now calculated at 582.0782 mm^3, this multiplication operation is now limited by the contributor with the least number of significant figures, in this case, two significant figures from the diameter. *Technically*, the volume should be expressed as 5.8E2 mm^3 or 580 mm^3, both of which indicate the limiting measurement with two significant figures. *Practically*, scientific notation is not always the most suitable choice, and whole numbers ending in zero are considered ambiguous significant digits. The better choice is to present the number as accurately as possible at 582 mm^3. By including the uncertainty calculation with the result, the lack of

DOI: 10.1201/9781003329237-20

precision will be clear. Finally, none of these details state anything about the *accuracy* of the results. If the lab technician did not correctly zeroize the calipers prior to measurements, all of the precise results are inaccurate. Either the lab technician catches and fixes the inaccuracy or it goes unnoticed; either way, accuracy (or lack thereof) has no direct impact on how the numbers are reported. Accuracy is an error analysis and statistical issue, not a reporting issue.

- Number classifications
  - Exact: discrete, countable items with no ambiguity
  - Inexact: any and all measured values
- Significant digits and decimal precision
  - Do not indicate more precision or certainty in the result than is known
  - Follow the rules of significant digits when values propagate through mathematical operations
  - Multiplication and division results are limited by the fewest significant digits in operation
  - Always carry full precision in the calculator
- Significant digit rules:
  - Non-zero digits are always significant
    - Four significant digits: 2.365E–3
  - Any zeros between two significant digits are significant
    - Four significant digits: 2305
  - A final zero or trailing zeros *after* the decimal point are significant
    - Four significant digits: 2.500
  - Leading zeros are not significant
    - Three significant digits: 0.00500 ←*trailing* zeros are significant (not leading zeros)
  - Using scientific or engineering notation clarifies the number of significant digits
    - 0.00500 → 5.00E–3 (three significant digits)
    - 200 → 2E2 (one significant digit)
    - 200.0 → (four significant digits)
- Decimal precision (also, just precision)
  - Numbers associated with physical measurements have limitations in precision
  - Do not indicate more precision than actually exists in a numerical result
    - If length measures 53 mm, writing 53.000 mm indicates an incorrect higher precision
  - A length of 3.00 mm has a two-decimal precision or a precision of hundredths of a millimeter
  - Addition and subtraction results are limited by the lowest precision in the operation
    - Adding a 53.0 mm measurement and a 10.02 mm measurement results in 63.0 mm, not 63.02 mm
  - The last digit is an imprecise estimation, so any additional digits are inappropriate

UNCERTAINTY

Uncertainty, significant digits and precision work together to ensure results are accurately represented in numbers. While decimal precision ensures *numbers* express proper limitations, uncertainty accounts for limitations in the *measurement devices.* Following the uncertainty calculation rules ensures results are within the limits of the measurement equipment while following the rules of significant figures and precision ensures these results are properly written without indicating greater precision than the uncertainty.

- Uncertainty, u, quantifies the limits of precision in the measurement process (source of error)
  - Uncertainty values are cumulative as measurement errors combine and propagate
- Instrument uncertainty is a manufacturer published value associated with a device
- Uncertainty is the *plus or minus* in the final result, written as *value ± uncertainty*
  - Example of 0.1 kg measurement uncertainty: mass = 2.0 kg ± 0.1 kg
- Uncertainty should be written to the same decimal precision as the associated result
  - Decimal precision will be limited by significant figure rules for computational results
- Uncertainty can be expressed as either an absolute value or relative value
  - Relative uncertainty is written as a percentage: 100 Ω ± **5%** ($u = \pm 5$ Ω)
  - Absolute uncertainty is specified in the same units as value: 100 Ω ± **0.5 Ω**
- Two formulas for calculating uncertainty
  - Root mean squared (RMS) (easy): addition or subtraction are the only mathematical operations
  - Total derivative (difficult): everything else
- Formulas for uncertainty, u
  - RMS is given in Equation 20.1

$$u = \sqrt{{u_1}^2 + {u_2}^2 + {u_3}^2 + \cdots + {u_n}^2} \tag{20.1}$$

where: $u_1, u_2, \ldots u_n$, are individual uncertainties for each measurement

  - Total derivative (known as the Kline and McClintock method)[1] per Equation 20.2

$$u = \sqrt{\left[\left(\frac{\partial F}{\partial x}\right)u_x\right]^2 + \left[\left(\frac{\partial F}{\partial y}\right)u_y\right]^2 + \left[\left(\frac{\partial F}{\partial z}\right)u_z\right]^2 + \cdots} \tag{20.1}$$

where: F is a function of x, y and z, or as many independent variables in the function

u_x is the uncertainty in the x-measurement

- The RMS formula is a simplification of the total derivative formula
- Units are always a guide on which formula to use
  - If individual measurements and results have the same units, RMS is acceptable.
  - If units in the overall result (area, m^2) are different from individual measurement units (length, m), total derivative formula must be used.

Example 20.1

Consider a beam's cross-section using height and width measurements. The RMS and total derivative formulas give different results for the same problem.

- GIVEN: Following measurement parameters of a rectangular beam cross-section
  - Lab measurements: $W = 8.0$ cm and $H = 5.0$ cm
  - Published measurement device uncertainty ±1%
    - Uncertainty in width: $u_W = 0.08$ cm
    - Uncertainty in height: $u_H = 0.05$ cm
- FIND: **Perimeter** of beam cross-section
  - The perimeter result requires *adding* measurements: $P = H + W + H + W$
  - Consistent units: individual measures = cm, resultant perimeter units = cm
    - Use the easy RMS formula since units are consistent
  - Perimeter: $P = (5.0\text{ cm}) + (8.0\text{ cm}) + (5.0\text{ cm}) + (8.0\text{ cm}) = 26.0\text{ cm}$
    - Adding (or subtracting) tenths precision, the result should also be tenths precision
  - Uncertainty:
    $u = \sqrt{(0.05\text{ cm})^2 + (0.08\text{ cm})^2 + (0.05\text{ cm})^2 + (0.08\text{ cm})^2} = 0.133\text{ cm}$
    - Reduce precision to the lowest measurement precision of tenths: $u = 0.1$ cm
  - Result: $P = 26.0$ cm ± 0.1 cm
- FIND: cross-sectional **area** of beam
  - Area requires *multiplying* measurements: $A = H * W$
  - Inconsistent units: individual measures = cm, resultant area units = cm^2
    - Total derivative formula required since units are not consistent
  - Area: $A = (5.0\text{ cm})(8.0\text{ cm}) = 40\text{ cm}^2$
    - Multiplying (or dividing) two significant digit numbers should result in no more than two significant figures in the final computation
  - RMS (**INCORRECT**) $u = \sqrt{(0.05\text{ cm})^2 + (0.08\text{ cm})^2} = 0.09\text{ cm}$ ←wrong units
  - Total derivative formula (**CORRECT**):
    - $$u = \sqrt{\left[\left(\frac{\partial F}{\partial x}\right)u_x\right]^2 + \left[\left(\frac{\partial F}{\partial y}\right)u_y\right]^2 + \left[\left(\frac{\partial F}{\partial z}\right)u_z\right]^2 + \cdots}$$
    - $F = HW$ (Area), variable #1: H, variable #2: W

- $u = \sqrt{\left[\left(\frac{\partial(HW)}{\partial H}\right)u_H\right]^2 + \left[\left(\frac{\partial(HW)}{\partial W}\right)u_W\right]^2}$
- $u = \sqrt{[(W)u_H]^2 + [(H)u_W]^2}$
- $u = \sqrt{[(8.0\ \text{cm})(0.05\ \text{cm})]^2 + [(5.0\ \text{cm})(0.08\ \text{cm})]^2}$
- Ensure unit consistency by including units in computation
- $u = 0.56\ \text{cm}^2$ ← correct units for area

- Area = $40\ \text{cm}^2 \pm 1\ \text{cm}^2$
  - Round up the uncertainty value for consistent precision with the result

EXERCISES

A ball is dropped from rest at unknown height. A stopwatch is used to time the fall, and the time is used to calculate the unknown height. The kinematics formula for height as a function of time: $h = \frac{1}{2}a_g t^2$, where $a_g = 9.81\ \text{m/s}^2$. Timing instrumentation has an uncertainty of $\pm 1.0\ \text{s}$. Time to fall is measured at $t = 6.5\ \text{s}$. Determine the following.

1. Is $t = 6.5\ \text{s}$ an exact or inexact value?
2. How many significant digits are in 6.5 s?
3. What is the height of the drop? ($h = ?$)
4. Write the result with appropriate significant digits.
5. Write the most appropriate practical value without using scientific notation.
6. How many independent variables are in the height equation?
7. How many significant digits are in the gravitational constant value, a_g?
8. Can the RMS formula be used for uncertainty? Why or why not?
9. Should the total derivative formula be used for uncertainty? Why or why not?
10. What function, F, should be used in the total derivative formula?
11. Perform the partial derivatives in the total derivative formula without entering values.
12. Write the height and uncertainty in height using appropriate precision.

NOTE

1 Holman, J. (1984). *Experimental methods for engineers*. McGraw-Hill Book Company. p.50.

21 Active Writing

Engineers are good at solving engineering problems; writing, not so much. Technical reports tend to be wordy and exceedingly passive. Fixing only these two issues will greatly improve written reports.

Somewhere between grade school and graduate school, engineers often "unlearn" how to write. A well-written report should be clear and easy to understand. The material itself may be complex, but an effort should be made to present complex material in a straightforward manner. Convoluted sentence structure might seem impressive, but most people are far more impressed with a complex topic simply presented.

Engineers could take a lesson from the literary arts, where a very complex novel can be written with simple sentence structures. A well-written report should read like a *story*. Not in the sense of fictional characters and fictional data, rather as a compelling document. The document should clearly introduce the relevant material, move into the plot, avoid extraneous and boring material and wrap up with a nice, cohesive and consistent conclusion.

Confusion often exists over whether a data table belongs in the body of a report or in an appendix. If the data table is a *boring* plot element, causing a reader to skip over it for the more interesting material, then it belongs in an appendix. A good story should keep the reader interested, just as should a well-written experimental report.

An active sentence is the simplest sentence form that follows the subject-verb-object (SVO) structure. This structure concept is also relatively independent of language. Differing languages may have different orders, such as subject-object-verb, but the basic components are the same. If a sentence is built around this simple structure, it will be easy to understand.

PLAIN WRITING ACT

In 2010, the U.S. federal government signed into law the Plain Writing Act of 2010 to "*use clear government communication that the public can understand and use.*" [1] These guidelines are also suitable for research and experiment report writing.

- *Write for your audience*
  - Does not mean to "dumb down"
  - Do your research on the audience
  - Write to "one person," not the many
  - Use language your audience is comfortable with
  - People only want to know what applies to them
- *Organize the information*
  - State your purpose
  - Give the Bottom-Line Up Front (military-style: BLUF)

DOI: 10.1201/9781003329237-21

    - Most important info at the beginning with background later
    - Use topic sentences to indicate what the reader is about to encounter
    - Logical
- *Choose your words carefully*
    - Do not complicate with jargon, technical terms or unknown abbreviations
    - No need for "literary flair"
    - Use common words versus obscure words
    - Use specific words versus ambiguous words
    - Use short words versus long words
- *Be concise*
    - Cut excess wordiness
    - Avoid redundancy
    - Avoid modifiers (adverbs with -ly endings)
- *Keep it conversational*
    - Use active voice
    - Use present tense when possible
    - Use examples as appropriate
- *Design for reading*
    - Make it "easy on the eye" and inviting (visual appeal)
    - Use short sentences
    - Organize with headings
    - Use bullets, lists and tables as appropriate
    - Use simple fonts and layouts
- *Follow web standards (not discussed here)*
- *Test your assumptions*
    - Seek peer reviews or edits suggested by others
- *Plain Writing Act Example*[2]
    - *Wordy: "The Dietary Guidelines for Americans recommends a half-hour or more of moderate physical activity on most days, preferably every day. The activity can include brisk walking, calisthenics, home care, gardening, moderate sports exercise, and dancing."*
    - *Clear: "Do at least 30 minutes of exercise, like brisk walking, most days of the week."*

ACTIVE VERSUS PASSIVE WRITING

Difficulties in technical writing arise since authors are usually encouraged to write passively to avoid the appearance of bias. Active writing requires a direct subject such as *I*, *We* or *The Team*. "We used a gravitational constant of 9.81 m/s^2," is more direct and compelling compared to the passive version of "A gravitational constant of 9.81 m/s^2 was used."

Both active and passive writing styles have their place, and journals may recommend one style over the other. Even if passive writing is necessary, keeping the rules of active writing close at hand can prevent writing styles from becoming too passive and wordy.

Aside from the S-V-O construct, the easiest way to distinguish active and writing styles is by identifying the main verb in the sentence. An active sentence requires only a single verb. A passive sentence requires two verbs and often has three verbs. The second verb in passive writing is usually a form of the "*to be*" verb.

- Active writing
  - Most technical documents are *too* wordy and *too* passive with difficult sentence structure
  - Active writing is always more powerful and usually improves writing quality
  - Subject performs the action of the verb
  - Includes a direct subject: I, We, The researcher, …
  - Fewer words are required
  - Can be written with *only one verb*
  - Follows the S-V-O construct
    - Example: *"We verified results."*
    - Succinct: three words
    - One verb: *verified*
    - Begins with a direct subject: *We*
- Passive writing
  - Passive styles are considered less biased, and technical journals may require passive styles
  - Passive styles always contain two or more verbs: usually with an "-ed" verb form
  - Does not follow the S-V-O construct and contains an auxiliary verb
    - Wordy example with six words: *"The results were verified by us."*
    - Two verbs: verified (main verb) + were (auxiliary verb form of *to be*)
    - People ("*us*," convert to "*we*") should be directly performing the action of *verification*
    - Active revised example with four words: *"We verified the results."*
- The case for passive writing[3]
  - *The performer (subject) is unknown, irrelevant or obvious*
    - "Thermocouples are used extensively in engineering temperature measurements."
    - Passive structure is easily identified by two verbs: *are* + *used*
    - In this case, the direct subject is wordy: "As engineers, we use thermocouples..."
  - *The performer (subject) is less important than the action*
    - "Thermocouple wires are twisted together to form a junction."
    - Passive structure is easily identified by two verbs: *are* + *twisted*
    - As a statement, active writing does not help: "We twist thermocouple wires together."
  - *The recipient (object) is the main topic*
    - "Thermocouples were used for temperature measurements."
    - Passive structure is easily identified by two verbs: *were* + *used*
    - Thermocouples are the focus, rather than the people

REVISING A PASSIVE SENTENCE

If passive writing is necessary or more appropriate, know the difference between active and passive to *improve* the passive style. Use the following process to revise a passive sentence into an active sentence:

1. Identify the *one main verb* in the sentence
2. Identify or create the noun that *best precedes* main verb and is capable of performing the verb action
   - If no direct subject exists, begin the sentence with "We," "I," "The team" or other subject
3. Rewrite the sentence with noun subject and only one main verb; everything else follows

Example 21.1

Revise the following sentence from passive to active: *"Temperature was verified by a thermocouple."*

- Passive: two verbs: *was + verified*
  1. Identify main verb: *verified*
     - Auxiliary verb *"was"* does not convey action or context (*was*, a form of *to be*)
     - Revision step one: *"—— verified ——"*
  2. Identify or create noun best preceding verb capable of action
     - NO: "Temperature *verified ——*" ← Temperature cannot perform verification act
     - OKAY: "A thermocouple *verified ——*" ← Loosely, thermocouples *might* verify
     - CORRECT: "We *verified ——*" ← People are directly capable of verification act
  3. Everything else follows
     - *"We verified temperature with a thermocouple"*

OTHER WRITING PITFALLS

Regardless of active or passive writing styles, always seek to use fewer, more concise words, correct grammar and correct spelling. The best strategy for improving writing is to keep sentences short, with no more than two clauses. When in doubt, use a period and begin a new sentence. Avoid the overuse of conjunctive adverbs (*however, furthermore, thus* and *although*) and prepositions (*in spite of* and *despite*). Always check if the sentence reads well without the preposition or conjunctive adverb and remove it if possible. Do not use superlatives; point out the facts, and allow the information to speak for itself.

- Keep sentences short and use fewer, concise words
- Avoid *-ion of* and *-ment of*
  - "*A spreadsheet was created in preparation of data collection*" ← Passive and wordy
  - "*A spreadsheet was prepared for data collection*" ← Passive but more succinct
  - "*We prepared a spreadsheet for data collection*" ← Active
- Diction: Use higher quality words
  - Poor choices: *stuff* and *things*
  - Better: refer to objects directly by name
- Precision: Use specific references
  - NO: "*The results were good.*"
  - YES: "*Experimental results agreed within 5% of theoretical results.*"
  - NO: "*Various materials were tested.*"
  - YES: "*We tested aluminum, steel and copper.*"
- Remove unnecessary words
  - "*That*" can often be removed from a sentence
    - "*We verified the fact* that *aluminum was more conductive than steel.*"
    - "*We verified aluminum was more conductive than steel.*"
  - "*in order to*" → "*to*"
  - "*for the purpose of*" → "*for*"
- Avoid ambiguous subjects
  - "*It is...*" or "*There is...*"
  - "*It was established*": *What* was established? Is "*it*" truly clear from the preceding sentence?
- Do not confuse and misuse common homophones
  - *Their, there and they're*
    - *Their*: Possessive
    - *There*: Adverb or pronoun
    - *They're*: Contraction of "They are"
  - *Affect* and *effect*
    - *Affect*: Verb (*I affect the result*)
      - To confirm it is a verb, a past tense is available: *affected*
    - *Effect*: Noun (*I dislike the effect*)
      - To confirm it is a noun, a grammatical article should be present: (the, a or an) *the effect*
  - *Its* and *it's*
    - *Its*: Possessive
    - *It's*: Contraction of "*it is*"
    - Neither is preferred since both words are often ambiguous
    - Note *that* publisher requirements will likely dictate certain writing styles that vary from the above recommendations (as was true for this textbook during editing to conform with the publisher's style).

INAPPROPRIATE REPORT PHRASING

Report writing should tell the story of what happened. It should not direct the reader *what to do* or make extraneous statements. By the time the discussion and conclusion of the report begin, no additional content should be included that has not already been introduced. Avoid each of the following constructs.

- "For this experiment, *be sure to wear safety glasses* at all times."
  - Avoid telling the reader what to do, unless a unique hazard requires highlighting
  - List equipment used and inform the reader what is necessary to recreate the experiment
- "In this experiment, *we will be trying to find* allowable current."
  - Use fewer words, "*We will determine allowable current.*"
- *"The wire that we will be using for this test is called* Nichrome wire."
  - Too verbose, use fewer words, "*We will test Nichrome wire.*"
- "Finally, we *should* compare results that are experimental to those of theoretical results."
  - Make a clear statement without ambiguity, "*We will compare...*"
  - Alternatively, "*Experimental results are compared to theoretical results.*"
- "For our results, *we expect to have successful results...*"
  - Everyone expects success, but what are the expected values?
- *"You can see* in Figure 5."
  - Better: "As presented (or *shown*, or *depicted*) in Figure 5..."
- "When doing this lab, *we learned a huge amount of information about thermocouples.*"
  - Readers do not care about what <u>you</u> learned; this does not belong in a conclusion
- "*...we were able to record our data.*"
  - This is a superfluous statement of no use
- "...values were *extremely close* to the actual published values."
  - How close is "extremely close"? (provide a specific percentage of error)
- *"We knew we were doing the experiment correctly."*
  - Lead the reader to the conclusion; do not be presumptive with frivolous statements
- How to simplify convoluted writing
  - Convert long sentences into shorter ones
  - Work on one sentence at a time
  - Use just a few words to summarize the sentence
  - Replace pronouns (*it* and *they*) with nouns; be specific
  - Remove unnecessary words
  - Reduce the number of verbs used in a single sentence
  - Write a simple outline or bulleted version
  - Rewrite the text in a concise but suitable narrative

EXERCISES

Consider the following report statement and answer questions 1 through 3: "*The wing aerodynamic drag coefficient before adding winglets and the drag coefficient after adding winglets can be obtained based on the Navier-Stokes equations.*"

1. Identify the main verb.
2. Does the statement have a direct subject?
3. Rewrite the sentence in active form with only one verb.

   Consider the following report statement and answer questions 4 through 6: "*The new algorithm of the electrical conductivity assessment method has been applied to measurements performed on several conductive wires.*"
4. Identify the main verb.
5. Does the statement have a direct subject?
6. Rewrite the sentence in active form with 14 words or less.

NOTES

1 PlainLanguage.gov. (2011 March). *Federal plain language guidelines.* U.S. General Services Administration. https://www.plainlanguage.gov/media/FederalPLGuidelines.pdf.

2 PlainLanguage.gov. (n.d.). *Examples>Reports, Dietary Guidelines: Losing Weight Safely.* U.S. General Services Administration. https://www.plainlanguage.gov/examples/brochures/hhs-brochure/.

3 Every, B. (n.d). *The value of passive voice.* BioMedical Editor. http://www.biomedicaleditor.com/passive-voice.html.

22 Presentations

Many opportunities exist to present technical information. Accepting the challenge of public speaking is an opportunity for recognition and advancement in any professional organization.

The ability to give a proper presentation is a learned skill. Some people are naturally gifted speakers, and the task of speaking comes easy to them. For the majority of people, public speaking creates a range of anxieties. While the emotional stress may never be completely overcome, the ability to deliver a quality presentation can always be enhanced with sufficient preparation and practice.

Technical presentations do not need to be, and should not be, boring. The presenter should consider themselves as actors on a stage, ready for a performance. Actors have lines to memorize and deliver in a manner where the performance and content are interesting enough to watch. Every technical presentation, large or small, should take a similar approach. Scale the delivery appropriate to the audience, but practice the material and give a flowing and interesting presentation.

Just as preparation and practice are essential to delivery, audio and timing are critical to success. Public speaking requires a consciously louder voice than normal conversational levels. If no amplification is available, the natural voice must be loud and clear to fill the room. Do *not* ask the all-too-common awkward question, "*Can you hear me in the back?*" The audience is not warmed up to the presentation and will remain mostly silent. It is *the speaker's duty* to ensure his or her voice can be heard ahead of time. Visit the forum and practice. Bring in an assistant and have that person sit at the back of the room and provide feedback. If a microphone and amplification are available, test everything ahead of time with sufficient time to correct technical issues prior to the presentation. Poor audio is a failed presentation, period. Timing is likewise critical. Presentations must be ready to go on time with no delays in logging on to computers, finding files or fixing audio issues. The presentation should be up and running within 15 seconds, and it should absolutely end within the designated time limit. If no time limit is clearly specified other than a one-hour block of time, presentations with questions must be limited to 50 minutes. People expect at least a 10-minute break between activities and should be afforded that opportunity. The audience always has expectations, even if they are not clearly defined.

Giving a well-prepared and well-practiced presentation is key to success in any organization. Key personnel and stakeholders will take note of good presenters who will be called upon for future opportunities. Stepping up to the challenge and investing an appropriate level of effort for a proper presentation leads to more visibility and advancement within any organization.

DOI: 10.1201/9781003329237-22

PREPARATION AND PRACTICE

Good presentations do not require natural speaking ability. Preparation and practice are the two most important requirements for proper presentations.

- Preparation
  - Know the audience expectations in terms of content, delivery style and timing
    - *Every* audience has a set of expectations, especially for timing
    - When possible, confer with key individuals to confirm expectations ahead of time
    - Conferring with key individuals also increases their vested interest in the outcome
  - Creativity and appropriate humor are often welcomed, even for a technical presentation
    - Deliberate humor can be a risk; confirm the content and delivery of humor with others
  - Have a hard copy backup and provide copies to key personnel, if appropriate
  - Always plan suitable attire
    - Minimum standards are collared shirts, slacks, dresses or skirts
    - Jeans, non-collared shirts, sports attire and sports shoes are generally unacceptable
  - Review the presentation forum ahead of time, preferably the day prior
    - Go through the *entire* process of turning on equipment and bringing up electronic files
    - Check the audio; visual issues can be forgiven, but *audio failures are unacceptable*
    - Presentations must always be ready to go at the start time without fuss
  - Have a recovery plan such as note cards to quickly recover if lost during presentation
- Practice
  - Always ensure at least one complete, *out-loud* practice
    - It is nearly impossible to meet time constraints without actual, out-loud practice
    - Practicing alone and then with at least one live person is always recommended
    - Always pay close attention and avoid filler words
      - Typical filler words include: *um, uh, you know, like, as you can see*
      - To correct, *awareness* is first required, either from a recording or another person
      - Practice and attentiveness are then required to remove filler words
      - Simple pauses in speech are far preferable and effective compared to filler words

- Do not create presentations with continuously running animations or videos
  - Animations and motion.gif images distract the audience from the speaker
  - Videos should be short, deliberate and supplement, not compete, with spoken words
- Images and diagrams MUST be practiced!
  - Presenters often think diagrams are easy and self-explanatory and gloss-over during practice
  - Presenters get bogged down with unpracticed explanations and exceed time limits
  - Presenters get "*sucked into*" looking only at slide diagrams and never look at the audience
  - Get up-close and personal as necessary on slides with graphics to adequately explain
    - Use pointer or hand to work through graphics explanations
    - Do *not* stand at a distance, waving hands around and saying "*as you can see*"
    - "*No, I cannot follow what you are trying to convey as you pointlessly wave your hands around!*" (Practice and use a pointer to work through diagrams.)

- Delivery
  - Always introduce yourself upon speaking, both as a team (if applicable) and individually
  - NOT: *"I'll give you a few moments to write down my name,"* just introduce and begin
  - Stick to the script; ad-libbing almost always results in busting the time limit
  - Speak to audience, not slides or cards, especially when explaining diagrams
  - Always speak sufficiently loud and clearly
    - Proper presentation volume is much louder than normal conversation volumes
  - Maintain a coherent, organized delivery
  - Maintain professionalism if something goes wrong
  - Always re-introduce yourself and the topic at the end of the presentation
  - Offer an opportunity for questions
- Team Presentations
  - The team lead should introduce the topic and all team members
    - Individuals should still introduce themselves when speaking for the first time
  - Everyone in the team must speak an equitable portion
  - Team practice is required to prevent confusing transitions between speaking parts

ONLINE PRESENTATIONS

Although online presentations are likely less stressful than standing in front of an audience in real-time, technical issues can make delivery more challenging. Online presentations are never an excuse for *anything goes* or failure to set up a suitable presentation environment. Technical issues must be completely checked ahead of time, including logging on to the presentation forum software. Lighting, framing and background are the three essentials to consider for setting the proper visual environment. Laptops require additional consideration since they usually sit below eye level and must be adjusted for proper framing. Refer to Figures 22.1 and 22.2 for online environment considerations.

- All technical issues MUST be resolved and checked ahead of time
- Audio is vitally important and must be loud and clear
  - Have a good quality microphone
  - Ensure microphone is sufficiently CLOSE to mouth
- Lighting must illuminate the face properly, as shown in Figure 22.1
  - If necessary, set up an extra light source, such as a lamp, to illuminate the face
  - Do not rely on the changing light from the computer monitor during the presentation
  - The light source should be diffuse and located at eye level or slightly above eye level
  - Ensure eyes show through eyeglasses
    - Tilt glasses slightly down to direct monitor reflections downward
    - Use an external webcam off-axis from the monitor to avoid direct reflections from monitor
- Proper framing
  - The camera should be at the correct distance such that *you* are the main item of interest
  - The face must be clearly seen and directed at the camera

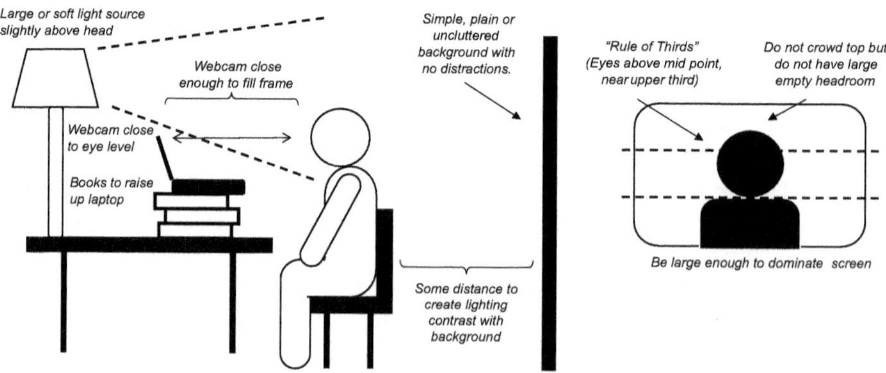

FIGURE 22.1 Proper laptop configuration for online presentations.

- Do not have the camera oriented low to see typing or mouse clicks while presenting
- Raise the camera position near eye level to avoid *up-the-nose* camera angles
  - Laptops with integral cameras can be temporarily elevated on stacked books
- Consider the "*rule of thirds*" to align the eyes *about* one-third down from the top of the video image

- Distracting backgrounds must be corrected, as shown in Figure 22.2
  - Use lighting to illuminate the face compared to the background
  - Strong backlighting from windows in background should have shades closed
  - Keep other people and pets out of the background and reduce clutter
  - Keep backgrounds simple with solid colors, solid walls or simple, non-descript elements
  - Open rooms in the background should be darker compared to the presenter to reduce distractions
  - Avoid *gimmicky* virtual backgrounds during presentations; use simple schemes
  - Do not have moving elements such as virtual animations or operating ceiling fans
- Demeanor
  - All professional aspects of in-person presentations still apply for online presentations

Dark, small in frame, poor lighting andno contrast with background

Small, bottom of screen, bright background lights, spinning ceiling fans

Ball caps on walls or shelves, pictures or other clutter competing for attention

Open doors, bright unshaded windows, unrestrained pets

FIGURE 22.2 Proper background considerations for online presentations.

- Avoid distracting behaviors such as rocking in chairs while speaking
- Online audience attendees also require professional, non-distracting on-camera behavior

SLIDE PREPARATION

Presentation slides must be clear and readable. Avoid low-contrast color schemes. Keep font sizes large, and certainly no smaller than 20 points. Keep bulleted statements short at one or maybe two lines. Avoid excessive verbiage since bullets are key points and not sentences to read.

- Organize topics into a few key points, clearly presented
  - Keep key points to a minimum, usually only three for short presentations
- Overall structure requires slides with the following information
  - *Title slide* with topic and presenter name with team members as appropriate
  - *Overview* outlining each of the key points
    - Do not get bogged down discussing details during the overview; keep it succinct
  - Main body of presentation with each of the key points
  - Include a *conclusion* only if it is suitable for the type of presentation
    - Making a case or defending a thesis should have a conclusion
    - Informative presentations do not have a conclusion
  - Always include a *summary* slide that is a repeat of the overview slide
  - Include any necessary *references* (fonts will be much smaller)
  - Include a *closing title slide* similar to the title slide with topic, name and "Questions"
    - Always offer an opportunity for questions on the last slide
    - Ensure the name of the presenter(s) is included on the closing slide (with the topic)
  - Save additional supporting material for backup slides
- Slides must be clear and legible for a large audience
  - Use large fonts, 24-pt or larger and no less than 20-pt
  - Avoid large, pointless graphics in the style
  - Keep bullets short as keywords and memory joggers, not full sentences
  - One-line bullets are best; remove unnecessary grammar articles and filler words
  - Ensure consistent punctuation: bullets do not have periods since they are not sentences

DATA PRESENTATION

Charts imported from spreadsheets must be cleaned up from their default settings. The most common problems to address are axes, significant digits and titles to correctly label all information and avoid clutter. All data tables require formatting for an appropriate number of significant digits with labeled columns.

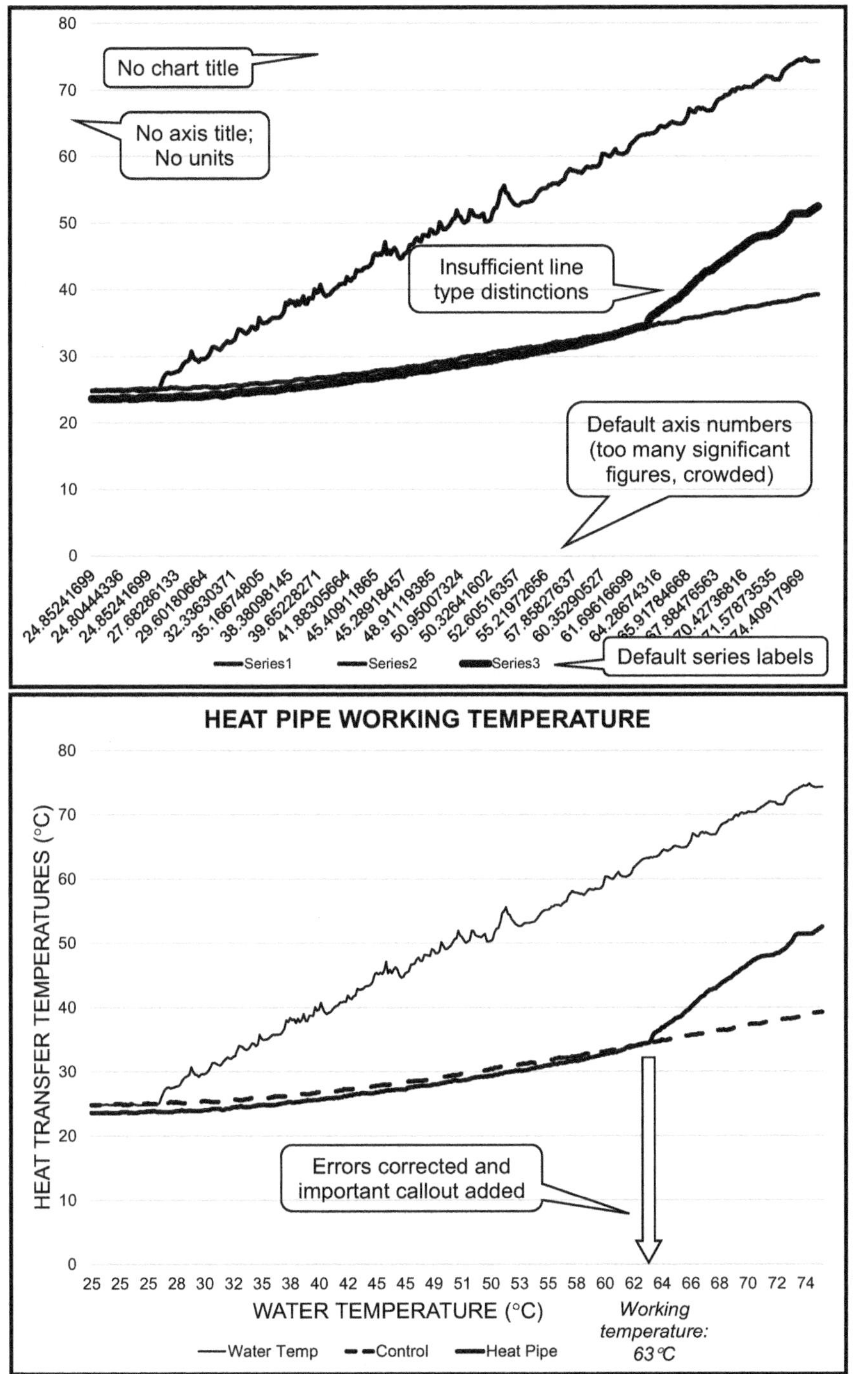

FIGURE 22.3 Common data chart issues and fixes. The top chart has many common errors that are corrected in the lower chart.

- Do not cut and paste tiny equations into a slide
  - Equations should identify key parameters with a legend
  - Recreate equations as necessary rather than using a small cut and paste image
- Charts must be clear and legible, with labels and uncluttered axes
- Do not just cut and paste busy spreadsheet charts straight into presentations
  - Rebuild and simplify charts as required with large fonts and clear graphics
- Do not use spreadsheet defaults with multiple significant digits
  - Format numbers to suitable significant digits and reduce axis divisions for readability
- All charts need titles, axis labels and units
- An example of common chart issues and corrections is presented in Figure 22.3

EXERCISES

1. TRUE or FALSE. Good speaking ability cannot be learned.
2. TRUE or FALSE. All presentations should have a conclusion slide.
3. TRUE or FALSE. Failure to completely practice talking through graphics often leads to exceeding time limits.
4. Where should laptops be placed for proper on-camera presentations?
5. Why are laptop microphones unsuitable for presentations?
6. Identify all the errors associated with the chart in Figure 22.4.

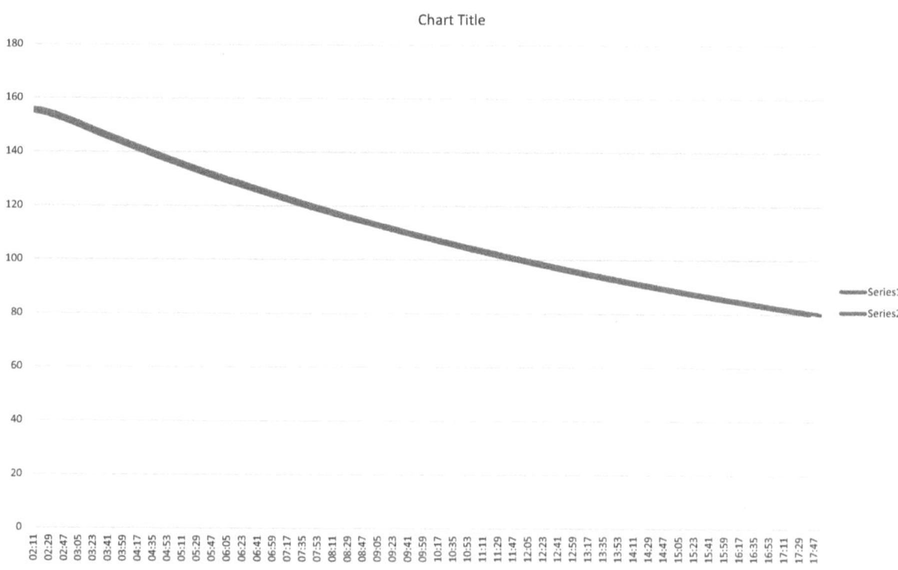

FIGURE 22.4 Chart errors.

LAB 0
Teams

Safety Requirements

No biting, scratching or fighting

Procedural Requirements

Choose lab partners
- Total of three or four individuals per team
- Permanent once chosen
- Only attend sessions with your own lab team

Choose team lead
- Each team has one lead who serves as the team's point of contact
- Responsible for the conduct of team activities
- Responsible for submitting finalized documents

Submit a properly formatted lab report
- Reference and use the report format template with a cover page
- For the report, directly copy the sections below from this manual
  - Introduction, Theory, Method, Results, Conclusions and Appendix L0.A
  - No rewording required; no reference required; no additional appendices
  - All team member information must be completed
- Only the team lead submits a single report for the lab team

Clean-up Requirements

Fist bumps

INTRODUCTION

Most career and research activities require some form of team collaboration. Team collaboration has both positive and negative attributes. A team with motivated individuals usually results in an overall better product through the exchange of ideas and review of individual work. A team with unmotivated individuals usually results in an increased workload for the responsible individuals while the unmotivated individuals receive unwarranted benefit from the team's overall success.

Choose team members based on similar motivations as much as possible. However, choices are usually constrained. Expect to have low-contributing team members. This may also be true of future career projects, so the challenge is learning how to work through these issues. Honest reporting is required, and individuals are assessed accordingly. Never 'cover' for weak contributors.

Identify the team lead. The team lead receives an extra credit for the burden of managing team activities. The team lead is always the first point of contact for lab team activities.

OBJECTIVES

Select lab teammates and the team lead. Practice submitting a lab report in the correct format.

THEORY

Group dynamics and resolution techniques are discussed in Chapter 2. No additional theory is presented in this section.

EXPERIMENTAL METHOD

Choose lab partners from among the section students who are available. Enter all the required information into Table L0.1 in the results section. Format the lab report according to the available template guidelines. Only the team lead submits the completed report for the team.

RESULTS

Table L0.1 details team member information. No additional discussion is required.

DISCUSSION

No discussion is required.

CONCLUSIONS

This lab is self-explanatory. No further conclusions are required.

TABLE L0.1
Team Member Information. First Entry with Asterisk Is Team Lead

Name (Last, First)	Email	Alternate Contact
*		

APPENDIX L0.A: NOTES ON CONFIGURATION AND IMPLEMENTATION

It is imperative to establish team and team member lines of communication as soon as possible. All large groups have competing time and schedule constraints. Difficult situations can only be resolved when the bulk of the effort is completed as quickly as possible. These labs are designed for three or four personnel. More thanfour results in under-employment while less than three results in over-employment.

LAB 1
Discrete System Vibrations

Safety Requirements

Normal safe lab practices
Lab coats and safety glasses required

Equipment Requirements

Tracks and carts with springs, magnets and ultrasonic sensor
Computer with data-logging software
Stand to hang spring vertically
Small weight for spring
Steel ruler
Mass or weight scale

Procedural Requirements

No unique considerations

Clean-up Requirements

Return all equipment to proper locations
Unplug all lab equipment

Emergency Actions

No unique considerations

INTRODUCTION

A second-order system is any physical system that can be mathematically described by a second-order differential equation. These systems involve inertia with the ability to exchange energy between states. A classic example is a pendulum, which continuously exchanges potential energy for kinetic energy and back again. The system has inertia in the pendulum's bob mass. Any such system can oscillate or *vibrate* if the system is excited by an external force or displaced from equilibrium. The frequency of this oscillation is the *natural frequency,* since the system naturally moves at this frequency without being otherwise forced or constrained by an external function or force.

The most basic oscillatory system is a discrete, single degree-of-freedom (1-DOF) system composed of a mass, a spring and a damper. The physical characteristics of these three components determine the natural frequency of the system. Stiffer springs tend toward higher frequencies, while higher damping tends to reduce frequency. Many systems, simple and complex, are often modeled as a basic 1-DOF system to help understand the characteristics of the system in simplest form.

In this lab, a discrete mass-spring-damper system is studied. A wheeled cart of a given mass is placed on an inclined track and attached to a spring. A magnetic attachment can be affixed to the cart as a damper component. An ultrasonic sensor is attached to a data-logging computer to measure the position of the oscillating cart. The key parameters of the system (mass, spring constant and damping coefficient) are individually measured to calculate the theoretical natural frequency. The experimental frequency is then measured and compared to the theoretical value.

OBJECTIVE

Assess the validity of experimentally obtained natural frequencies of a one-degree-of-freedom system relative to the theoretical model. An accepted theoretical model is available for a discrete, single degree-of-freedom vibrating system. The objective is to replicate a 1-DOF experiment and assess whether the experiment yielded satisfactory data. Plot and compare experimental and theoretical system responses for undamped, under-damped and over-damped conditions.

THEORY

Mass-spring-damper systems. In their most fundamental form, oscillatory systems can be modeled as a simple mass-spring-damper system, as shown in Figure L1.1.

The mass, m, has inertia, and its position is displaced in only one direction. In this case, the one-dimensional movement is vertical, using a y-coordinate in the positive up direction. The mass is subject to the force of gravity, the spring force and the force of the damper.

The spring exerts force on the mass based on its spring constant, k. Hooke's law provides the relationship between force and the spring constant as $F_{SPRING} = kx$, where "x" is displacement from the neutral position of the spring. Given the coordinate system established in Figure L1.1 using "y," force can be written as $F_{SPRING} = ky$.

The damper impedes any motion based on a damping constant (damping coefficient), c. Dampers are similar to any relatively viscous liquid that resists movement. Water in a swimming pool acts as a damper as it impedes efforts to walk along the bottom of

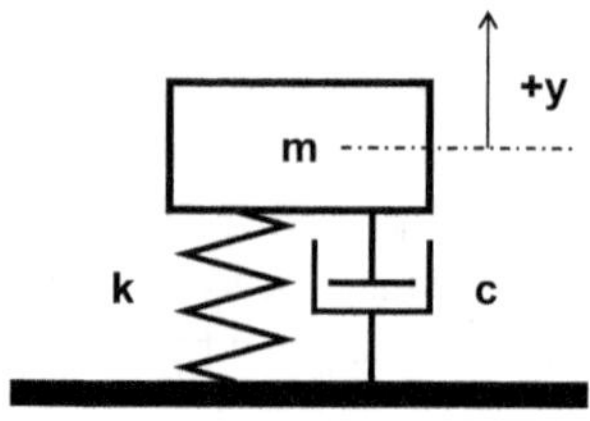

FIGURE L1.1 Mass-spring-damper physical model.

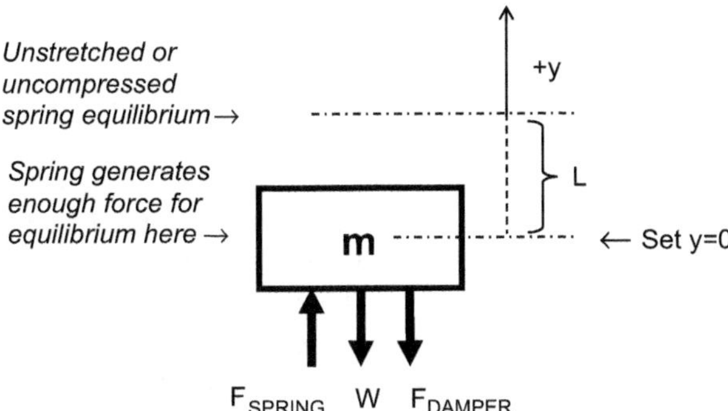

FIGURE L1.2 Vertical free-body diagram of mass-spring-damper system, displaced a distance "*L*" due to weight, *w* (mg). (All forces are assumed to act *in line with each other* through the mass CG and are only shown as above for clarity.)

the pool. However, without movement (velocity), there is no resistive force. A damper only exerts force if the system is in motion, so the damping force, *F*, is a function of velocity, *v*, where $F_{DAMPER} = cv$. As a one-dimensional system, velocity can be written simply as a change in position with respect to time, *dy/dt* or $\dot{y}$. For this system, the damping force is $F_{DAMPING} = -c\dot{y}$. It is negative since the force will *always* act opposite of the displacement direction. This is in contrast to the spring force, where the spring may exert either a pushing or pulling force on the mass, even if the system is at rest.

A free-body diagram (FBD) can be drawn from Figure L1.1 with forces in place of physical components as shown in Figure L1.2. If the entire system was rotated horizontally, the force of weight, *mg*, would not enter into the equation (assuming no friction), since weight would act perpendicular to the direction of motion and the spring will be at its neutral position. However, in a vertical orientation, the spring is displaced from neutral by weight due to mass. The model setup is therefore different for a horizontal versus a vertical orientation, but *the result is the same* where weight as a force does not affect the dynamics of the system. Mass affects the system, but the combined term "*mg*" will not affect the system as discussed below. As a result, the orientation of the mass-spring-damper model will not alter the system response under idealized conditions.

Mathematical model. Using Newton's second law, $F = ma$, the forces in the *y*-direction can be summed and set equal to the product of mass times acceleration, *ma*. Since this is a one-dimensional system (*y*-direction only), acceleration is the second time derivative of position in the *y*-direction, d^2y/dt^2, or $\ddot{y}$. Using the selected coordinate system with up as the positive direction, summing forces in accordance with Newton's second law results in the *preliminary* Equation L1.1, with all forces on the LHS, and mass times acceleration on the RHS. (Note that Equation L1.1 is not quite accurate until fully discussed and rewritten as Equation L1.2 and ultimately Equation L1.3.)

$$\boldsymbol{k}(\boldsymbol{displacement}) - c\dot{y} - mg = m\ddot{y} \quad \text{(L1.1)}$$

where: $k(displacement) = F_{SPRING}$
$c\dot{y} = F_{DAMPER}$
$-mg = \text{Weight } (W)$
$m\ddot{y} = \text{Force due to acceleration } (ma)$

All forces are accounted for in Equation L1.1. However, the *k(displacement)* term requires extra attention to be properly defined. This analysis is a *mini-FBD* of just the mass and spring. The mass weight (gravitational force) is already captured in Equation L1.1 with the "$-mg$" term, so that portion is complete.

Focusing only on *k(displacement)*, we know a spring force acts on the mass to push or pull the mass back to the equilibrium position, where all forces are statically balanced. If the mass is displaced in the positive y-direction, the spring force is "$-ky$," since the spring force will act in the *negative y-direction* to pull the mass back to equilibrium. If the mass is displaced in the negative y-direction, this spring force is "$k(-y)$" since displacement is negative in the assigned coordinate system. Therefore, spring force *always* results in "$-ky$."

Since the spring is vertical, there is also a static displacement force acting on the mass that would not be present if the system were entirely oriented in the horizontal. A large mass will compress the spring until the spring generates enough force to hold the mass in equilibrium. This static displacement force due to gravity is "kL," as depicted in Figure L1.2. "kL" is separate consideration from the "$-ky$" force. By conveniently choosing the coordinate system origin of $y=0$ at this resting equilibrium point of spring compression (or tension), "kL" has a positive sign since it is acting upward from zero. The mass equilibrium is chosen at $y=0$, but the spring was compressed by a length of "L" to get there. Therefore, "kL" must be positive since it is both acting upward and the distance L is on the positive side of the chosen origin of $y=0$.

Choosing this convenient $y=0$ origin at static equilibrium simplifies analysis since the amplitude of vibration is always about this neutral point. Mathematically, the choice of origin location does not matter, but a wise choice can greatly simplify the analysis. The mini-FBD analysis of *k(displacement)* is now complete, and both the spring force and equilibrium displacement force are entered into Equation L1.2.

$$(kL - ky) - c\dot{y} - mg = m\ddot{y} \tag{L1.2}$$

At equilibrium, the kL term must equal the mg term, since only the kL force is holding the mass in a static position. Since $kL=mg$, then kL-mg in Equation L1.2 cancel each other out to zero for Equation L1.3a. *Weight is therefore removed from the equation.* Without this *displacement analysis*, the mg term might erroneously remain in the equation developed from an FBD. A correct analysis results in Equation L1.3a, where weight as a force, mg, does not affect the dynamics of the system. Mass, however, as an inertial element does affect the system. This also means the system behavior is the same whether it is horizontally oriented (no friction) or vertically oriented, and only the equilibrium position changes. Equation L1.3a can be rearranged into Equations L1.3b and L1.3c for a ready solution technique.

$$-ky - c\dot{y} = m\ddot{y} \tag{L1.3a}$$

$$\ddot{y} + \left(c/m\right)\dot{y} + \left(k/m\right)y = 0 \tag{L1.3b}$$

$$\ddot{y} + a\dot{y} + by = 0 \tag{L1.3c}$$

With reference to static equilibrium, the spring force is always negative in Equation L1.3a, as discussed. The damping force is also *always* negative since it always opposes direction and acceleration. Equation L1.3a can be rearranged into Equations L1.3b. Equation L1.3c is the generic equivalent for a linear, ordinary differential equation (ODE) with constant coefficients. This ODE form has a readily available solution.

No damping. For simplicity, consider only the case of Equation L1.3b, where there is no damping term such that Equation L1–3b can be written as Equation L1.4.

$$\ddot{y}+\left(k/m\right)y=0 \leftarrow \text{No damping term} \tag{L1.4}$$

Use the characteristic equation solution technique to solve for the roots of the ODE as shown in Equation L1.5a, using the omega symbol, ω, in the characteristic equation.

$$\omega^2+\left(k/m\right)=0 \leftarrow \text{Characteristic form of L1.4} \tag{L1.5a}$$

$$\omega = \pm i\sqrt{k/m} \tag{L1.5b}$$

Omega, ω, is the characteristic equation root and identifies the *natural frequency* of the system, defined by the spring constant and the mass as shown in Equation L1.6.

$$\omega_n=\sqrt{k/m} \tag{L1.6}$$

Equation L1.5b is an imaginary term (complex number with no real part) since it contains "*i*." If a system is oscillatory, the mathematical solution will have an imaginary term. Therefore, by dropping "*i*," Equation L1.6 *is* the oscillatory frequency based on stiffness and mass.

For sinusoidal functions, the natural frequency solves out as the *angular frequency*, ω, in terms of radians per second. For a discrete, oscillatory system, it is also useful to *count* system cycles in a second, which is known as the *ordinary* frequency, or just *frequency*, f. The number of times a mass passes by the same point (in the same direction) per second is its frequency. The relationship between frequency and angular frequency is given in Equation L1.8. Angular frequency should be used for all mathematical evaluations.

$$\omega = 2\pi f \tag{L1.7}$$

Damping. If damping is present in the system, the quadratic equation can be used to solve the roots of Equation L1.3b. Equation L1.3b can be written in characteristic equation form in Equation L1.8, and the quadratic equation is used for the roots in Equation L1.9.

$$\omega^2 + (c/m)\omega + (k/m) = 0 \tag{L1.8}$$

$$\omega = \frac{-\frac{c}{m} \pm \sqrt{\left(c/m\right)^2 - 4(1)\left(k/m\right)}}{2(1)} \tag{L1.9}$$

Depending on the values of stiffness, mass and damping (k, m and c), the **radical portion** of Equation L1–9 can be **imaginary, zero** or a **real** number.

Consider the unique case where the *radical* in Equation L1.9 is zero. Only one set of parameters will create this situation, which results in repeated real roots for the solution. This is the critically damped case where the oscillatory system drives to steady state in the minimum time with no oscillatory overshoots. This condition only occurs when Equation L1.10 is true, which is why "c" is defined as critical for this unique case in Equation L1.10.

$$c_{CRITICAL} = 2\sqrt{km} \tag{L1.10}$$

If system damping is less than the critical value, the system will oscillate as it drives to steady state. If damping is greater than the critical value, the system will drive to steady state without oscillation but slower than the critical case. Damping also affects the natural frequency, where ω_d is the damped natural frequency.

ODE solution. The most general solution root for the quadratic equation is the complex number in Equation L1.11, where alpha is the real part and beta is the imaginary part. Mathematically, alpha might be zero or beta might be zero, which provides insight into the oscillatory (or non-oscillatory) nature of the system.

$$\omega = \alpha \pm i\beta \tag{L1.11}$$

Equation L1.11 is a repeat of Equation L1.9. When mass, stiffness and damping are known, the values can be entered into Equation L1.9 and the result is a complex number, written in general terms in Equation L1.11. If $\alpha = 0$, the system has no damping and β can be entered directly into Case 0 in Table L1.1 for a plot. If Equation L1.9 solves out to two real numbers, the system is over-damped and the two real numbers, ω_1 and ω_2, can be entered into the Case I equation in Table L1.1. The same scheme can be followed for the remaining two root cases.

Table L1.1 shows how ODE solutions may be written in more than one form as exponential or trigonometric equations with or without a phase angle, ϕ. Recognize that different forms are possible to represent the exact same solution and one form may be easier to work with than the other. Table L1.1 summarizes the ODE solutions

TABLE L1.1
ODE Solution Cases

Case	Damping	Solution	Equations
0	None	$y = A\cos(\beta t + \phi)$	(L1.12a)
I	Over	$y = c_1 e^{\omega_1 t} + c_2 e^{\omega_2 t}$	(L1.12b)
II	Critical	$y = c_1 e^{\alpha t} + c_2 t e^{\alpha t}$	(L1.12c)
III	Under	$y = e^{\alpha t}\left[\, c_1 \sin(\beta t) + c_2 \cos(\beta t) \,\right]$ **OR** $y = e^{\alpha t}\left[A \cos(\beta t + \phi) \,\right]$	(L1.12d)

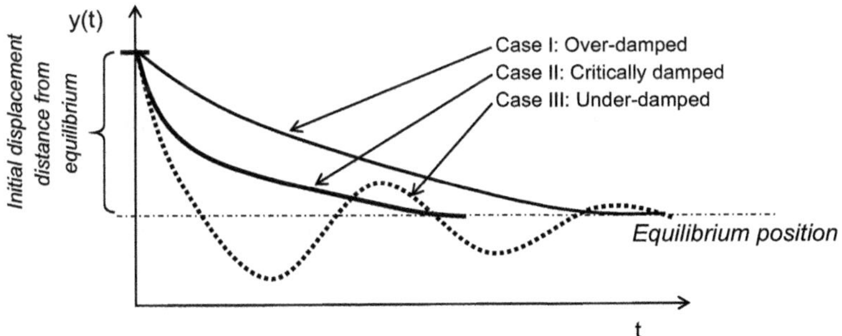

FIGURE L1.3 Damping cases.

based on the roots of Equation L1.11, where time, t, is the independent variable ($y(t)$ = *function of time*).

Consider the three damping cases from Table L1.1 in their mathematical forms. Case I is over-damped because both exponential functions superimpose and significantly delay getting to equilibrium. For a stable system, ω must be negative for the exponentials to decay. This can be seen in Equation L1.9, where the first negative term dominates the result due to the high damping coefficient, c.

Case II has repeated roots for a combined exponential curve that cannot be driven any faster to its settling value without an oscillatory overshoot. Neither Case I nor Case II has an imaginary term, so they are both non-oscillatory.

Case III imposes an exponential function on an oscillatory function. This solution is under-damped and will oscillate until the exponential eventually drives it to a settled value for a stable system. Figure L1.3 summarizes the three damping cases.

Case 0 is a pure oscillatory function. There is no damping since there is no exponential component causing the function to die out. This is an idealized case since real-world systems always lose some energy and eventually dampen out.

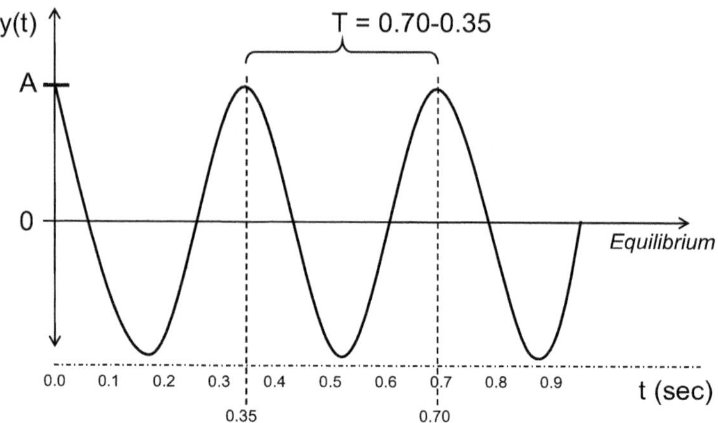

FIGURE L1.4 Determining system frequency based on wavelength measurement.

Period and frequency. The natural frequency of a highly oscillatory system can be determined through a variety of techniques. The most straightforward method is to plot the time domain data or curve as shown in Figure L1.4, and measure any peak-to-peak time for one full wavelength.

Total time for one wavelength is the period, T, in seconds, which is related to frequency, f, in Hertz (s^{-1}) with Equation L1.13.

$$f = {}^{1}\!/_{T} \tag{L1.13}$$

Amplitude and initial conditions. The maximum achievable amplitude, A, in Figure L1.4 is the initial displacement of the mass from its equilibrium position. If the system is displaced 15 cm, then $A = 15$ cm in Figure L1.4. For Case 0 in Table L1.1, the constant, A, also equals 15 cm for a 15 cm displacement since the cosine function will oscillate with a maximum of $A = 15$ cm.

Solving the constants for the Case I through III equations requires more effort. In each case, at $t=0$, the function will evaluate to the maximum amplitude, just as with Case 0. When oscillations commence at $t=0$, position, y, is equal to displacement amplitude. If the displacement is 15 cm, then $y(0) = 15$. Also, at $t=0$, initial velocity is zero, so $y'(0)=0$. The derivatives of the position function equations in Table L1.1 provide the velocity equations. Both the position and velocity equations must be used with the initial conditions for position and velocity to solve the undetermined constants in Table L1.1. The constants can be checked by setting $t=0$ and solving for "y" and confirming the originally planned amplitude.

EXPERIMENTAL METHOD

Prepare a table or spreadsheet similar to Table L1.2 in the results section to record the system parameters of mass, spring constant, and damping coefficient for both theoretical and experimental values. For the lab plan, include this table in the expected results section with a notional cart mass of 0.75 kg and a spring constant of 2.0 N/m. Update these notional values with actual values during the lab for correct theoretical calculations in the final lab report results and discussion section.

Theoretical plots. Prior to the lab, generate theoretical response plots using Equations L1.9 and L1.12(a–d) for *each* of the four oscillation conditions (undamped, critically damped, under-damped and over-damped). In both the lab plan and lab report, use topic sub-headings to clarify each of the damping cases. Complete the following steps to generate the plots:

1. For the lab plan, use the provided notional values of mass and spring constants.
2. For the lab report, regenerate plots as necessary with actual measured values.
3. Calculate the critical damping constant, c_{cr}, with Equation L1.10.
4. For the over-damped case, choose any desired value for the damping coefficient, c, above the critical value. For the under-damped case, choose any desired value less than the critical value.
5. Choose a displacement amplitude in the range of 10–25 cm.
6. Take derivatives of Equations L1.12(b–d) and solve for the unknown constants. Follow the instructions in the **Amplitude and initial conditions** paragraph.
7. Use any desired plotting program to create the theoretical plots.
8. Place the plots in the expected results section of the lab plan and the results section of the lab report. Ensure each plot has a title and each axis is labeled.
9. Clearly include the equation used with each theoretical plot. This will be one of the equations from Table L1.1 with solved values in place of the unknown constants.
10. Clearly include the natural frequency with the oscillatory plots. For the undamped case, use Equation L1.6. For the under-damped cases, use Equation L1.9. Do not compute a natural frequency for the critically damped and over-damped cases.

For the final lab report, ensure these theoretical plots are updated with actual values. The experimental plots must also be included in the final lab report for comparison. Theoretical and experimental plots can be side-by-side or overlayed as desired to provide a comparison.

System parameters. Use a scale to determine the mass of any suitable weight that can hook onto a spring. Hang the spring vertically, as shown in Figure L1.5, from a stand or other fixture with no weights attached and measure the unstretched spring length from any convenient reference point. Hook the weight onto the free end of the spring and record the extended spring length from the same reference point.

Use a simple FBD to **equate spring force, kL, to weight, mg** ($k\Delta L = mg$), and solve for the spring constant, k (units of kg/s^2 or N/m). Be sure to use the stretched spring *delta length*, ΔL (difference in lengths), as indicated in Figure L1.5, since it is the stretch force $k(\Delta L)$ opposing the weight, mg. Use a scale to determine the mass of the carts used in the system, along with their weights and damping magnets.

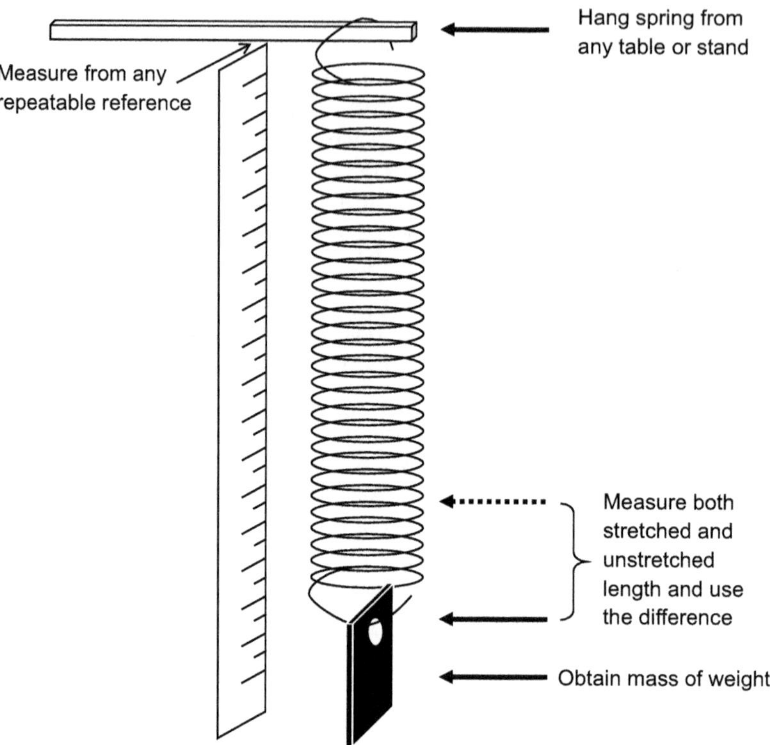

FIGURE L1.5 Determining the spring constant.

Setup. Place rolling carts on an inclined track. Attach one end of the spring to the cart and the other end to a fixed point at the upper end of the track, as shown in Figure L1.6. Place an ultrasonic distance measuring device at the lower end of the track, connected to computer data-logging software to constantly record the position of the cart as it moves. Ultrasonic sensors may have a limited effective range, so the cart position relative to the sensor may require adjustment to ensure proper detection.

Software. Use any suitably provided data-logging software to record ultrasonic distance. Use a minimum sampling rate of no less than 50 Hz. Confirm the software is measuring the correct distance to the cart. Ensure the software captures and saves enough significant digits. For the quickly oscillating cart, three-decimal precision in thousandths of a second is required.

No damping. The cart should move freely, with no damping magnets near the metal track to interfere with movement. Ensure the correct mass is used, depending on whether magnets are included. With the cart in place on the track, note the equilibrium position at the front end of the cart. Pull the cart down approximately 10–15 cm below its equilibrium position. Ensure the software is in data acquisition mode, capturing data, and release the cart so it begins to oscillate. Allow the cart to oscillate for several cycles until there is sufficient data. Repeat the process as desired

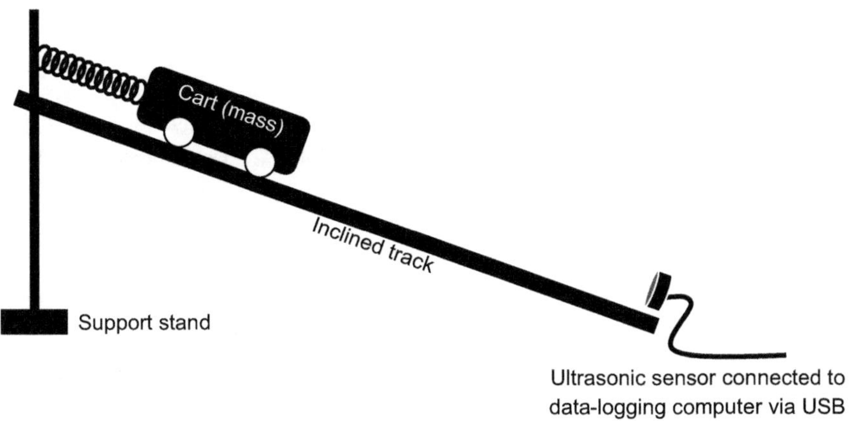

FIGURE L1.6 Oscillating cart system setup.

to ensure sufficient data acquisition. Since friction is present, the result will have a slightly damped behavior. Use only the first few stable cycles to determine the natural frequency.

Under-damped. Although the aluminum track is non-magnetic, moving magnets near a conductive metal create eddy currents. This process extracts energy from the system to create the damping effect. Re-orient or attach the magnet assembly on the cart such that the magnets are in very close proximity to the track but not touching. Repeat the data acquisition process by displacing the cart and recording data as it oscillates. If the carts were not originally weighed with the magnets in place, ensure an updated mass measurement is taken.

Over-damped. The cart magnets in conjunction with a thin aluminum track may not provide enough damping for the over-damped case. To improve the simulation, an additional metallic component can be placed below the track. A metallic straightedge or other length of metal can be held or taped into position under the track. Steel will create a static magnetic force, which might interfere with the experiment. Aluminum will create eddy currents only during movement, which may prove more controllable. Attempt a couple of test runs to verify the correct behavior before collecting data. The cart should not oscillate in this case. Repeat the data collection process as before.

Critically damped. This condition is not tested. There is no capability within this experiment to "*dial in*" the critical damping coefficient. For the lab plan and lab report, only the theoretical values and plots should be provided for this case.

RESULTS

All system parameters must be summarized in Table L1.2. While three cases are experimentally tested (undamped, under-damped and over-damped), only the undamped case is compared between theoretical and experimental results in Table L1.2. Mass and spring constant can be physically measured in the lab for theoretical

TABLE L1.2
System Parameters

Parameter	Theoretical (Expected Values Pre-Lab)	Experimental	% Error
Mass, m	*(0.75 kg)*		
Spring constant, k	*(2.0 N/m)*		
Critical damping, c_{cr}			
Undamped natural frequency, ω_n			

calculations and experimental comparison, but the damping coefficient cannot be directly measured or set for theoretical calculations. All theoretical and experimental plots should be included in the results section of the report (expected results in the lab plan).

Data processing. Import the saved experimental data into analysis software, such as a spreadsheet with graphing capability. Depending on the data capture software, two columns of data should be of interest. One column is for increments of time based on the sampling frequency, and the other column is for position. It is also possible the data-logging software may return only a position data column with no direct timing information. If this is the case, each position data point is based on the sensor sampling frequency. Merely create a new spreadsheet column to convert *line entries* into timing based on the sampling frequency. The first data point should be when time equals zero. If the sampling frequency is set at 10 Hz, data is logged every 0.1 seconds (1/10 Hz) and timing for each position entry occurs at 0.1 s, 0.2 s, 0.3 s… If 20 Hz was used in the experiment, use a time factor of 0.05 (1/20 Hz). Select both the time and position columns and create a *scatter chart with smooth lines* based on this data. The result should be an oscillatory waveform with time on the x-axis.

Determine frequencies. Determine the experimental frequency for both the undamped and under-damped cases only. Frequency can be determined from the waveform manually or automatically with the spreadsheet data analysis tools. To perform a manual analysis, enlarge the waveform chart sufficiently to discern details, and use Figure L1.4 as a guide. Choose any discernible point on the plot, such as max or min amplitude points or horizontal axis crossing points. Hover the mouse over the point to determine the x-axis time value and write down the number. Move the mouseover one full wavelength and hover for the time value at that point. Subtract the two values to determine the period. Use Formula L1.13 to determine the frequency, f. Use Equation L1.7 to convert between angular frequency and regular frequency when comparing theoretical and experimental frequencies.

Additional considerations. The experimental waveform plots can be analyzed, if desired, to determine the actual damping present within the experimental setup. The damped natural frequency solves according to Equation L1.14.

$$\omega_d = \omega_n \sqrt{1 - \left(c / c_{cr} \right)^2} \tag{L1.14}$$

Several resources are available describing how to analyze a decaying waveform amplitude due to damping. This concept is also discussed in the chapter on first-order models and decaying exponential functions. These techniques can be used in conjunction with Equation L1.14 to solve the damping coefficient. Still, there is no ability to *set a damping value, c,* with the experimental carts, since the magnets can only be moved closer or farther to the track to increase or decrease damping for qualitative results.

Present results. Provide an overlay or side-by-side comparison of theoretical and experimental plots for each of the appropriate cases. Title each chart and label all axes with units. Format each axis value for appropriate significant digits and readability. Include the appropriate theoretical equation with each chart from Table L1.1. Include the experimentally determined frequency values for the oscillatory cases. Compute the percentage error only for the undamped case between theoretical and experimental frequencies and enter this percentage error into Table L1.2. Provide figure numbers and captions for each plot within the results section of the report. All plots should be introduced, explaining *what* the plots are and *what* is being presented in the plots.

DISCUSSION

Describe what is occurring with the data for each case. Identify significant differences between theoretical and experimental results and values for each case. Discuss possible reasons for the differences between theoretical and experimental results. Discuss whether the undamped experimental case appears to have damping. Mention any unique challenges encountered during the experiment that may have impacted results and how these challenges appear in the data. Mention sources of error that were either controlled or potentially uncontrolled. Do not mention human error as a source of error unless it is necessary to re-accomplish or redefine the experiment due to a failed result.

CONCLUSIONS

Based on the objective, cite the percentage errors between the experimental results and theory and assess whether the experimental data appears valid. All assessments and conclusions should be a logical result of the presented data. Offer any ideas on how *your* experimental setup could have been improved for better results. Do not mention whether you "*learned a lot*" from the experiment or whether it was "*fun*" or not; stick only to the context of the experiment.

APPENDIX L1.A: ADDITIONAL NOTES

Experiment configuration and implementation. This lab was written based on commercially packaged vibratory cart and track equipment with dedicated software. Several solutions are also available for ultrasonic position sensors and data logging without the need for dedicated software. The experiment can be configured as a simple spring-hanging-mass system, but damping is more difficult to implement.

Regardless of the configuration, damping can be implemented through a few different mechanisms. Friction is simple, but associated stiction affects equilibrium return. Magnets with ferrous metals generate static forces that can affect results. Magnets moving relative to non-magnetic aluminum create drag through eddy currents only during movement, which is truer to the theoretical model. If the magnets are small or the aluminum thickness is insufficient, it is difficult to generate damping for the over-damped condition. Preparation and analysis are somewhat lengthy for this lab, but it is routinely completed within about an hour and a half.

LAB 2
Continuous System Vibrations

Safety Requirements

Normal safe lab practices
Lab coats and safety glasses required

Equipment Requirements

Function generator
Vibration generator
Ultrasonic sensor
Computer with data logging software for ultrasonic sensor
Polycarbonate beam (26 cm long, 2 cm wide, 2 mm thick or any similar dimensions)
Small ~0.125″ hole exactly in center for attaching to vibration generator
Cantilever support structure (any wood or 3D printed support with clamp)

Procedural Requirements

Avoid excess stress on polycarbonate beams to avoid cracks
Lock vibration generator prior to attachments and unlock prior to operation
Use a maximum of 5 V on the function generator

Clean-up Requirements

Return all equipment to proper locations
Unplug all lab equipment

Emergency Actions

No unique considerations

INTRODUCTION

As with a discrete oscillatory system, a continuous structure such as a cantilever beam also has a natural frequency of vibration. If the structure is displaced from equilibrium or otherwise energized by an external force, it will seek equilibrium as it vibrates at its natural frequency. In addition, similar to a discrete oscillatory system where the mass, stiffness and damping determine the natural frequency, these same attributes establish the vibration frequencies of a solid structure.

In mathematical terms, vibration is an oscillatory function best described in sinusoidal form. Sinusoidal functions are repetitive and have potentially infinite solution sets. Just as the sine of zero is zero, the sine of two-pi radians is zero and every such interval is equal to zero as the function repeats. In terms of vibration, the first solution to a repetitive function is the *fundamental* frequency, while *harmonic* frequencies are the subsequent solutions. These vibration frequencies (solutions) can be visually identified by the distinct *mode shapes* of the vibrating beam.

For this experiment, a cantilever beam is a continuous system. The first mode shape is mathematically determined and experimentally checked against the beam's freely vibrating fundamental frequency. Higher harmonic frequencies are then solved and checked against the appropriate mode shapes while the beam is excited with an external forcing function. While the beam will vibrate at any frequency set by the external vibration generator, only harmonic frequencies create the unique mode shapes associated with solutions to the function.

OBJECTIVE

Assess the validity of the experimentally obtained natural frequencies of a cantilever beam relative to the theoretical model. An accepted theoretical model is available for a vibrating cantilever beam. The objective is to replicate a vibrating beam experiment and assess whether the experiment yielded satisfactory data.

THEORY

The natural frequencies and mode shapes of a vibrating cantilever beam require developing and solving the dynamic beam equation. This equation is a time-based (dynamic) mathematical model specifically developed for a solid cantilever beam. It begins first with static beam bending (no time component) and then introduces time via acceleration with Newton's second law. A full derivation of the dynamic beam equation is rather involved and not presented in complete detail here. Instead, an overview of the approach is presented, beginning with classic Euler–Bernoulli beam theory.

The Euler–Bernoulli model neglects shear and rotational deformations, which differs from the Timoshenko beam model. As a result, the Euler–Bernoulli model results in a theoretically stiffer beam, which also tends to overestimate natural frequencies, especially in higher modes.

Static beam displacement. Several approaches are available to develop the Euler–Bernoulli model. The following approach emphasizes simplicity with basic Cartesian geometry, minimum notation and one-dimensional strain along the x-axis. The following Euler–Bernoulli assumptions require emphasis:

1. Rotational effects are neglected, so all angles are right angles.
2. Deformations are small, so small angle approximations apply ($\sin\theta \approx \theta$).
3. The neutral axis endpoints remain vertically aligned between the bent and unbent beam.

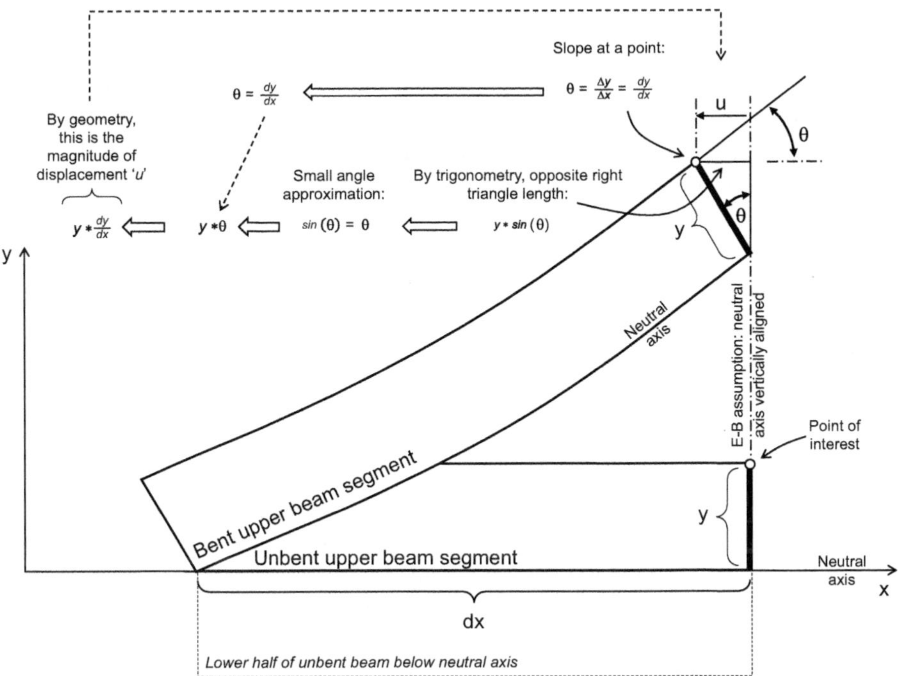

FIGURE L2.1 Geometry of the Euler–Bernoulli beam.

For most structural applications, these three assumptions are satisfactory since deformations are usually quite small. However, for a vibrating cantilever beam, the displacements are greater, and these assumptions begin to break down. As a result, these assumptions are a likely source of error in the vibration frequencies.

Figure L2.1 depicts the comprehensive geometry used in the derivation. The geometry is greatly exaggerated to show detail. To reduce clutter, only the upper half of the beam is depicted above the neutral axis.

The goal of derivation is to obtain a general expression for strain, ε, in the x-direction of interest. Strain in the x-direction is a function of both x and y. Strain is proportional to stress, and stress changes along the length of the cantilever beam from maximum at the fixed root to nothing at the free end. Deflection in the y-direction is also changing with x.

The first key feature to note in Figure L2.1 is the upper right point on the unbent beam that moves up and to the left on the bent beam. This is the displacement to define. The next important feature in Figure L2.1 is the beam-half thickness, y. The point of interest is defined by this distance along the y-axis. However, this same exact length, y, rotates up with the bent beam, which is important in defining the geometry. As the beam bends up, the upper right point of interest is displaced along the x-axis by a distance defined as "u." By convention, u, is the position function of both x and y to locate the point displaced by bending. For this analysis, all that is required is to note that the point moves "*negative u*" as defined by the Cartesian coordinate directions. The magnitude of u is the base leg of the right triangle shown in Figure L2.1, formed by the hypotenuse, y, and the angle, θ. Theta is also defined by the slope at the displaced point in the global x-y coordinate system, $\theta = \dfrac{dy}{dx}$.

The base of the right triangle in Figure L2.1 is $y*\sin(\theta)$ by simple trigonometry. Both y and θ are already defined to create the relationship. A small angle approximation is made to reduce $\sin(\theta)$ to just θ, and the triangle base leg, which is also the displacement, u, becomes $y\theta$. Since θ is also defined as a slope, a substitution can be made for the result in Equation L2.1. A negative sign is included to show the u-displacement direction in the coordinate system for the x-component of interest.

$$u = -y\frac{dy}{dx} \quad \text{(L2.1)}$$

Strain is defined as a change in length over the original length. In terms of displacement in the x-direction of interest, this relationship is expressed in Equation L2.2.

$$\varepsilon_x = \frac{du}{dx} \quad \text{(L2.2)}$$

Equation L2.2 calls for the derivative of the displacement function, Equation L2.1, with respect to the segment, dx, as shown in Equation L2.3a. The operation is complete in Equation L2.3b. As discussed in the development, y, in Equations L2.3a and L2.3b is a fixed length and can be pulled out of the differential operator.

$$\varepsilon_x = \frac{du}{dx} = \frac{d}{dx}(u) = \frac{d}{dx}\left[-y\left(\frac{dy}{dx}\right)\right] \quad \text{(L2.3a)}$$

$$\varepsilon_x = \left[-y\left(\frac{d^2y}{dx^2}\right)\right] \quad \text{(L2.3b)}$$

Material considerations. Up to this point, the analysis has been independent of material properties. The beam is in static equilibrium, so Newton's second law in Equation L2.4a applies. Stress (units of pressure) times area gives force, so integrating stress over differential units of area in Equation L2.4b is the same as Newton's law from Equation L2.4a.

$$\Sigma F_x = 0 \quad \text{(L2.4a)}$$

$$\int \sigma_x dA = 0 \quad \text{(L2.4b)}$$

Next, use Hooke's law as provided in Equation L2.5 to relate stress to strain.

$$\sigma_x = E\varepsilon_x \quad \text{(L2.5)}$$

Two more substitutions are required. Begin with Equation L2.4b and replace stress, using Equation L2.5 to arrive at Equation L2.6a. Next, replace the version of strain in Equation L2.6a with the version of strain in Equation L2.3b. The result of both

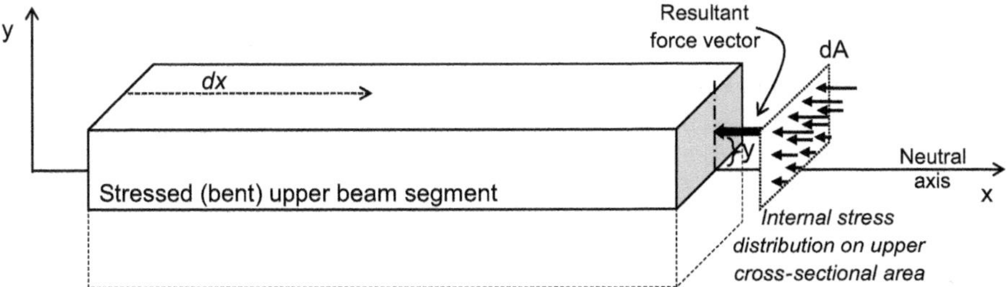

FIGURE L2.2 Internal stress with resultant force vector in stressed beam.

substitutions is Equation L2.6b, which is still just Newton's second law for static equilibrium.

$$\int (E\varepsilon_x)dA = 0 \tag{L2.6a}$$

$$\int E\left[-y\left(\frac{d^2v}{dx^2}\right)\right]dA = 0 \tag{L2.6b}$$

Equation L2.6b integrates stress along the length of the beam as it is applied to the cross-sectional area, as depicted in Figure L2.2.

Given the static equilibrium, moments at any point can be summed to zero per Equation L2.7a. Since force times distance is a moment, Equation L2.7a can be rewritten as Equation L2.7b using the distance, y, to the force vector in Figure L2.2 and summing moments about the neutral axis point.

$$\Sigma M_{NA} = 0 \tag{L2.7a}$$

$$\Sigma F_x y = 0 \tag{L2.7b}$$

Since both Equations L2.7a and L2.7b are equal to zero, they can be set equal to each other for Equation L2.8.

$$F * y = M \tag{L2.8}$$

The left-hand side of Equation L2.6b is force, so it can be substituted for F_x in Equation L2.8 to arrive at Equation L2.9.

$$\int E\left[-y\left(\frac{d^2y}{dx^2}\right)\right]dA * y = M \tag{L2.9}$$

Equation L2.9 can be re-arranged, recognizing that $\int y^2 dA$ is the definition for moment of inertia, I, to arrive at the second-order differential equation for static Euler–Bernoulli beam bending in Equation L2.10.

$$EI\left[\frac{d^2y}{dx^2}\right] = M \tag{L2.10}$$

Dynamic beam displacement. To move from static to dynamic bending, apply Newton's second law for dynamics in Equations L2.11a and L2.11b.

$$\Sigma F = ma \tag{L2.11a}$$

$$\Sigma F = m\frac{d^2y}{dt^2} \tag{L2.11b}$$

Since it is a fundamental relationship that the second derivative of moment is load (force), take the second derivative of the LHS in Equation L2.10 and place that into the LHS of Equation L2.11b. Convert mass in the RHS of Equation L2.11b to density times area per unit length for integration. Change the regular differentials into partial differentials to account for the newly created function of two variables, x (position along the x-axis) and t (time). The result of these substitutions and adjustments is Equation L2.12.

$$\frac{\partial^2}{\partial x^2}\left(EI\left[\frac{d^2y}{dx^2}\right]\right) = -\rho A\frac{\partial^2 y}{\partial t^2} \tag{L2.12}$$

All of the necessary elements of a cantilever beam are now accounted for in Equation L2.12. Equation L2.12 is a fourth-order partial differential equation, which can be solved through the separation of variables into Equation L2.13a and L2.13b. Equation L2.13 is a solution for only the separated position variable, x, where beam amplitude, y, is described for any point along the length of the beam.

$$y = A\cosh(\beta x) + B\sinh(\beta x) + C\cos(\beta x) + D\sin(\beta x) \tag{L2.13a}$$

$$\textit{where } \beta^4 = \frac{(\rho A)y^2}{EI} \tag{L2.13b}$$

Applying the boundary conditions of a fixed-end and free end for a cantilever beam resolves the coefficients A, B, C and D. The roots of Equation L2.13 return the natural frequencies, ω_n, of the continuous system. Including the relationship $\omega = 2\pi f$, the first three natural frequencies of the beam are given in Table L2.1. Plotting solutions to Equation L2.14 gives the corresponding beam mode shapes in Table L2.1.

The *nodes* are locations of zero amplitude (no vibration). For a cantilever beam held fixed at one end, a node will necessarily be present at this fixed location.

TABLE L2.1
First Three Harmonic Frequencies and Shapes for a Vibrating Cantilever Beam

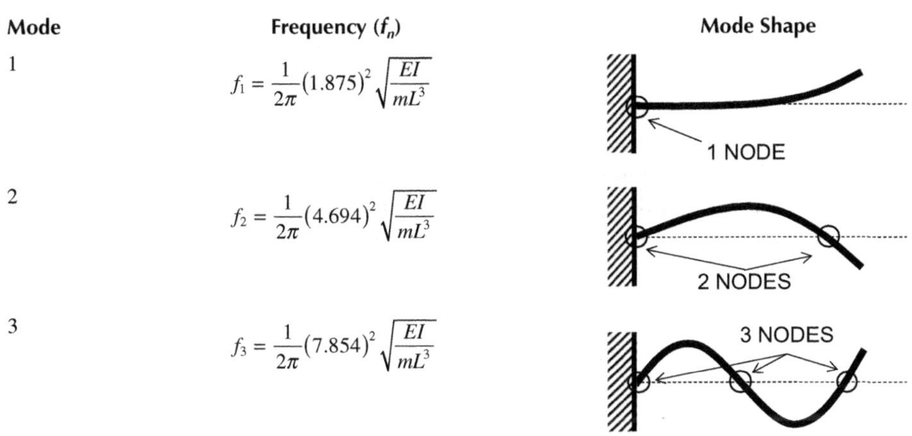

Mode	Frequency (f_n)	Mode Shape
1	$f_1 = \frac{1}{2\pi}(1.875)^2\sqrt{\frac{EI}{mL^3}}$	1 NODE
2	$f_2 = \frac{1}{2\pi}(4.694)^2\sqrt{\frac{EI}{mL^3}}$	2 NODES
3	$f_3 = \frac{1}{2\pi}(7.854)^2\sqrt{\frac{EI}{mL^3}}$	3 NODES

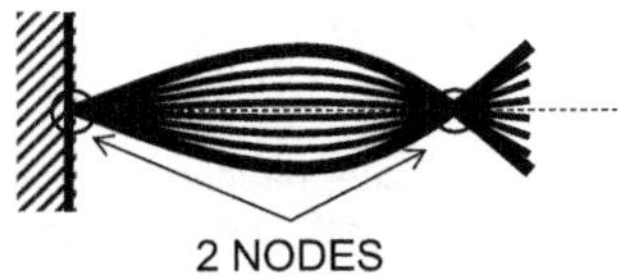

FIGURE L2.3 Nodes of the second mode shape in real time.

A cantilever beam vibrating at its first fundamental frequency will vibrate in its first mode shape with only one node at the fixed location. The second mode shape will have two nodes and so on for higher mode shapes. Figure L2.3 repeats the image of the second mode shape from Table L2.1 as it tends to look while vibrating in real time.

EXPERIMENTAL METHOD

Calculations. Consider a small cantilever beam made from polycarbonate plastic. Create a table to record values similar to Table L2.2 in the results section. For the lab plan, use the expected values from Table L2.2 for computing the first three natural frequencies of the cantilever beam. The physical beam is actually twice as long as indicated in Table L2.2, so use the *cantilever length* (half of the polycarbonate beam's actual length) provided in the table. Update all lab plan expected values with *actual* values during the lab and recompute any necessary properties. For the final lab report, do not include a column for *expected* values. Expected values from the lab plan are superseded and replaced by actual values from the lab.

Prepare another table and spreadsheet similar to Table L2.3 in the results section to record both theoretical and experimental system parameters. All theoretical parameters in Table L2.3 must be completed in the lab plan prior to the lab session using the expected values from Table L2.2. Update all parameters with actual values once they have been verified in the lab.

Calculate the cantilever beam's natural frequencies using the formulas from Table L2.1. Review the basic formulas for mass as a function of density and moment of inertia, I, for a rectangular cross-section. The moment of inertia is the resistance to bending and is based on the cross-sectional area of the beam, as shown in Figure L2.4. Record all intermediate material parameter calculations in Table L2.3.

Free vibration (Mode 1). Secure the polycarbonate beam into a cantilever support fixture as shown in Figure L2.5. Place an ultrasonic sensor on the table, pointed upward toward the end of the beam.

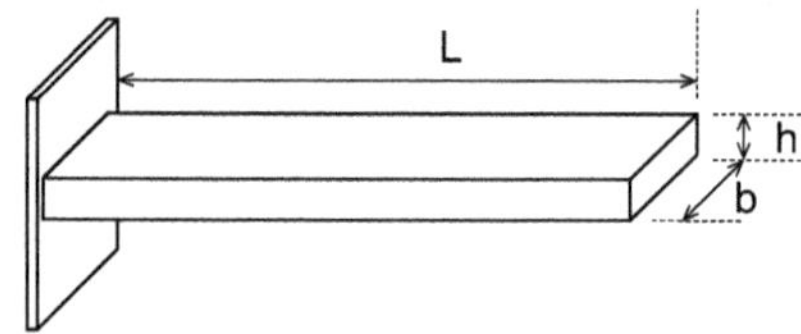

FIGURE L2.4 Cantilever beam dimensions.

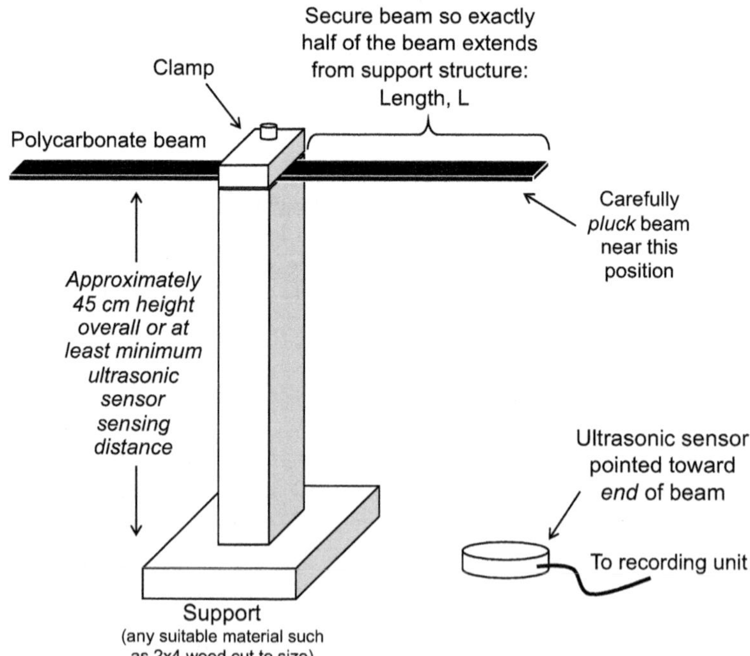

FIGURE L2.5 Free vibrations setup.

Set up the ultrasonic recording software according to the instructions. Set the sampling frequency rate to a minimum of **100 Hz**. Turn on the ultrasonic data recording unit and gently *pluck* the beam to initiate a free vibration. Move hand quickly away to avoid interfering with the ultrasonic sensor. Record and save the oscillation data for later analysis.

Forced vibration (Modes 1 and 2). Remove the polycarbonate beam from the cantilever support and secure the beam onto a vibration generator as shown in Figure L2.6. Electrically connect the function generator to the vibration generator. Connection polarity does not matter. If the vibration generator has a lock feature, be sure to lock the generator when affixing and removing the beam and unlock it prior to energizing. Prior to energizing, ensure the function generator is set to 5 V with a sine wave, if the waveform is selectable.

The goal is now to experimentally determine the natural frequencies of the first two mode shapes under forced vibration. Since no feedback sensors are used in this setup, achieving the exact mode shape frequency will require some qualitative assessment, both visually and tactilely. This is possible to a few Hertz, but this may be a SOURCE of error. This is not *human error.* This is a random TYPE of error from an observational source since the error will change somewhat randomly between experiments or personnel, subject to a limit in the ability to accurately measure the result.

Once the beam is secured in place and the vibration generator is in the unlocked position, turn on the function generator. The function generator usually defaults to standby mode. Ensure the voltage is about 5 V. Adjust the frequency close to the computed first natural frequency of the beam. Begin adjusting the frequency up and down in 1-Hz increments to achieve the **first mode** vibration frequency. This mode can be recognized by minimum to *no oscillation at the center attachment point* of the beam, where the node should be present. Realize this setup is meant to simulate a cantilever beam, so consider the center attachment as a fixed wall, as shown in Figure L2.7.

A near maximum deflection will be present at the end of the beam, but this deflection is also a superposition of oscillations at the attachment. Therefore, accurate

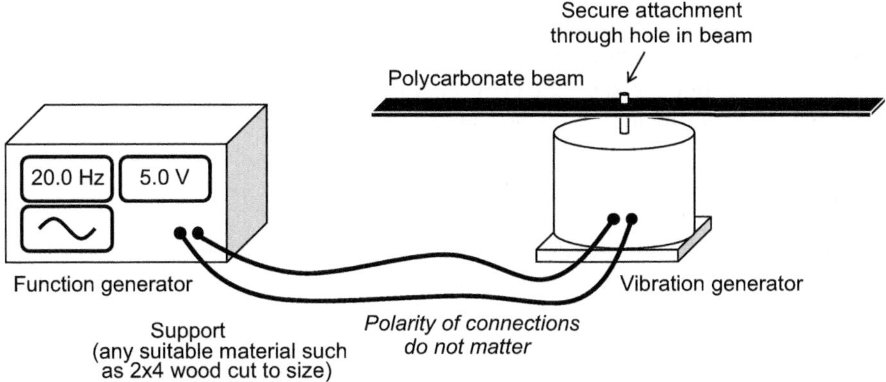

FIGURE L2.6 Forced vibrations setup.

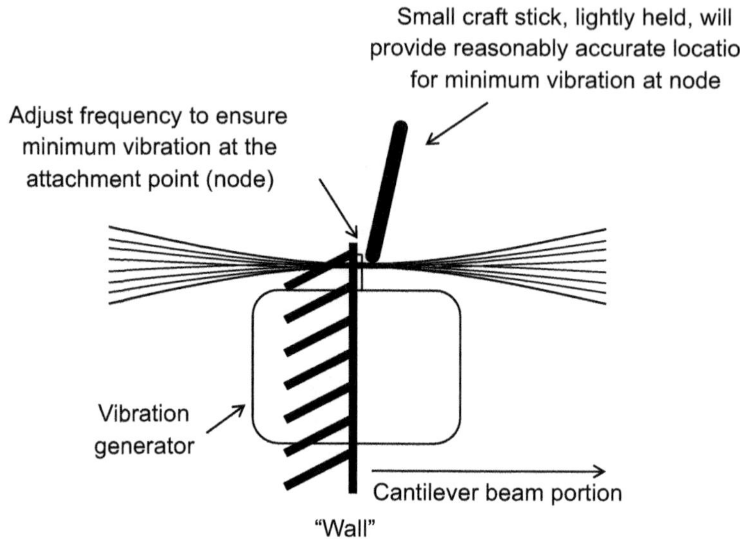

FIGURE L2.7 Using a craft stick to accurately detect first mode shape frequency.

results can be best achieved by ensuring a minimum to no-vibration node occurs at the attachment point.

To help confirm a node at the center attachment, *lightly hold and rest* a small craft stick on or near the attachment of the beam to feel for minimum vibration at the attachment point, as shown in Figure L2.7. Adjust the function generator frequency up or down to ensure minimum vibration occurs at the center attachment, as felt by the craft stick *vibrating loosely in your hand*. Once the first mode natural frequency has been identified, record the frequency value displayed on the function generator.

Place the function generator in standby to suspend vibrations. Adjust frequency to near the **second mode** shape theoretical frequency. Press the frequency generator standby button again to resume vibrations. Repeat the previous process to identify the second natural frequency mode shape.

For the **third mode** shape frequency, only compute the theoretical value. Do not determine the experimental frequency since it is too difficult.

RESULTS

The polycarbonate beam's physical and material properties are summarized in Table L2.2. Theoretical and experimental results for each mode shape are recorded in Table L2.3.

For the first mode shape, calculate the error only between the theoretical and experimental results for free vibration (not forced vibration). Also, do not attempt to calculate the error between the experimental free and forced values with the first mode shape.

TABLE L2.2
Experimental Values

Parameter	Expected	Actual
Length, L	0.18 m	
Width (base), b	2 cm	
Height (thickness), h	2.4 mm	
Material	Polycarbonate	
Density, ρ	1.2 g/cm^3	
Modulus, E	2300 MPa	

TABLE L2.3
System Parameters Summary

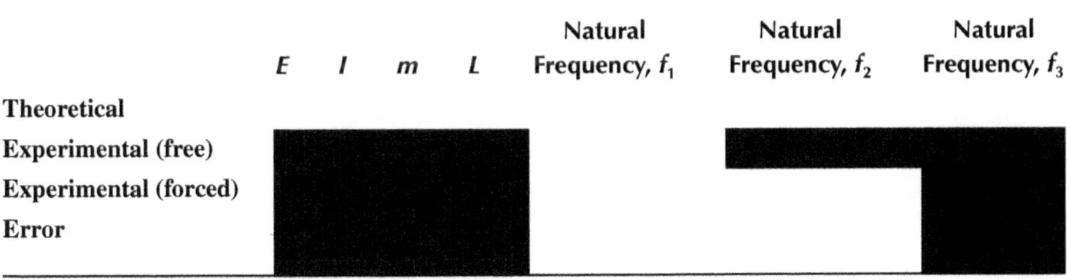

	E	I	m	L	Natural Frequency, f_1	Natural Frequency, f_2	Natural Frequency, f_3
Theoretical							
Experimental (free)							
Experimental (forced)							
Error							

DISCUSSION

Describe what is occurring with the data for each case. Note any significant differences between theoretical and experimental values and offer any potential sources of error. Sources of error might include observational errors, beam stiffness due to theoretical assumptions and differences in actual and published material properties. Refer the theory section about assumptions with the Euler–Bernoulli beam theory contributing to a theoretically stiffer beam. Offer any factual reasons why the errors may be present, such as the observational difficulty in precisely determining the forced mode frequency. Avoid *speculative* conclusions about the impact of errors by assessing the data and specifically indicating how the data shows the impact of error.

CONCLUSIONS

Based on the objective, cite the percentage errors between the experimental results and theory and assess whether the experimental data appears valid. Offer any ideas on how *your* experimental setup could have been improved for better results. Do not mention whether you "*learned a lot*" from the experiment or whether it was "*fun*" or not; stick only to the context of the experiment.

APPENDIX L2.A: ADDITIONAL NOTES

Experiment configuration and implementation. This lab was written based on commercially packaged vibration equipment with dedicated software. The experiment has also been completed using individually sourced components. Several solutions exist for ultrasonic sensors with data logging capability. Polycarbonate beams have consistently demonstrated satisfactory results. Thin aluminum beams of uncertain alloy composition have had less satisfactory results. When properly configured and prepared, this lab is routinely completed within about an hour and a half.

LAB 3
Waves

Safety Requirements

Normal laboratory concerns
Lab coats and safety glasses required

Equipment Requirements

Oscilloscope, function generator, vibration generator
Power supply, sound sensor (amplifier)
Spreadsheet program
Tuning fork, block of wood
Open tube (various lengths, approximately 4 cm diameter)
Large tube with capped end (approximately 4″ × 3 m)
Ruler, tape measure, bucket with water, string, weights, mass or weight scale
"Clicker" or other sharp noise source

Procedural Requirements

Ensure the power supply is set to 4.5–5.0 V before connecting the sound sensor
Do not strike tuning forks on tables (use the bottom of the shoe or a block of wood)

Clean-up Requirements

Dispose of water down drains
Return all equipment to proper locations
Unplug all lab equipment

Emergency Actions

No special considerations

INTRODUCTION

Waves surround us. Sound waves carry our voices and allow us to communicate. Electromagnetic waves carry music and data to be picked up by our devices. Light travels as a wave and allows us to see and perceive color.

There are three main types of waves: transverse, longitudinal and surface. These types are distinguished by the direction the *wave itself* travels versus the direction of motion of the *particles* excited by the wave. Throw a rock into a pond of water, and the wave travels away from the splash to the edges of the pond. However, the water just moves up and down, so this is a transverse wave. The sound of the splash travels to your ears as the air molecules push and pull back and forth parallel to the same direction the sound wave is also moving toward you. This is a longitudinal wave. Once a strong wind kicks up, the water molecules get kicked up and dropped back down in a somewhat circular motion as the windswept waves travel across the surface of the pond. This is a surface wave.

Standing or stationary waves are not a *type* of wave but rather the result of interference patterns set up within either transverse or longitudinal types of waves. When a wave is reflected, such as the ripples of water from a splash against a large partially submerged rock, the waves will ripple back in the opposite direction. The reflected waves will interact with the oncoming waves and either constructively or destructively interfere with each other. If the interference pattern results in a wave that does not *appear* to move, this is a standing wave.

All waves have properties such as amplitude, wavelength, frequency and velocity. These particular properties of a standing wave give direct insight into the same properties of the transverse or longitudinal waves creating the standing wave. This lab determines the fundamental properties of both transverse and longitudinal waves in different media by creating and analyzing standing waves.

OBJECTIVE

Assess the validity of experimentally obtained longitudinal and transverse wave properties of wavelength, frequency and velocity relative to theory. An accepted theoretical model is available for waves, and properties can be measured from standing waves. These wave properties have theoretical inter-relationships.

THEORY

Transverse waves. A transverse wave travels away from a source of disturbance or excitation while the particles of the medium carrying the wave are displaced perpendicular to the travel of the wave. Consider only one ripple created by throwing a rock in calm water, as shown in Figure L3.1.

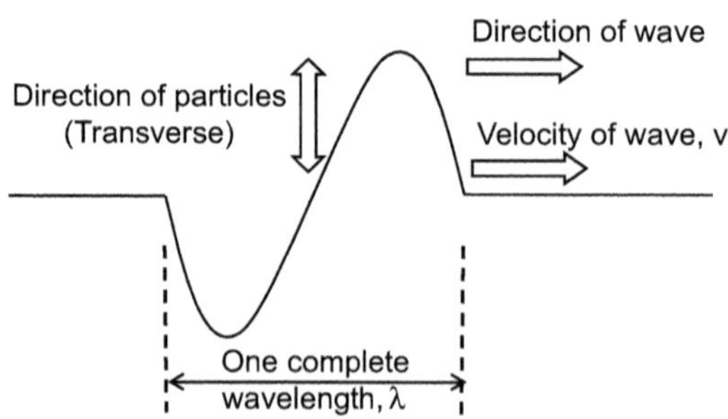

FIGURE L3.1 One transverse wavelength.

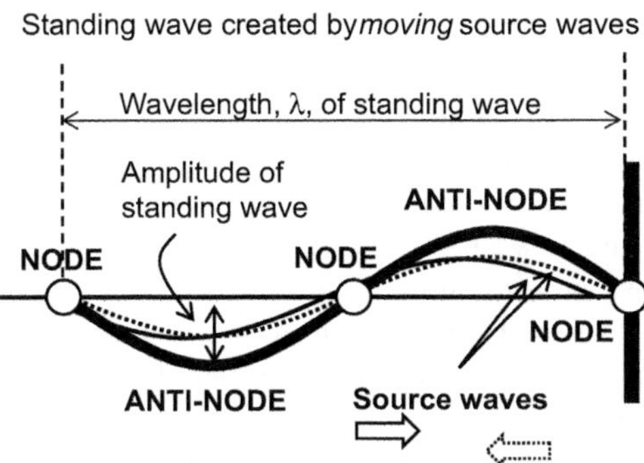

FIGURE L3.2 Standing wave properties with interference pattern from underlying traveling waves.

This transverse wave will travel onward until it runs out of energy or hits an obstruction. If it encounters a reflective surface, such as a large rock face or wall, the wave will reflect back upon itself to some degree. Since there are actually several ripples or waves created by throwing a rock in the water, the reflected waves will interact with the additional oncoming waves. Depending on the phase of the waves, particles at the crest of a reflected wave may get a boost upward from the crest of an oncoming wave. Particles on the crest might also happen to see a trough and be pulled back down. This interaction creates an interference pattern, as shown in Figure L3.2.

If the source of waves is continuous, then the interference pattern continues indefinitely based on the length of the medium and the energy in the system. The resulting *interference pattern* appears as a stationary *standing wave*, even though the source and reflected waves creating the standing wave are constantly traveling back and forth. The standing wave has the same properties of amplitude, wavelength, nodes and anti-nodes as shown in Figure L3.2. Ripples or waves in water are transverse waves. Waves in vibrating strings are also transverse waves.

Longitudinal waves. A longitudinal wave also travels away from a source of disturbance or excitation, but the particles of the medium are displaced parallel to the travel of the wave instead of perpendicular. Consider a single, loud clap in otherwise still air, as shown in Figure L3.3. The air molecules are initially pushed away from

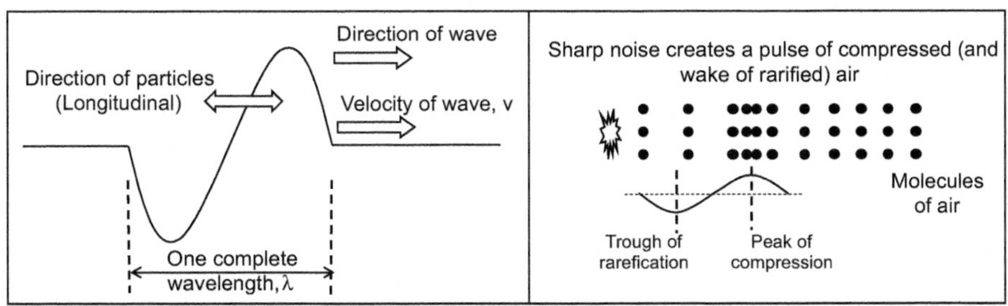

FIGURE L3.3 One longitudinal wave.

the source of the noise as they continue to compress molecules in the same direction the wave travels. As the wave of compressed molecules moves forward, a wake of lower pressure air follows to create what can be represented as one wavelength.

Just as with transverse waves, the particles of a longitudinal wave can hit against a reflective surface and begin moving in the opposite direction as the wave reflects back into the oncoming wave particles. An interference pattern is created, which again may or may not result in a standing wave.

Standing waves. The interference pattern of a source and its reflected wave do not necessarily create a standing wave. The wavelength of the source wave must be a specific length relative to the length of the confined space creating the reflected wave. If it is not at the correct wavelength, the resulting interference wave is irregular and does not appear stationary. However, if the source wavelength can *properly fit* into the confined space, a stationary wave will develop where the nodes (and anti-nodes) remain in fixed locations. Although the standing wave remains stationary, it still has vibratory motion in amplitude at its anti-nodes.

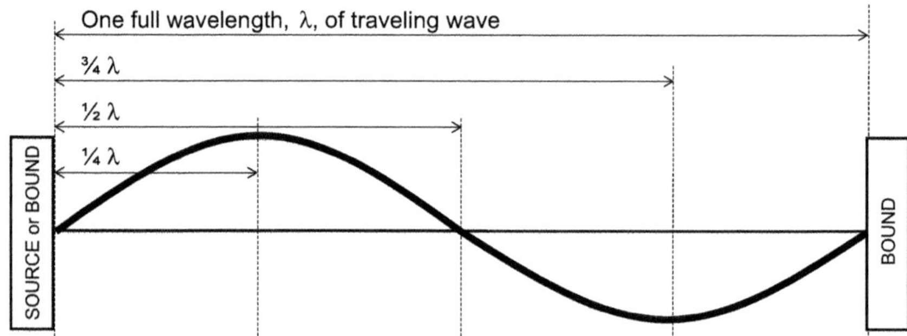

FIGURE L3.4 Wavelength and fractions of a wavelength within fixed bounds.

The traveling source wave must be bounded at particular lengths so its wavelength (or fractions of wavelength) fits into the length of the bounds. This fit must be such that a minimum or maximum amplitude occurs at the bound, corresponding to ¼, ½, ¾ or one full wavelength (or multiples thereof), as shown in Figure L3.4, depending on the *type of bound*. A correct fit permits the maximum amplitudes of inbound traveling waves to constructively interfere exactly with the maximum amplitudes of the outbound reflected waves and vice versa with minimum amplitudes. If the wavelength does not perfectly fit, a *messy in-between* pattern results without creating a standing wave.

A bound or endpoint can be either fixed (closed) or free (open), creating three end-condition scenarios as shown in Figure L3.5: fixed-fixed, free-free or fixed-free (same as free-fixed). In the first case with two fixed ends (fixed-fixed or closed-closed), only a node (not an anti-node) can exist since movement is forcibly constrained. As a result, any *integer multiple of half-wavelengths* can meet the fixed-fixed boundary conditions and can create a standing wave. In the second scenario, with both ends free, any bounding length of *half-wavelength multiples* can create a standing wave.

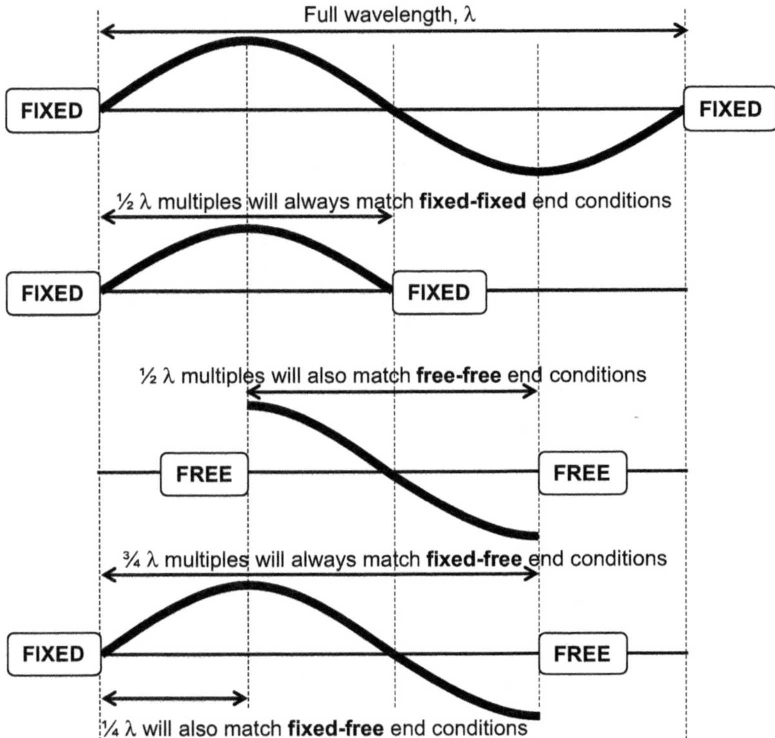

FIGURE L3.5 End conditions.

For the fixed-free condition, multiples of three-fourths wavelengths will match up with the boundary conditions. Also, in the fixed-free condition, the unique case of a one-fourth wavelength will match the fixed-free boundary conditions and can create a standing wave.

Generally, it's easy to envision a fixed-fixed boundary condition such as the vibrating string on a guitar after it is plucked. It is also possible to create the free-free boundary condition by grabbing the free end of a length of rope and shaking it up and down, if hand movement successfully creates an anti-node both where it is held and at the other freely moving end.

A more common utilization of a fixed-free or free-free boundary condition is with sound waves. Wind instruments are designed with lengths that correspond to either one or both ends being free. Sound amplification is at its maximum when a standing wave develops with an anti-node at the free end. This is an idealized condition, however. Sound waves within real tubes experience a variety of effects, such as viscosity and pressure differences at the free end, which means a real tube requires a *corrected length*[1] to match the theoretical conditions given in Figure L3.6. Empirical evidence indicates resonance of the standing wave within an open-end tube occurs at a corrected length given by Equation L3.1. This is an empirically accepted relationship, so no theory is offered.

$$L_{CORRECTED} = L_{WAVELENGTH} + 0.3(TUBE\ DIAMETER) \quad \text{(L3.1)}$$

Since a standing wave is a function of the underlying traveling waves, it possesses the same fundamental frequency, velocity and wavelength of the source waves. As a standing wave, it is also *much easier* to determine the properties of the source waves.

Wave equation. The one-dimensional wave equation is often derived by modeling a string under tension with Newton's second law. The derivation is not offered here, but several good resources exist. The result of the derived mathematical model is a second-order partial differential equation (PDE), known as the wave equation given in Equation L3.2. The equation is *one-dimensional* since the wave is described along a single x-dimension path. The amplitude displacement, y, in Equation L3.2 varies both as a position along the wave path, x, and at any moment in time, t.

$$\frac{\partial^2 y}{\partial x^2} = \left(\frac{T}{\mu}\right)\frac{\partial^2 y}{\partial t^2} \quad \text{(L3.2)}$$

The constant, T/μ, Equation L3.2 is a result of using Newton's second law in the derivation to account for string tension, T, and the string's mass per unit length, μ. It is possible to envision a lightweight string with low tension behaving differently from a heavy cable with higher tension. The tension per unit mass is a constant for the string or cable and dictates the parameters of the wave propagated by the string or cable. These are physical parameters introduced via Newton's second law in the derivation.

The PDE wave equation can be solved by differing techniques, including the separation of variables. Equation L3.3 is a general solution to the wave equation, which can be checked for validity by taking derivatives and plugging them into Equation L3.2. The constants, A, k and ω, are included to keep Equation L3.3 as general as possible.

$$y(x, t) = A\sin(kx - \omega t) \quad \text{(L3.3)}$$

Wave velocity. Equations L3.4a and L3.4b are the second derivatives of Equation L3.3.

$$\frac{\partial^2 y}{\partial t^2} = -\omega^2 A\sin(kx - \omega t) \quad \text{(L3.4a)}$$

$$\frac{\partial^2 y}{\partial x^2} = -k^2 A\sin(kx - \omega t) \quad \text{(L3.4b)}$$

Equations L3.4a and L3.4b can be substituted into Equation L3.2 to arrive at Equation L3.5.

$$\frac{T}{\mu} = \left(\frac{\omega}{k}\right)^2 \quad \text{(L3.5)}$$

Returning to Equation L3.3, the sine argument must be a dimensionless number for mathematical evaluation. If omega is angular velocity, the ωt can be divided by wavelength to become dimensionless, such that $\omega = 2\pi v/\lambda$. Similarly, if $k = 2\pi/\lambda$, the x-distance units cancel. The 2π in both terms is the repeating sine solution for the repeated wavelength. Solving Equation L3.5 with these substitutions results in the velocity given in Equation L3.6.

$$v = \sqrt{T/\mu} \quad \text{(L3.6)}$$

Equation L3.6 is the velocity of a traveling wave in a string or cable based on the physical properties of string tension and mass per unit length. The wave's velocity is a function of the medium carrying the wave. If the wave is a sound wave instead of a "*string wave*", then the velocity will be a function of the air medium. A sound wave will travel at the speed of sound in air, while a wave in a string will travel at the "*string speed*," which is the square root of tension per unit mass. This wave propagation speed is called celerity (not *string speed*).

Wave frequency. Based on the discussion so far, waves possess properties of both velocity and wavelength. These two properties can be combined for a third property, frequency. The frequency, f, of a wave is defined as the number of times the wavelength repeats itself per unit of time, so it has units of *per time* (Hertz), s^{-1}. Simply divide the velocity property by the wavelength property to obtain frequency, given in Equation L3.7, with units of Hertz. If two wave properties are known or can be determined, Equation L3.7 can be used to determine the third wave property.

$$f = v/\lambda \quad \text{(L3.7)}$$

EXPERIMENTAL METHOD

PART I. *WAVELENGTH* of longitudinal traveling waves. Sound propagates as a longitudinal wave. If a sound wave's frequency and velocity are known, then wavelength can be computed by solving Equation L3.7. The flow of activities is as follows:

1. Determine the velocity of the sound wave in a large echo tube with an oscilloscope.
2. Calculate the wavelength of sound based on velocity and a tuning fork frequency.
3. Experimentally determine the wavelength of sound by hearing resonance in a small tube.

Prepare a spreadsheet similar to Table L3.1 in the results section. The first row of Table L3.1 must be populated with expected results for the lab plan. Choose a common tuning fork frequency from Appendix L3.A and update this frequency in the lab as necessary if the actual tuning fork frequency is different from the planned frequency (Table L3.A1). For the lab plan, use a published value for the velocity of sound in air prior to actual velocity determination in the lab. Once in the lab, record the lab room temperature and update Table L3.1.

Velocity. Prepare a single spreadsheet or three tables similar to Tables L3.A2–L3.A4 in Appendix L3.A for recording data and computations. These tables are the supporting data for Table L3.1 in the results section. An expected tube length for the echo chamber is provided for lab plan computations in Table L3.A2 of the appendix. This length must be confirmed and updated in the lab. Enter a published value for speed of sound, v, for the lab plan. The speed of sound will be measured in an *echo tube*, as shown in Figure L3.6. Since velocity is distance divided by time, calculate the expected *time of flight* for the sound wave, Δt, using *twice* the tube length as distance for Table L3.A2. The sound pulse will travel down the tube and echo back, traveling *twice* the distance of the tube during measurement. This factor of two must be corrected. The easiest approach is to use twice the tube length, D, as the actual distance, as shown in Table L3.A2 along with Equation L3.8.

$$V = D / \Delta t \tag{L3.8}$$

Solve Equation L3.8 for time of flight in the lab plan to assist in setting the oscilloscope to the proper interval measurement range.

Connect a sound sensor to both a power supply and an oscilloscope, as depicted in Figure L3.7. A sound sensor has a microphone and an amplification circuit to create a measurable signal for the oscilloscope. Do not exceed 5 V on the power supply. Check the voltage level below 5 V before connecting the sound sensor wires to the power supply. Approximately 4.5 V is sufficient. Refer to Appendix L3.C for oscilloscope operation.

Place the sound sensor flush with the opening of the tube as shown in Figure L3.6, and create a sharp sound with a "*clicker*" to record on the oscilloscope. The sound will travel the length of the tube, bounce off the sealed end, and return to the sound sensor as a second peak. With the equipment connected and operating, create the sound pulse and determine the time interval between the original sound peak and the return sound peak, as described in Appendix L3.C. Determine the speed of sound using Equation L3.8 with Δt from the oscilloscope. Several measurements may be conducted as desired for an average value. Record sound velocity in both

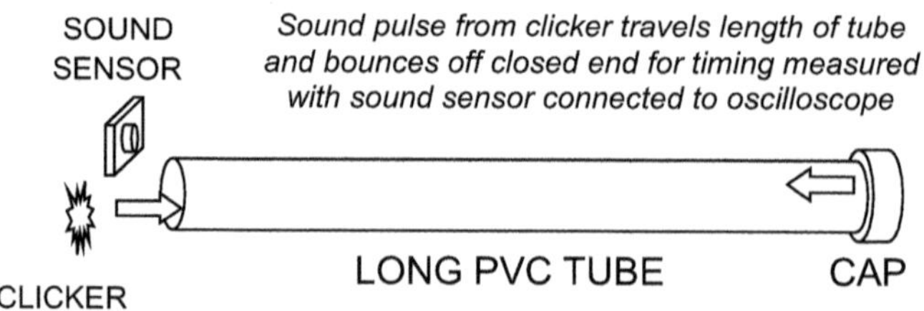

FIGURE L3.6 Sound velocity echo tube.

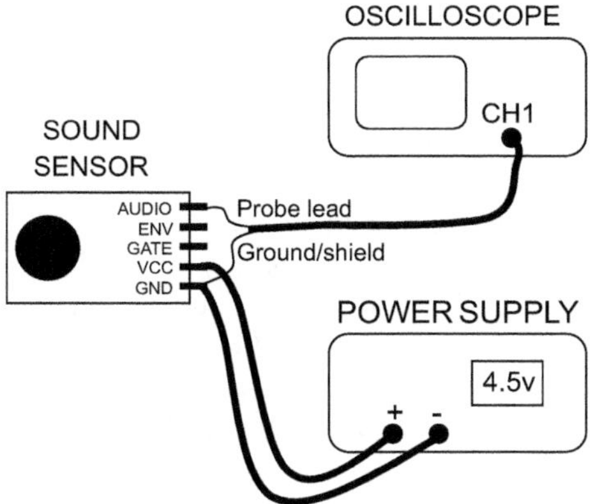

FIGURE L3.7 Sound sensor configuration.

Table L3.A2 and Table L3.1. Update the wavelength value in Table L3.1 based on the actual speed of sound and tuning fork frequency.

Wavelength. Ideally, one full wavelength can be experimentally measured and compared to the theoretical value in the first row of Table L3.1. However, harmonic fractions of a wavelength can also be measured and used to calculate the full wavelength. Sometimes, it is more convenient to work with shorter wavelength harmonics rather than the longer, full wavelength. For a tube closed at one end and open on the other, a quarter-wavelength or three-quarter wavelength will create a maximum amplitude anti-node at the open end, according to Figure L3.5. If a quarter-wavelength or three-quarter wavelength standing wave can be created in the closed-open tube, simple algebra can be used to determine the full wavelength for Table L3.1.

Prepare a spreadsheet and tables similar to Tables L3.A3 and L3.A4 in Appendix L3.A. Enter the appropriate tuning fork frequency in the table (@ f =____). Enter the speed of sound in all cells of the velocity column, v. Use the expected (published) velocity for the lab plan and the experimentally determined value for the lab report. Calculate the theoretical wavelength, λ_{THEORY}, using Equation L3.7 for each cell in the wavelength column (each cell should have the same value). Be sure to update the theoretical wavelength, λ_{THEORY}, in real time during the lab once the actual speed of sound is determined.

The wavelength column, λ_{THEORY}, will contain the full wavelength value based on the actual frequency and the actual speed of sound. Now it is necessary to calculate fractional portions of this full wavelength to fit the boundary conditions. Refer to Figure L3.5 and look for three possible options to fit the closed-open (fixed-free) boundary conditions. These three options will include two multipliers less than 1.0 and one multiplier greater than 1.0. Enter these three multipliers into the multiplier column, k, from smallest to largest. Calculate the theoretical harmonic tube lengths in the last column of Table L3.A3, L_{THEORY}, as shown.

Table L3.A3 is merely the application of Figure L3.5 to plan for a standing wave within an open-ended tube (fixed at the opposite end). Listening for the resonant peak during the experiment is nearly impossible without starting close to this fixed-free wavelength tube length.

Use Table L3.A4 to apply a small correction factor to the theoretical tube length and record the experimental results. Copy the theoretical tube length from Table L3.A3 into the appropriate cells of Table L3.A4. Use Equation L3.1 to apply the small correction factor as shown in the table for column $L_{CORRECTED}$.

Refer to Table L3.A4 and choose only *one* harmonic length to test, either $n=1$, $n=2$ or $n=3$. Fill a small bucket with water. Choose a small tube of approximately 1.5 inches in diameter with a length at least a few inches *longer* than the desired harmonic length, $L_{CORRECTED}$ from Table L3.A4. The tube should be able to dip one end into the bucket of water, as shown in Figure L3.8. The tube length above the water should be approximately $L_{CORRECTED}$ with the ability to adjust slightly up or down while still keeping the end of the tube in the water. This creates the fixed-free (closed-open) boundary condition for a standing wave.

Have one lab teammate hold the tube and be ready to adjust it up and down in the water. Have another teammate ready to measure the length of the tube above the water. Strike the tuning fork on an appropriate surface (block of wood, a shoe heel, but *not* a hard table surface) and hold it right in the top of the tube. Place your ear next to the open end of the tube to listen for subtle changes.

Slowly adjust the tube up and down in the water until the tuning fork sound is sharply amplified by the creation of a standing wave. Several attempts may be required. At the correct tube length, an anti-node will form at the open end of the tube, which is the maximum amplitude of the standing wave created from the constructive interference of the traveling waves. This phenomenon can be heard as a resonating, sharp increase in sound at just the correct length. Hold the tube position steady and have a lab partner measure the height of the tube above the water. Enter this value in the appropriate cell of the $L_{EXPERIMENTAL}$ column of Table L3.A4. Failure to set up close to the expected length will make this task nearly impossible.

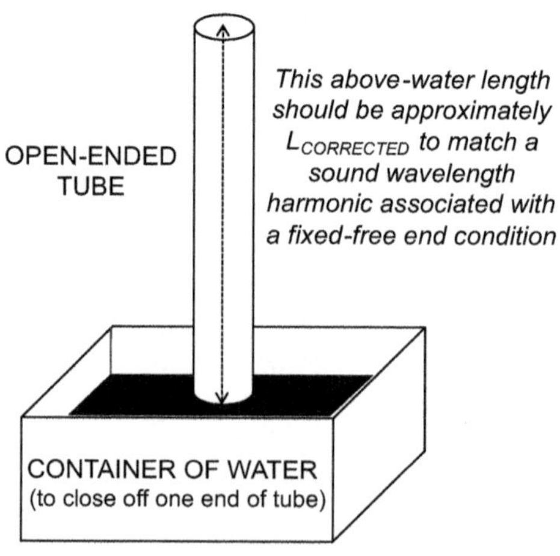

FIGURE L3.8 Open tube in water to measure wavelength of sound in air.

After entering the experimentally determined wavelength harmonic, $L_{EXPERIMENTAL}$, in Table L3.A4, it is necessary to recover the experimental wavelength from this value for the final column in Table L3.A4. According to Equation L3.1, the tube should appear a bit longer than the theory suggests. To account for this, subtract 0.3D from $L_{EXPERIMENTAL}$ (Equation L3.1, solved for $L_{WAVELENGTH}$). Ensure consistent units in the calculation. After this correction factor is resolved, use Equation L3–9 to convert the fractional portion of the wavelength back into the full wavelength.

$$\lambda_{EXP} = L_{EXP(AFTER\ CORRECTION)}/k \quad \text{(L3.9)}$$

After resolving the experimental wavelength from Equation L3.9, enter the result into Table L3.1 and compute the experimental wavelength error relative to the theoretical value.

PART II. *FREQUENCY* of transverse traveling waves. A wave in a cable, string or water propagates as a transverse wave. In the first part of the experiment, both frequency and velocity were established as fixed values, and Equation L3.7 was used to calculate the wavelength of a sound wave. For this second part of the experiment, velocity and wavelength will have fixed values, and frequency is resolved with Equation L3.7. The flow of activities is as follows:

1. Determine the velocity of a wave in a string based on the string's properties.
2. Establish any desired fixed-fixed bound length for the string.
3. Calculate an appropriate frequency to fit the string wavelength into the available bounds.
4. Experimentally determine the frequency that fits the string wavelength into the bounds.

Prepare a table or spreadsheet similar to Table L3.2 for the results section. The first row of Table L3.2 must be completed for the lab plan with expected results. Refer to Appendix L3.B to complete the calculations in Table L3.2. Nominal expected values are provided in Table L3.B1 for expected results. These values must be updated during the lab.

Prepare a spreadsheet and tables similar to Tables L3.B1 and L3.B2 in Appendix L3.B. Refer to Figure L3.9 for the fixed-fixed boundary conditions on the string. Knowing these planned string boundary conditions, refer to Figure L3.5 and select the first three wavelength multipliers that meet the fixed-fixed string boundary conditions. Note that any half-integer multiple of a wavelength will match the fixed-fixed boundary conditions. Enter these multipliers in the appropriate column of Table L3.B2 in Appendix L3.B.

Complete Tables L3.B1 and L3.B2 for the lab plan using the available information. All values must be updated with actual values during the lab for the final report. The L_{λ} column in Table L3.B2 is the portion of wavelength that will be *captured* in the bounding length based on the fixed-fixed end conditions. Compute each of the frequencies in the f_{THEORY} column for the lab plan and update these values during the lab based on actual measurements.

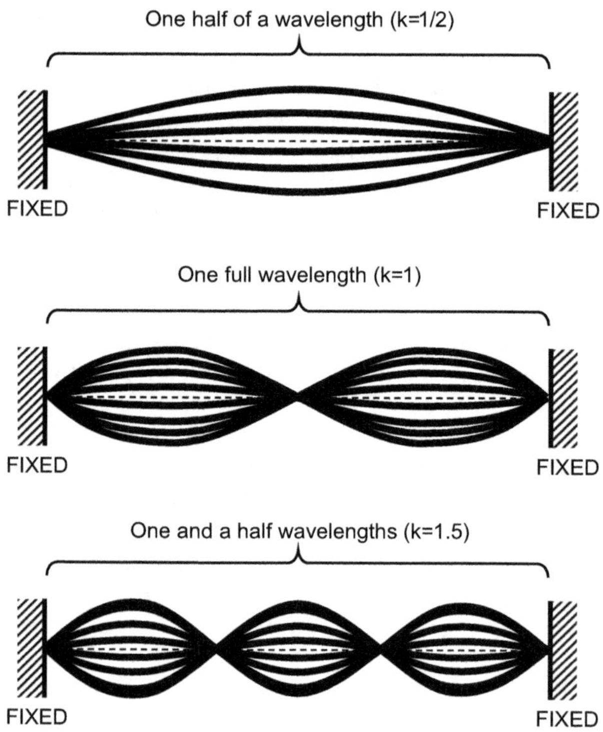

FIGURE L3.9 First three mode shapes and wavelength multiples for fixed-fixed end conditions.

Obtain a section of stout string, place it on a scale and record the string's mass into Table L3.B1. Next, obtain a tensioning weight for the string. Weigh the tensioner and record the weight in Table L3.B1. Be sure to record the mass for the string and the weight (force) for the tensioner. A simple free body diagram will show this weight *force* is also the string's tension force, T, when the weight is suspended on the string per Figure L3.10.

Configure the experimental setup as shown in Figure L3.10. Place a vibration generator near one end of a long table and attach one end of the string to the generator. Attach a C-clamp or other support at the opposite end of the table at approximately the same height as the generator. The free end of the string should hang over the C-clamp handle (or other support). Attach the tensioning weight to the free end of the string. *After* the setup is configured, measure the total stretched length of the string as shown in Figure L3.10. Add both the horizontal and vertical portions of the stretched string to get the total length. Enter the total stretched string length into the $L_{STRING(TOTAL)}$ cell of Table L3.B1, and use this total stretched string length for μ.

Measure only the horizontal string length distance between the vibration generator and the C-clamp support, and enter this value as the wavelength, λ, in Table L3.B1. This is the fixed-fixed wavelength or portion of wavelength that the standing

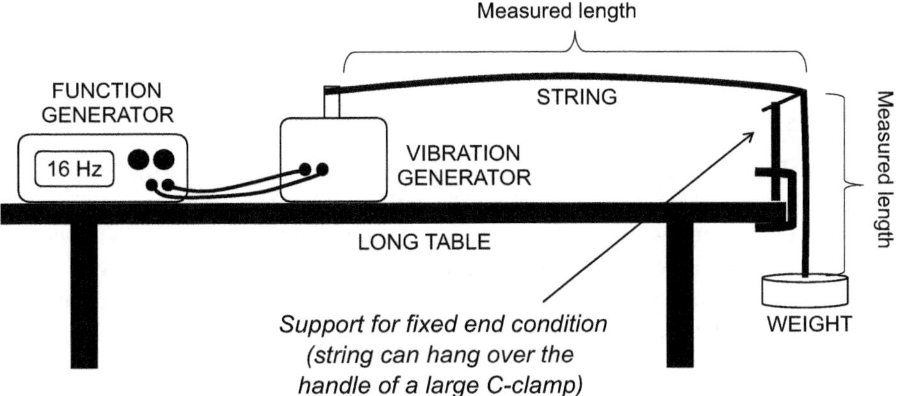

FIGURE L3.10 Transverse wave string setup.

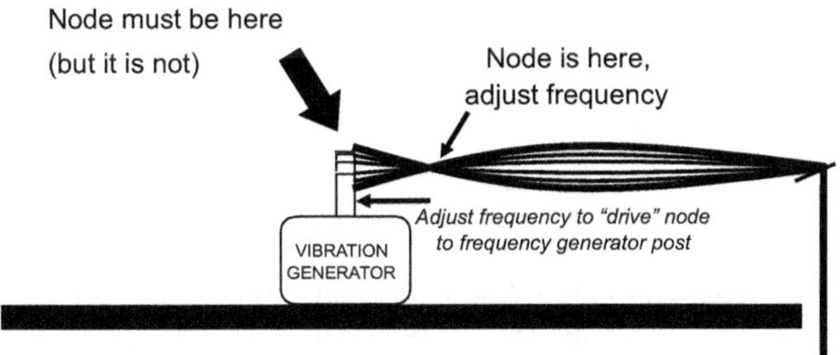

FIGURE L3.11 Seeing and adjusting the half-wavelength standing wave frequency ($n = 1$).

wave must fit within. The goal is to find the frequency that will fit a wavelength (or portion) within this available bounding length.

Each of the frequencies at $n = 1$, 2 and 3 in Table L3.B2 will be checked and also computed for error. However, only the full wavelength will be entered into Table L3.2 along with the error.

Set the function generator at the first theoretical frequency for the half-wavelength condition. Ensure the vibration generator is unlocked and energize the function generator. Adjust the frequency until the first standing wave mode shape is observed for $n = 1$ as indicated in Figure L3.9. Record the frequency in the $f_{EXPERIMENTAL}$ column. Increase the frequency to observe the next two mode shapes and enter those values in the experimental frequency column. Compute the error between theoretical and experimental frequencies.

Some individuals have difficulty seeing the node for the first fundamental frequency. Adjust the frequency *below* the expected value, and look for a stationary node as shown in Figure L3.11. Once the node is identified, increase the frequency to *drive* the node toward the vibration generator post. Once the node is located at the vibration generator, the vibration generator post should appear to have very little movement. It may also be easier to begin with a higher fundamental frequency to see the standing wave develop.

RESULTS

PART I. Longitudinal wavelength. For the first part of the experiment, sound is used as the longitudinal wave medium. Equation L3.7 relates the properties of frequency, velocity and wavelength. Frequency is provided with a tuning fork; velocity is measured with an oscilloscope; and wavelength is computed. The wavelength is then experimentally measured by creating a standing wave within a sound tube and compared with the theoretical value. These values are summarized in Table L3.1.

TABLE L3.1
Longitudinal Wavelength of Air Data Summary

@ Temp=______ °C	V	f	λ
Theoretical	(Published/o'scope)	(Tuning fork)	(Equation L3.7)
Experimental			(Use tube)
% Error			(Calculate)

PART II. Transverse frequency. For the second part of the experiment, a transverse wave is propagated in a string. The wave velocity is set by the physical properties of the string, and the wavelength is fixed by the string length. The frequency for this wavelength is theoretically computed with Equation L3.7. The frequency is then experimentally determined by adjusting a frequency generator until a standing wave is created within the wavelength bounds. These values are compared and summarized in Table L3.2.

TABLE L3.2
Transverse Frequency of String Data Summary

T = N	v	λ	f
Theoretical	(Equation L3.6)	(Bound length)	(Equation L3.7)
Experimental			(Freq gen)
% Error			(Calculate)

All supporting data for Tables L3.1 and L3.2 are included in Appendices L3.A and L3.B.

DISCUSSION

Describe and summarize significant information about the data. Identify potential sources of error. One obvious source of error is the ability to hold the tube in bucket of water while taking a measurement of length. This is an observational measurement source of error. It is a random type of error since the effect will be somewhat different for each measurement taken, without a particular trend. Another source of error is the speed of sound, which is determined from the oscilloscope measurements. The measurement peaks may be difficult to precisely determine due to noise in the signal. These noise peaks in the oscilloscope signal are inherent to the setup, especially if the noise source is not sufficiently brief. They may result in a systematic type of error as well as a random type, depending on how measurements are taken. Discuss at least one more possible source of error not mentioned here. Present assessments and possible sources of error as objectively as possible based on the data and notes taken during the experiment.

CONCLUSIONS

Based on the objective, results of the experiment and errors, assess whether the experimental data appears valid. Offer any improvements or adjustments to *your* experiment for anyone attempting to replicate the experiment or for a follow-up study.

APPENDIX L3.A: SPEED OF SOUND SUPPORTING DATA TABLES

TABLE L3.A1
Tuning Forks

Tuning Fork	Frequency, *f*
C	256 Hz
E	320 Hz
G	384 Hz
C	512 Hz

TABLE L3.A2
Velocity Echo Time of Flight Parameters

Measured (Expected) and Computed Values	
L (Echo chamber)	(8.0 ft)
$D = 2*L$	(16.0 ft)
V	
Δt	
λ (Equation L3.7, v/f):	

TABLE L3.A3
Harmonic Length Calculations

@*f* =____	*v*	λ_{THEORY}	*Multiplier, k*	L_{THEORY}
$n = 1$	(Table L3.A2)	(Equation L3.7)	(First possibility from Figure L3.5)	$(\lambda*k_1)$
$n = 2$	↓	↓	(Second) possibility from Figure L3.5)	$(\lambda*k_2)$
$n = 3$			(Third possibility from Figure L3.5)	$(\lambda*k_3)$

TABLE L3.A4
Experimental Wavelengths

@*f*=____	L_{THEORY}	$L_{CORRECTED}$	L_{EXPMTL}	λ_{EXPMTL}
$n = 1$	(Table L3.A3: $\lambda*k_1$)	$(L_{THEORY1} - 0.3*D)$	(Lab result)	(L_{EXP1}/k_1)
$n = 2$	(Table L3.A3: $\lambda*k_2$)	$(L_{THEORY2} - 0.3*D)$	(Lab result)	(L_{EXP1}/k_2)
$n = 3$	(Table L3.A3: $\lambda*k_3$)	$(L_{THEORY3} - 0.3*D)$	(Lab result)	(L_{EXP1}/k_3)
Small tube diameter, *D*		(1.5 in)		

APPENDIX L3.B: SUPPORTING DATA TABLES FOR STRING CELERITY

TABLE L3.B1
String Properties

Measured (Expected) and Computed Values	
m_{STRING}	(15 g)
$L_{STRING(TOTAL)}$	(250 cm)
μ_{STRING} (m/L)	
T_{STRING} (Weight of tensioner)	(2 N)
λ (String bounding length)	(200 cm)
v_{THEORY} ($\sqrt{T/\mu}$)	
f_{THEORY} (v/λ)	
$f_{EXPERIMENTAL}$	

TABLE L3.B2
String Wave harmonic frequencies.

	Multiplier, k	λ	L_λ	v_{THEORY}	f_{THEORY}	f_{EXPMTL}	% *Error*
$n=1$	(First possibility from Figure L3.5)	(Bound length)	$(k_1*\lambda)$	$(\sqrt{T/\mu})$	(v/L_{λ_1})		
$n=2$	(Second possibility from Figure L3.5)	↓	$(k_2*\lambda)$	↓	(v/L_{λ_2})		
$n=3$	(Third possibility from Figure L3.5)		$(k_3*\lambda)$		(v/L_{λ_3})		

APPENDIX L3.C: OSCILLOSCOPE OPERATION

Refer to Figure L3.C1 and Table L3.C1 for detailed oscilloscope operation.

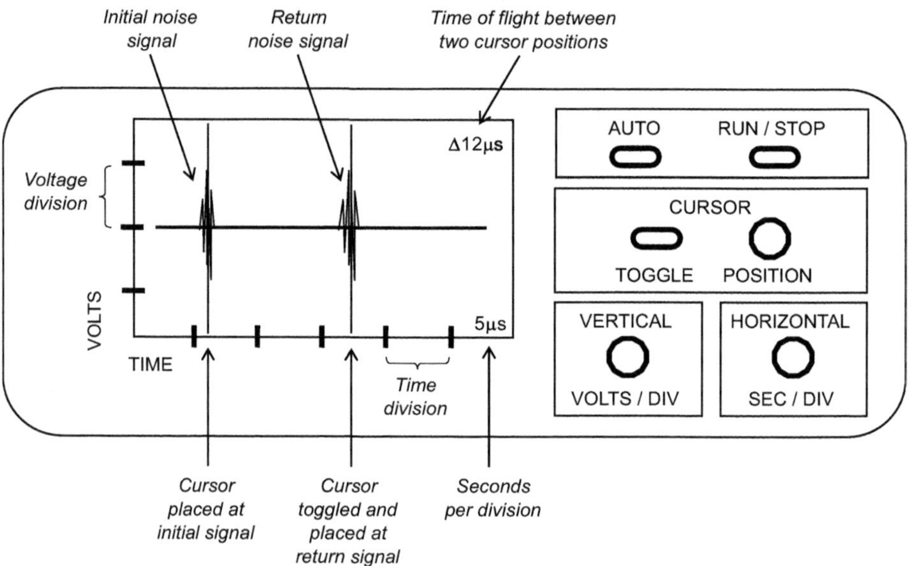

FIGURE L3.C1 Typical oscilloscope display and primary controls.

TABLE L3.C1
Typical Oscilloscope Steps for Measurement. Controls Vary with Oscilloscope Models, but Basic Adjustments and Procedures Are Similar

	SETUP
1.	Connect appropriate cables (per setup in Figure L3.8) and turn on oscilloscope.
2.	Push AUTO (AUTOSET) to reset previous settings and enter acquisition mode.
3.	Turn the large HORIZONAL SEC/DIV to 5 ms per division (or 1 ms/div for tubes <3 m). The SEC/DIV are displayed at bottom of the oscilloscope as shown in Figure L3.C1.
4.	Adjust the VERTICAL POSITION knob to center up the signal.
5.	Create a noise source and view the signal. Two very distinct signals should be visible.
6.	Quickly press the RUN/STOP button to capture the signal for analysis. The oscilloscope is in an auto trigger mode and will only momentarily hold the triggered noise signal.
7.	Adjust the HORIZONTAL POSITION knob to get the first main pulse toward left of oscilloscope screen as shown in Figure L3.C1.
8.	Adjust the HORIZONAL SEC/DIV to either up or down to see the two distinct signal pulses.
9.	Adjust the VERTICAL VOLTS/DIV knob to increase or decrease amplitude of the signal.
10.	Repeat the process as necessary. Press the RUN/STOP button as necessary to release the current signal and recapture a new signal.
	MEASUREMENTS
1.	Press the CURSOR or CURSOR TOGGLE button to view vertical bars on the screen. One vertical bar should be solid and the other should be dashed to indicate it can be repositioned.
2.	Adjust the cursor knob to move the single dashed vertical cursor line to the left and right on the display screen. Center this cursor over a prominent peak of the initial main pulse signal.
3.	Press the TOGGLE button to keep the cursor in position and move the other cursor.
4.	Adjust the cursor knob again to move the second cursor line to the prominent peak on the return signal. Choose a same relative peak position as with the main signal pulse.
5.	The "Δ" in the upper right corner of the screen now indicates the time of flight between the two cursor positions which corresponds to the main initial pulse and the return pulse.

NOTE

1 *Journal of Physics*, Vol 5, Issue 1 (January 2011). Pipe Diameter and End Correction of a Resonant Standing Wave. Taylor Boelkes and Ingrid Hoffmann.

LAB 4

Stress-Strain

Safety Requirements

Do not touch a hot soldering iron tip with your fingers or other nearby equipment
Lab coat, safety glasses and nitrile gloves must be worn during all activities

Equipment Requirements

Rectangular aluminum and polycarbonate test specimens
Strain gauge, strain gauge prep and bonding chemicals
Calipers, benchtop multimeter and power supply
Wheatstone bridge (breadboard, jumper wires and temperature stable resistors)
Load cell test apparatus with 4-point bending accessory and photoelasticity accessories
Computer with data logging software finite element method (FEM) capability
Soldering iron, solder, wire cutter, wire stripper, scrap wire, and scrap breadboard

Procedural Requirements

Pay close attention to maximum loads during all tests – do not break test samples
All load values must be checked for validity prior to the experiment
Apply only light loads to the photoelastic polycarbonate bending specimen
Follow the lab manual and do not jump ahead to a test planned for the following session

Clean-up Requirements

Dispose of wire and solder remnants
Unplug all equipment and return to proper locations

Emergency Actions

BURNS: Run cold water on the affected area and call 911 as required

INTRODUCTION

Two of the most fundamental material properties for structural design are a material's yield stress and is its modulus of elasticity. A material's yield stress is the limiting design strength, and the modulus of elasticity is important for structural stiffness.

Hooke's law relates stress to elastic modulus via strain. Strain is the amount of material deformation (stretch) due to applied stress based on the material's elastic modulus. Each of these material properties can be verified through testing and experimentation prior to use in a structural design. Several tools are available for analysis, such as load cells, strain gauges, computational methods and photoelasticity. This lab explores each of these techniques to help develop a thorough understanding of key material properties and the distribution of stress within a structural component.

The goal of this experiment is a non-destructive verification of both elastic modulus relative to published data and stress relative to theoretical values. The lab begins with definitions of both stress and strain for a tensile test, since the tensile test provides direct measurements of both stress and strain. Hooke's law relates stress to strain via the elastic modulus, so the elastic modulus can also be verified. With a material's elastic modulus verified, a more complicated bending test is performed to verify stress using Hooke's law with the elastic modulus and a strain gauge for strain measurements. Finite element analysis (FEA) and photoelastic analysis of the bending test are included to provide additional insight into the overall stress analysis.

This lab takes place over two lab sessions. Thorough preparation is necessary to complete all tasks. The first session includes soldering practice, strain gauge bonding and finite element modeling. The second session includes the tensile test, 4-point bending test with a strain gauge and photoelastic analysis under 4-point bending.

OBJECTIVE

Verify elastic modulus via a tensile test relative to published data. With elastic modulus verified, verify bending stress relative to theoretical stress and computational stress. Perform a qualitative photoelastic bending stress analysis on an acrylic test specimen.

THEORY

Hooke's law and elastic modulus. Hooke's law is the linear-elastic relationship between stress and strain. The elastic modulus, also known as Young's modulus, is the stiffness constant of proportionality between stress and strain. Hooke's law is given in Equation L4.1.

$$\sigma = E\varepsilon \tag{L4.1}$$

where: σ is the stress

E is the Young's modulus, also elastic modulus

ε is the strain

By using a load cell to apply a known load while measuring strain, both stress and strain can be plotted. The slope created by the line of this plot is the elastic modulus as shown in Figure L4.1.

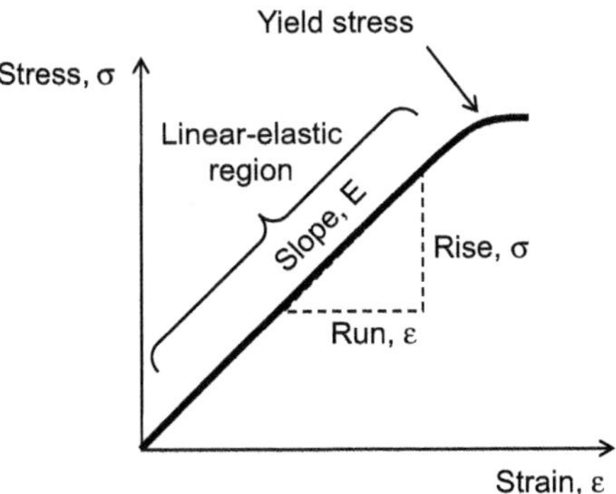

FIGURE L4.1 Hooke's law relating stress to strain via the elastic modulus.

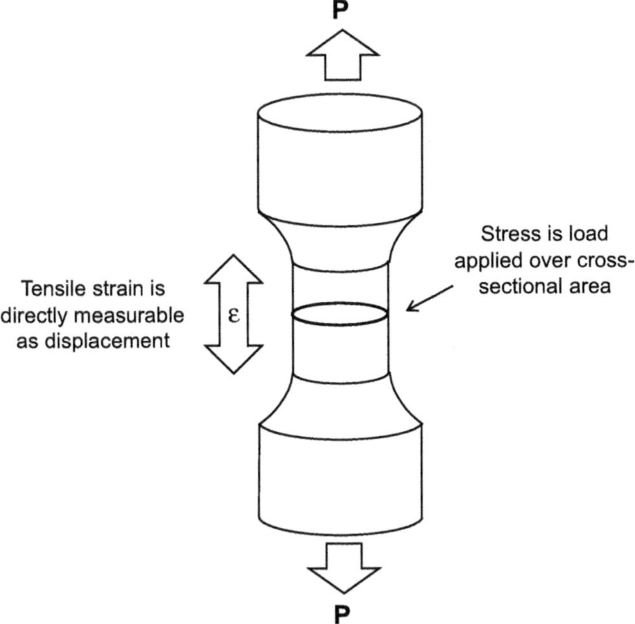

FIGURE L4.2 Tensile stress.

Tensile stress. Stress is *defined* as force applied over an area as given in Equation L4.2.

$$\sigma = {}^{F}\!/\!_{A} \text{ OR } \sigma = {}^{P}\!/\!_{A} \tag{L4.2}$$

where: σ is the stress
F or P is the applied load
A is the cross-sectional area

The highest stress of interest occurs at the smallest cross-sectional area, as shown in Figure L4.2.

Tensile strain. Strain is *defined* as the change in length over the original length as given in Equation L4.3. For a tensile test in a load cell, strain is easy to measure since any displacement measured by the machine is exactly the change in material length as indicated in Figure L4.2.

$$\varepsilon = {}^{\Delta L}\!/_{L_0} \tag{L4.3}$$

where: ε is the strain
ΔL is the change in test specimen length
L_0 is the original test specimen length

Bending stress. Bending stress is derived from the tensile stress definition of Equation L4.2. Consider the beam bending geometry in Figure L4.3. An external moment creates a distribution of compressive forces above the neutral axis and a distribution of tensile forces below the neutral axis.

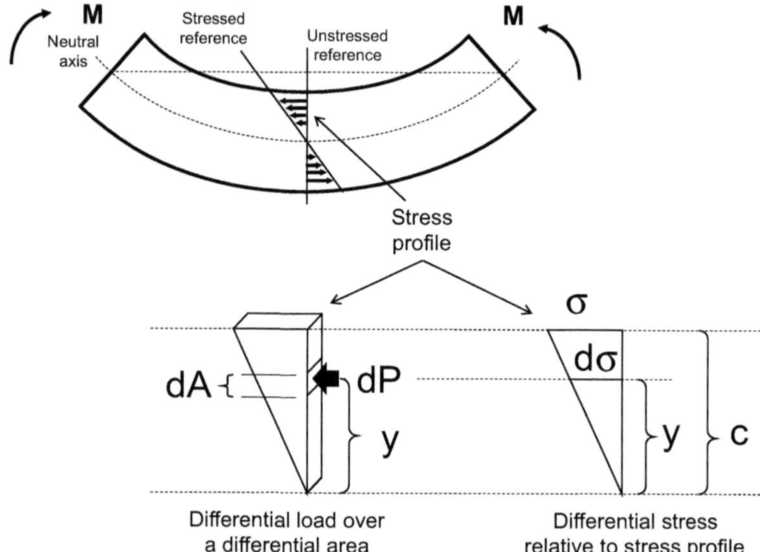

FIGURE L4.3 Bending stress analysis.

At some distance, *y*, from the neutral axis within the beam, a differential amount of force (load), *dP*, acts over a differential amount of area, *dA*, to create a differential stress, $d\sigma$, to arrive at a differential form of Equation L4.2 in Equation L4.4.

$$d\sigma = {}^{dP}\!/_{dA} \tag{L4.4}$$

As seen in the geometry of Figure L4.3, a proportional relationship is established with similar triangles to equate internal stress, $d\sigma$, to maximum stress at the outer surface, σ, since $d\sigma/y = \sigma/c$ (similar triangles). This relationship can be rearranged for Equation L4.5.

$$d\sigma = \sigma y/c \qquad \text{(L4.5)}$$

Equate the RHS of both Equation L4.4 and Equation L4.5 and rearrange for Equations L4.6a and L4.6b.

$$\sigma y/c = dP/dA \qquad \text{(L4.6a)}$$

$$dP = \left(\frac{\sigma y}{c}\right) dA \qquad \text{(L4.6b)}$$

The differential force, dP, in Equation L4.5b acts along the direction of the beam, parallel to the neutral axis. This is of little help since only an external moment is given for the beam and nothing is readily known about the internal force, dP. However, this differential force, dP, acting at a distance, y, is equivalent to the differential external moment, dM, such that $dM = dPy$. Both sides of Equation 5b can simply be multiplied by "y" to get the desired result of a moment in Equation L4.7a. Make the differential moment substitution and slightly rearrange and simplify the RHS of Equation L4.7a to arrive at Equation L4.7b.

$$(dP)y = \left[\left(\frac{\sigma y}{c}\right) dA\right] y \qquad \text{(L4.7a)}$$

$$dM = \left(\frac{\sigma}{c}\right) y^2 dA \qquad \text{(L4.7b)}$$

Equation L4.7b is now ready for integration, as shown in Equation L4.8.

$$M = \left(\frac{\sigma}{c}\right) \int y^2 dA \qquad \text{(L4.8)}$$

By definition, the integral, $\int y^2 dA$, is the cross-sectional moment of inertia, I, of the beam. This substitution into Equation L4.8 results in the final bending stress for Equation L4.9a and L4.9b.

$$M = \frac{\sigma I}{c} \qquad \text{(L4.9a)}$$

$$\sigma = \frac{Mc}{I} \qquad \text{(L4.9b)}$$

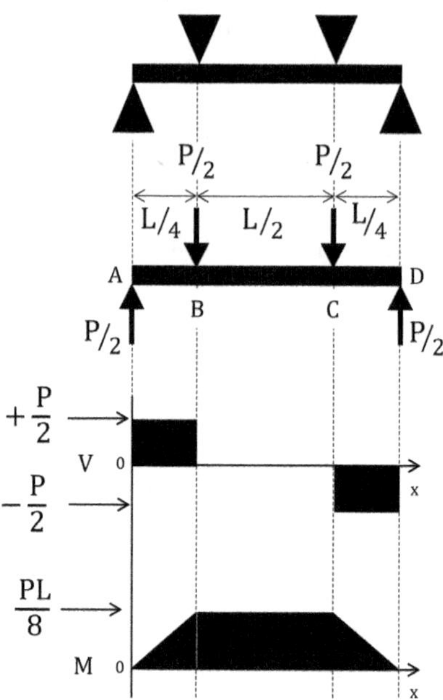

FIGURE L4.4 Shear and moment diagram for 4-point beam bending.

The maximum stress occurs on the surface of the beam where "c" is equal to the upper or lower thickness of the beam from the neutral axis, which is half the beam thickness.

Equation L4.9b is not adequate for most bending test experiments since load cells usually apply a load rather than a moment. To convert moment into load, the specific planned load condition must be examined. A shear-moment (*V-M*) diagram is a quick engineering tool to identify loads, moments and deflections at any point within the beam. The *V-M* diagram can provide the maximum bending stress moment for substitution into Equation L4.9b along with the location of maximum stress for proper strain gauge placement.

The bending beam external loads are first drawn as a free body diagram on the *V-M*, as shown in Figure L4.4. Note the convention to use $P/2$ for 4-point bending compared to P for 3-point bending. A tensile (or compression) tester measures a single applied load. When a 4-point bending anvil is used, the two point-loads each carry half the total applied test machine load. The two reaction forces likewise carry half the total load each.

Below the FBD in Figure L4.4, the first shear reaction force contributes to a positive CW tendency, so shear is drawn upward with a magnitude of $P/2$. There is no change in shear until point B when the $P/2$ magnitude drops shear back to zero. Shear remains zero until a similar situation between points C and D.

Since there is no moment reaction force at point A, the moment diagram begins at zero. The constant positive shear accumulates a positive linear slope for moment. The maximum moment occurs when the $P/2$ shear is fully accumulated (integrated) at a distance of $L/4$. For this simple load condition, the maximum moment is $PL/8$, as given in Equation L4.10.

$$M_{Max} = {PL}/{8} \tag{L4.10}$$

(Equation L4.10 is valid only for the 4-point bending configuration in Figure L4.4)

Bending strain. Measuring strain due to bending is more difficult since a load cell has no direct measure of the material *stretch* along the *surface* of the beam as the beam flexes. In this case, displacement measured by the load cell is *not* the same as strain within the material.

The most common method to make bending strain measurements is indirectly through changes in electrical resistance with a strain gauge as shown in Figure L4.5.

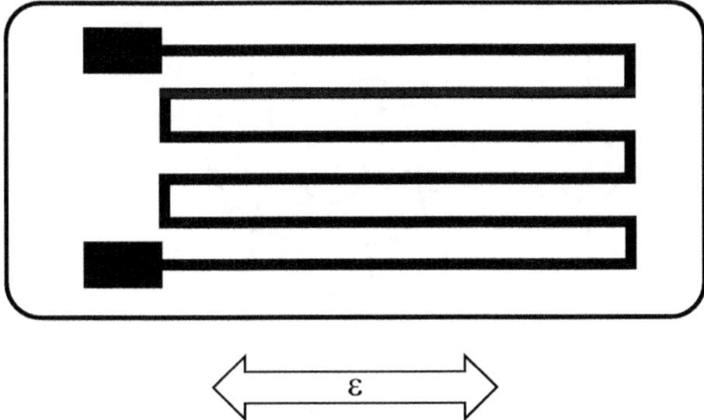

FIGURE L4.5 Strain gauge.

Strain gauges. A strain gauge bonds to a material's surface with an alignment such that, the thin lengths of wire in the gauge will stretch with the strain of the beam. As the thin wires in the strain gauge stretch, even slightly, electrical resistance increases. This change in resistance, ΔR, can be measured and related to strain as shown in Equation L4.11.

$$\varepsilon = \frac{\left(\Delta R_{GAUGE}/R_{GAUGE}\right)}{GF} \tag{L4.11}$$

where: ε is the strain

ΔR_{GAUGE} is the change in strain gauge resistance

R_{GAUGE} is the base strain gauge resistance

Strain gauges have a *gauge factor*, *GF*, which is provided by the manufacturer, and is usually around 2.0 for metallic strain gauges. The quantity ($\Delta R/R$) is the change in resistance from the initial resistance of the strain gauge. Strain gauges are usually produced with an initial resistance of either 120 Ω or 350 Ω.

Wheatstone bridges. If the change in resistance, ΔR, of a strain gauge could be easily measured, Equation L4.11 would readily provide a value for strain. Unfortunately, the changes in resistance are usually quite low and difficult to measure accurately. An ohmmeter (or multimeter) must first use its resources to measure the 120 Ω or 350 Ω strain gauge resistance, and *then* measure the very small milliohm change on top of this much larger initial value. Consider a digital display with only 5-digit accuracy; three digits are required to display 350 Ω with only two digits remaining to work on the milliohm change of the strain gauge.

A more accurate approach is to use a Wheatstone bridge to *zeroize* the initial strain gauge value and allow the instrument to measure the very small change with all of its resources. A typical Wheatstone bridge circuit is shown in Figure L4.6.

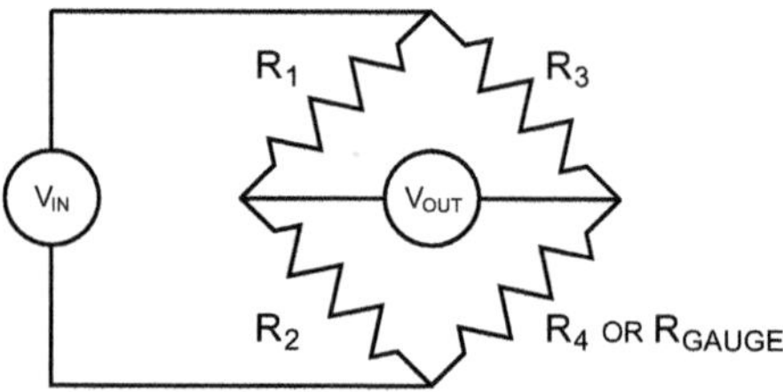

FIGURE L4.6 Wheatstone bridge circuit.

A Wheatstone bridge is a set of voltage dividers that provides a zero-voltage output when all resistors are equal, making the bridge *balanced*. As soon as any resistance changes, such as with a changing strain gauge, the bridge becomes unbalanced and the small voltage change is readily measured. Since Equation L4.11 requires resistance and a Wheatstone bridge returns voltage, a conversion is required with Equation L4.12.

$$R_{GAUGE} = \left[\frac{R_3}{\left(V_{OUT} / V_{IN} \right) + \left(\dfrac{R_1}{R_1 + R_2} \right)} \right] - R_3 \qquad \text{(L4.12)}$$

where all variables correspond to Figure L4.6.

Photoelastic stress. Photoelastic stress analysis is a technique to visually see the entire stress field within a transparent test specimen. To use photoelastic techniques for an opaque metallic part, a duplicate part is made from a suitable transparent material. Polarized light passing through the stressed transparent material will pass through some parts of the material while being blocked through other parts of the material based on the stress field. These alternating areas of light and dark are known as isoclinics and isochromatics. Lines of constant stress appear as dark-lined

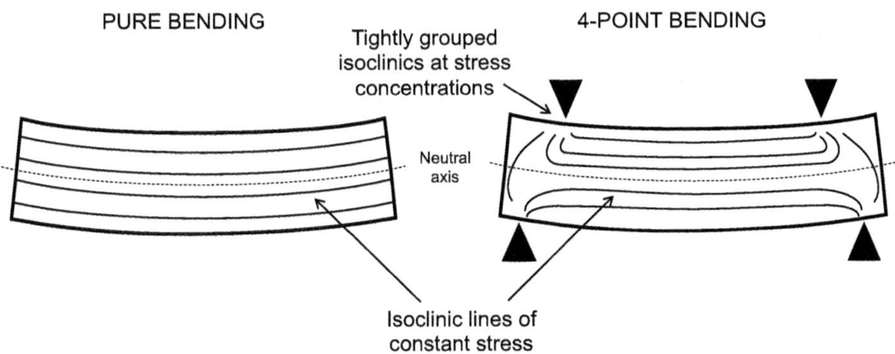

FIGURE L4.7 Photoelastic stress analysis of beam in pure bending and 4-point bending.

isoclinics. Although isoclinics and isochromatics can be used to quantitatively determine stress values, a mathematical analysis is not presented here. Instead, only a qualitative stress field analysis is considered, where tighter groupings of isoclinics indicate high stress concentrations. Any single, dark isoclinic line identifies a line of constant stress, which is typical of the bending stress in Figure L4.7. Isoclinics due to experimentally induced 3- or 4-point bending appear different from theoretically pure bending due to stress concentrations created by the bending anvils. The overall similarities in lines of constant stress are still present.

EXPERIMENTAL METHOD

The flow of lab activities includes the following:

1. Perform a tensile test and plot the stress (P/A) against load cell strain ($\varepsilon = \Delta L/L_O$).
2. Determine the elastic modulus from the slope of the tensile test plot.
3. Compare this experimentally determined elastic modulus with the published elastic modulus for the given material. Assume the published elastic modulus has now been experimentally verified for the next experiment, regardless of the materials used.
4. Attach a strain gauge to a bending test specimen.
5. Perform a bending test and record strain gauge data.
6. Use Hooke's law with strain gauge data and elastic modulus to determine bending stress.
7. Compare experimentally determined bending stresses with theoretical bending stresses.

Realize the materials are likely different between the tensile test and bending test, so the "*verified*" elastic modulus will not necessarily be the same value between tests. The first test *notionally* validates the published values.

TABLE L4.1
First Lab Session Tasks

	TASK
1.	Complete lab plan templates with all formulas prior to first lab.
2.	Verify all load value calculations with instructor or lab assistant.
3.	Verify test specimen materials and dimensions.
4.	Prep and bond strain gauge.
5.	Solder practice (each student).
6.	Build and test Wheatstone bridge (then disassemble for next lab team).
7.	Complete computational model and FEA simulation.
8.	Solder strain gauge leads, if required (last step while bonding agent cures).

PART I: SETUP

The tasks outlined in Table L4.1 should be completed during the first lab session. Organization and efficiency are required to complete all activities. Tasks are presented in the preferred order of accomplishment, and lab personnel duties should be decided prior to entering the lab.

Data tables and plots. Prepare all data collection tables, spreadsheets and plot sketches for the lab plan. All published and theoretical values that can be determined ahead of time must be included in the lab plan. Refer to the results section and appendices for example tables and plots with expected values for the lab. If exact dimensions are not known ahead of time, they can be estimated so that all tables and calculations are ready and approximated for the actual lab session. All lab plan plots must be populated with theoretical data as applicable. Experimental values will be added to these plots for the lab report, and theoretical values must be updated for the report.

The lab plan for the first session must be organized with the following minimum information. Information should be summarized in the *Expected Results* section of the lab plan. Long tables of data and complex tables should be placed in an appendix.

TENSILE TEST SUMMARY TABLE. Write a couple of introductory sentences in the results section to introduce a tensile test table similar to Table L4.2 for summarizing material properties and final results. Table L4.2 includes estimated values to complete the expected results for the lab plan. These values must be updated during the lab.

TENSILE TEST DATA TABLE. Create a table similar to Table L4.A1 in Appendix L4.A for recording stress and strain data. Enter the published values for the tensile test material into the table. The tensile test material may be aluminum, steel or brass. If the material is not known prior to the lab, use the material properties for AL6061T6 given in Table L4.2. At least five load value entries must be included, as shown in Table L4.A1. Clearly identify the load value for *point 5* **first** by calculating about 50% of the published material yield stress to use as the maximum total load during the experiment. Refer to Equation L4.2 for the tensile stress equation to calculate load from stress. Then, choose any four desired load values less than the maximum in point 5. Enter these load values for data point entries 1 through 4. Enter the associated stress values in the theoretical stress column. These load values will be recorded during the load cell tensile test. Point 0 is reserved for recording a pre-load of about 25 N to remove slack from the system.

TENSILE STRESS-STRAIN PLOT. Create a plot of stress versus strain similar to Figure L4.10 for the load values chosen in Table L4.A1. Use the published value of Young's modulus with Equations L4.1 and L4.2 to plot the theoretical values. This theoretical plot must be included in the lab plan's expected results. For the lab report, this same plot will be updated with actual experimental data.

BENDING TEST SUMMARY TABLE. As with the tensile test material, create a bending test summary table similar to Table L4.3.

BENDING STRESS DATA TABLE. Create a spreadsheet and tables similar to Tables L4.B1 and L4.B2 in Appendix L4.B for the lab plan. Enter the expected material properties in the table. To populate the load values in Table L4.B2, use Appendix L4.C to modify the bending stress Equation L4.9b in terms of **load** (not moment) and physical **measurements**. Do this by working out the algebraic steps in Appendix L4.C. Substitute Equation L4.10 into Equation L4.9b and solve for load. Replace the moment of inertia with the appropriate moment of inertia formula for a rectangular beam in terms of base and height. The value "c" is half the thickness of the beam height, which is the distance from the neutral axis to the outer surface. Simplify the result and clearly provide this equation in the "*Max Load Formula*" block of Table L4.B1. Use this *Max Load Formula* to calculate the maximum allowable load for point 5 in Table L4.B2. The maximum allowable stress for point 5 is 50% of the material's yield stress. Look up the material's yield stress property, multiply by 0.5, and use this value for stress in the *Max Load Formula*. Enter the values (or estimates) for beam base and height to calculate the point 5 load. Finally, choose any desired loads below this value to populate data points 1 through 4 in Table L4.B2. Enter all formulas into the spreadsheet table according to the notes.

BENDING STRESS-STRAIN PLOT. Create a plot of stress versus strain similar to Figure L4.11 for the load values chosen in Table L4.B2. Plot the five theoretical stress and strain values for each of the chosen loads. Use Equation L4.C1 from Appendix L4.C to calculate stress. Use this stress and the published elastic modulus with Equation L4.1 to solve for strain. Include this theoretical plot in the expected results of the lab plan. For the lab *report*, this same table will be updated with experimental values. The chart will also need to include at least one stress-strain point from the FEA results.

FEA STRESS IMAGE. Include a simple sentence or two in the lab plan stating that a numerical FEA will be performed for one load condition. Nothing more is required for the lab plan. An image of the simulated stress field will be included in the lab report.

PHOTOELASTIC STRESS IMAGE. A picture of the photoelastic analysis will be compared to the simulation image. Neither the photoelastic image nor the simulation image will be available for the lab plan, so a simple sentence or two can be used to describe the plan for including a qualitative photoelastic stress analysis.

The lab plan should also include enough information to utilize lab time efficiently. Specify team member duties as appropriate. This can be done by annotating team member names on Table L4.1 activities. All of the calculations for the tables must be set up in the spreadsheet prior to the lab for use in real time. Laptops are helpful for setting up calculations and performing all real-time calculations.

Update calculations. The first preparatory lab session provides the opportunity to update all estimates with actual values. Use calipers to measure both the cylindrical tensile test specimen and the rectangular bending test specimen. Confirm the material of the specimens. Update all values in the lab plan tables. The lab plan spreadsheets and tables must be updated prior to use in the second lab session.

Solder practice. Each team member must practice soldering with scrap wire and a scrap proto-circuit board. Most solders have a rosin core flux and do not require additional flux. However, extra flux can be used to facilitate challenging solder joints. To solder, attach the two parts to be soldered, heat the connection, and then apply the solder so it wicks into the joint. Soldering a strain gauge requires a delicate touch and is described in the following section.

Strain gauge preparation. Begin strain gauge prep early during the first lab session so the bonding agent can cure while other lab activities are performed. Refer to Appendix L4.D to prepare and bond a strain gauge to the rectangular test specimen made of aluminum or steel.

Refer to Figure L4.8 for the orientation of strain gauge placement. The prepared surface is the bottom center of the test specimen, where maximum tensile stresses occur. Do *not* bond the strain gauge upside down! **The gauge should be SHINY SIDE UP with dull side bonded to test specimen.** Allow the bonded strain gauge to cure while completing other lab activities. Strain gauges may or may not be provided with pre-soldered leads. If the strain gauge does not have pre-configured leads, finish up the first lab session by soldering leads to the strain gauge. If the strain gauge has pre-soldered leads, soldering directly to the strain gauge is unnecessary.

The strain gauge leads should include a strain relief loop to help prevent inadvertent detachment. Make a small, loose bend in the lead wires and tape the wires to the side of the strain gauge per Figure L4.8. Ensure the tape does not interfere with the strain gauge, and the wire leads come off the side of the test specimen to avoid interfering with the bending anvils. If lead wires are uninsulated, do not tape the wires directly to the metal test specimen or each other, or a short circuit will be created. Use an extra piece of tape as necessary to ensure the wires do not short out with the test specimen.

Wheatstone bridge. A Wheatstone bridge must be assembled on a breadboard according to the schematic in Figure L4.6. This schematic is recreated in Figure L4.9 for simple breadboard construction. Strain gauges typically have a resistance of 120

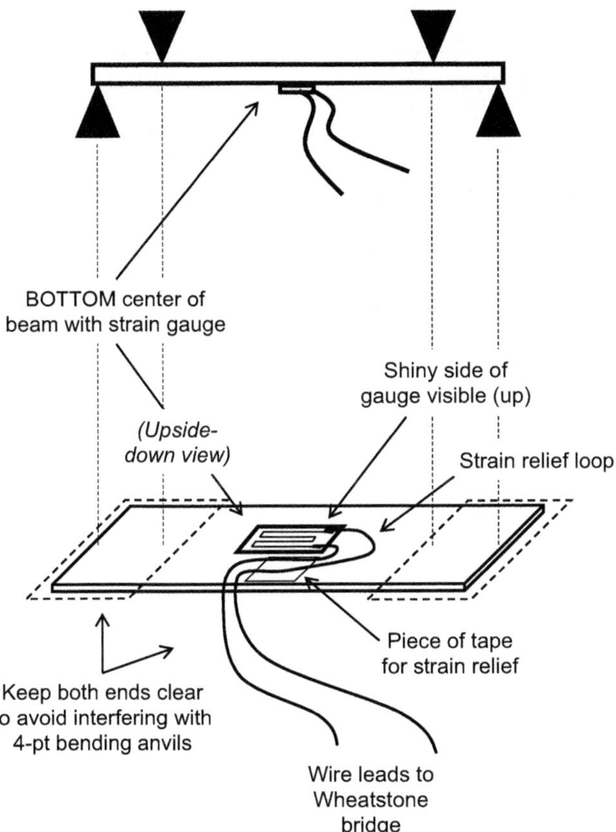

FIGURE L4.8 Strain gauge location with strain relief shown.

Ω or 350 Ω, so resistors must be chosen with the same value as the strain gauge to create a nearly balanced circuit. Either note the strain gauge value from the manufacturer's packaging or measure it directly with a multimeter. If the strain gauge is 120 Ω, then choose three 120 Ω resistors for the bridge. Construct and test the bridge as shown in Figure L4.9.

A variable resistor is often used to exactly balance the bridge at zero. However, only a "*nearly zero*" output of the bridge is required, so the bridge will probably not balance exactly to zero. Very low temperature coefficient resistors should be used so the bridge resistors do not change much with temperature during use of the bridge.

Connect the multimeter, power supply and strain gauge leads to the Wheatstone bridge according to Figure L4.9 to ensure the circuit is correct and strain gauge connections are reliable. The multimeter should indicate a VERY LOW millivolt value. With the strain gauge properly connected, physically flexing the test specimen

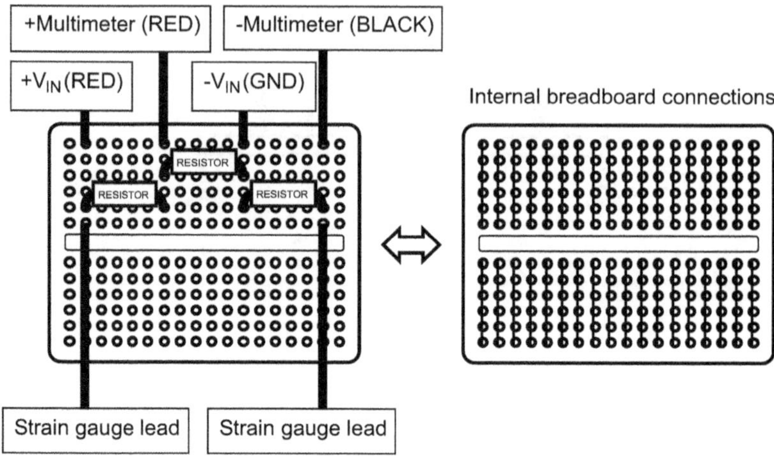

FIGURE L4.9 Wheatstone bridge configuration.

should register a changing voltage. If there is no change, recheck all connections carefully, including wire insertion into the breadboard. Correct any issues during the first lab session.

COMPUTATIONAL METHOD

Perform a numerical stress analysis of the bending test specimen at ONE of the five load data points from Table L4.B2. Several solutions are available for FEA stress simulations. The details vary between software packages, but most solutions follow a similar approach.

The general workflow approach to completing a numerical simulation is to first create a solid model using the bending test specimen parameters from Table L4.3. Create the solid model by (1) selecting or creating a ***plane***, (2) creating a 2D enclosed ***sketch*** on that plane, and (3) applying ***features*** such as extrusions or cuts of the sketch. To apply features, follow the general flow of (1) creating **solids**, (2) **cuts**, and then (3) **fillets**. All sketches should be dimensioned to fully define the sketch. When choosing a 2D contour to sketch, choose the contour with the most information; in other words, sketch a 2D shape that is closest to the final result.

When the solid model is complete, choose one of the load values from Table L4.B2. Run the stress simulation using the detailed steps outlined in Appendix L4.E for that load. Include a picture or plot of the stress simulation results in the lab report.

PART II: TESTING

Tensile test machines *may* require calibration before use. Without calibration, it is impossible to distinguish strain from the test specimen from flexure within the test machine. Follow any calibration steps as recommended.

Tensile test. Several solutions are available for tensile test machines with data logging capability. Test machines can also be constructed from individual load cells connected to a data acquisition device. Configure a material test apparatus for a tensile test using any pre-configured metallic test specimens (aluminum, steel or brass). Connect the load cell test machine to a computer or data acquisition device and launch the software. Choose a data display for both table and graph data, if available. Since a load cell test machine can only return time, force or position, select force and position as the data values.

Create a column using the force data [Force] provided by the device to calculate stress with Equation L4.2 where STRESS=[Force]/(area). Use the cross-sectional area of the test specimen and ensure consistent units. Create a column using the position data [Position] provided by the device to calculate strain with Equation L4.2 where STRAIN = [Position]/(length). Use the originally measured length of only the thinnest section of the test specimen. If using a data acquisition device, these formulas are sometimes entered as *engineering values* in the software.

Operation. Only small adjustments should be made on manually operated test machines. Small test specimens can quickly exceed their ultimate stress and break, depending on the device. Samples should be slightly *pre-loaded* for best results to remove anomalous data from slack initially in the system. To pre-load the sample, begin recording data and watch the load information until about 20 N is applied. Stop recording and reset the software.

When ready to begin the test, ensure data capture is running. Gently begin loading the test specimen while also manually recording the stress and strain at the pre-planned loads. Continue up to the maximum value determined in the lab plan (50% of yield). Upon reaching the maximum planned load, cease recording data. If data capture software includes analysis tools, measure the slope of the stress-strain plot to determine the elastic modulus. Be sure the software chart is plotting stress and strain rather than raw values. If no data analysis tools are available, export and save the data for analysis in a spreadsheet program.

Bending stress. Reconfigure the testing device for 4-point bending. Place the rectangular test specimen with bonded strain gauge into the 4-point bending test device as shown in Figure L4.8. Ensure the strain gauge is on the lower surface of the bending fixture. Attach the two strain gauge wire leads to the Wheatstone bridge as specified in Figure L4.9. Attach a power supply and multimeter to the Wheatstone bridge, also as specified in Figure L4.9. Turn on the power supply set to approximately 5 VDC.

Start a new session with the capture software, as before. Very slightly pre-load the bending test specimen, just enough to hold it in place in the fixture and record the pre-load force. As with the tensile test, use both the table and graph formats for data collection, if available in the data capture software. Choose force and position as the default measurements. Depending on the software, add a column or channel for stress. Use the *maximum stress formula*, derived as Equation L4.C1 in Appendix L4.C. Realize this is an analytic evaluation of stress based on the definition of stress using applied load and physical parameters. It does not evaluate stress based on actual strain deformation since strain is not measured directly by the test apparatus. Strain will be recorded by voltage readouts from the strain gauge.

Prior to the bending data run, ensure all circuit connections are correct and the multimeter is correctly reading in the millivolt range. Since the Wheatstone bridge is not perfectly balanced, the voltage will be low, but not zero. Record the initial voltage in Line 0 of the Table L4.B2 spreadsheet. Gently begin loading the test specimen to the first load value in Line 1 of Table L4.B2. Record the exact load and voltage readings from the multimeter. Continue this process for each of the five data points.

Photoelasticity. Remove the metallic bending test specimen and place a clear acrylic test specimen into the bending fixture in a manner similar to the 4-point metal bending test specimen. Use polarizing sheets with Velcro or tape and attach one polarizer sheet behind and one sheet in front of the acrylic test specimen on the test fixture. Place a lamp behind the test fixtures. There should be a clear optical path looking through the polarizing sheets to the lamp with the clear test specimen in place between the polarizing sheets. Turn on the lamp.

Very gently apply a load to the acrylic test specimen *only until* isoclinics and isochromatics become readily visible. It may be necessary to adjust the visual orientation to see the fringe patterns. Increase the load as necessary to see the photoelastic pattern, but do not permanently deform the test sample by applying too much load. Take a picture of the results for the lab report.

RESULTS

Tensile test specimen parameters and results are summarized in Table L4.2. Calculate and enter the percentage error from the experimentally determined modulus relative to the published elastic modulus.

TABLE L4.2
Tensile Test Specimen Parameters and Results

TENSILE TEST SPECIMEN PROPERTY	VALUE (*EXPECTED*)
GEOMETRY	CYLINDRICAL
MATERIAL	(*AL6061T6*)
LENGTH	(*35 mm*)
DIAMETER	(*3 mm*)
PUBLISHED MODULUS (E)	
EXPERIMENTAL MODULUS (E)	
% ERROR	

Figure L4.10 is a plot of key load stress and strain values for both published and experimental elastic moduli. The slope of the stress-strain curve is the elastic modulus.

Data analysis for experimental bending stress is provided by the formulas in the Table L4.B2 spreadsheet from Appendix L4.B. Follow the notes in Table L4.B2 and

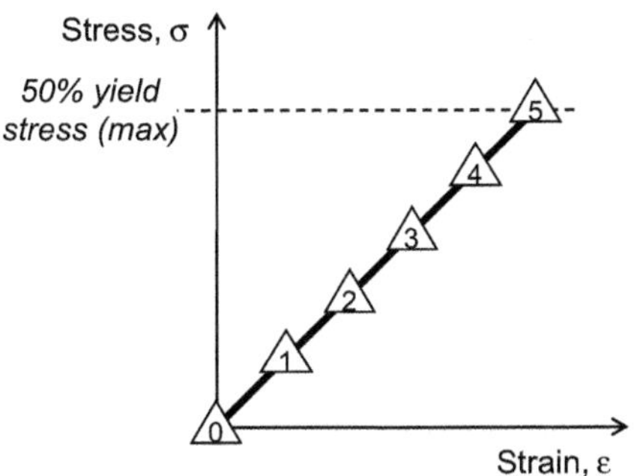

FIGURE L4.10 Tensile test stress-strain plot (published versus experimental).

ensure consistent units for calculations. V_{IN} from the power supply should be a constant value of about 5 V, and V_{OUT} from the multimeter is in millivolts ($\times 10^{-3}$ V). The change in resistance, ΔR_{GAUGE}, in each line is the R_{GAUGE} for that line minus R_{GAUGE} from the previous line. Strain, ε_{TOTAL}, is added together from each previous value of strain, ε.

Experimental stress is computed with Hooke's law, as indicated in the notes by Equation L4.1. For experimental stress, use a published value for elastic modulus in Equation L4.1, considering this modulus value was (notionally) verified by experiment in Part I. Use strain, ε, directly from the table for the appropriate load value point. Theoretical stress is computed with the equation derived in Appendix L4.C, Equation L4.C1, for the given load value. Compute the experimental stress error for each load point relative to the theoretical stress at each point. Include the FEA computational stress results for the one desired load point in Table L4.B2. Compute the error of the computational results relative to the theoretical results.

Bending test specimen parameters and results are summarized in Table L4.3. Calculate and enter the average percentage error from the five experimentally determined stress values relative to the theoretical stress values. Do not include percentages from computational results in this table.

TABLE L4.3
Bending Test Specimen Parameters

BENDING TEST SPECIMEN	
PROPERTY	**VALUE** (*EXPECTED*)
GEOMETRY	RECTANGULAR
MATERIAL	*(1018 STEEL)*
LENGTH	*(5 cm)*
WIDTH	*(2 cm)*
THICKNESS	*(2.4 mm)*
AVG THEORETICAL STRESS % ERROR:	

Refer to figure L4.10 to plot key load stress and strain values for both theoretical and experimental bending stress values. An additional stress point should be included for the computationally determined stress.

Include an image of the FEA stress distribution along with the photoelastic stress analysis picture.

DISCUSSION

Describe what is occurring with the data. Identify areas of high and low-stress concentrations and deviations from a pure bending case. Discuss the differences between the various results and approaches. Discuss the likely cause of the differences. Discuss the sources of error. Discuss how the strain gauge might contribute to errors. Assess whether experimental or computational results appear closer to theory and offer a potential explanation.

CONCLUSIONS

Based on the objective, cite the percentage error and assess whether the experimental results are consistent with published data and theoretical values. Provide a conclusion about the most accurate stress analysis tool. Discuss where any improvements could be made in *your* experimental setup.

APPENDIX L4.A: TENSILE TEST SUPPORTING DATA TABLES

TABLE L4.A1
Tensile Test Material Properties Summary

MATERIAL					
PUBLISHED YIELD STRESS					
PUBLISHED ELASTIC MODULUS					
		THEORETICAL		EXPERIMENTAL	
PT	LOAD (N)	STRESS	STRAIN	STRESS	STRAIN
0	Pre-load (~25 N)	0		0	0
1		$\left(LOAD / A_{XC} \right)$	*(Equation L4.1)*	$\left(LOAD / A_{XC} \right)$ *(From load cell)*	*(From load cell)*
2		↓	↓	↓	↓
3					
4					
5		*(0.5*YIELD STRESS MAX)*			

APPENDIX L4.B: BENDING TEST SUPPORTING DATA TABLES

TABLE L4.B1
Bending Test Specimen Properties

MATERIAL:	LENGTH:	$AREA_{XC}$:
MODULUS:	WIDTH:	RESISTORS:
YIELD STRESS:	THICKNESS:	MAX LOAD FORMULA (Equation L4.C2):

TABLE L4.B2
Bending Test Data

DATA POINT	0	1	2	3	4	5	NOTES
Load							
V_{IN}							← From power supply
V_{OUT}							← From voltmeter measurement
R_{GAUGE}							← Use Equation L4.12
ΔR_{GAUGE}							← Differences between R_{GAUGE}
E							← Use Equation L4.11
ε_{TOTAL}							← Accumulated totals of e
$\sigma_{EXPERIMENTAL}$							← Use Equation L4.1
σ_{THEORY}							← Use Equations L4.C1
% Error							← %Error for each and total average
$\sigma_{COMPUTATIONAL}$							← Choose only ONE load for FEA
% Error							← %Error for one point relative to theory

ADDITIONAL NOTES:

1. Record initial voltage for Point 0.
2. Enter initial value for R_{GAUGE} based on initial voltage.
3. ΔR_{GAUGE} has cumulative values such that $\Delta R_{G3} = R_{G3} - R_{G2}$.
4. ε_{TOTAL} has cumulative values such that $\varepsilon_{TOTAL3} = \varepsilon_0 + \varepsilon_1 + \varepsilon_2$.

APPENDIX L4.C: MAXIMUM STRESS FORMULA

TABLE L4.C1
Algebra to Convert Bending Stress as a Function of Moment into Load

Bending Stress in Terms of Moment (M):	$\sigma = \frac{Mc}{I}$	**Equation L4.9b**
Use substitutions for: • M (Use Equation L4.10, max moment) • c (in terms of beam height) • I (in terms of beam height and base)	↓ Algebraic steps ↓	
Bending stress in terms of applied load (P): • σ must be written only in terms of physical measurements in the lab • M and I are not directly measurable and must be substituted and simplified with in-lab measurable variables and applied load • Use P for load and L for length	$\sigma = f(P, L, b, h)$	Result becomes Equation L4.C1
Solve σ for P *Max Load Formula*	$P = f(\sigma, L, b, h)$	Result becomes Equation L4.C2

APPENDIX L4.D: STRAIN GAUGE PREP AND BONDING

See Tables L4.D1 and L4.D2 for detailed strain gauge bonding steps.

TABLE L4.D1
Steps to Prep Strain Gauge for Bonding to Bending Test Specimen

#	ACTIONS
1	PPE: Don a pair of nitrile gloves and safety glasses.
2	CLEAN: Degrease the test sample and the work surface with CSM-3 degreaser spray.
3	REMOVE OXIDES: Abrade the test sample bonding surface with 250 grit sandpaper. Follow up with wet sanding using higher grit (400 grit) paper. Use water, isopropyl alcohol or *bonding conditioner* for the wetting agent.
4	CLEAN: Use cotton ball or swab with alcohol or conditioner to clean the test specimen.
5	ETCH: Scribe center reference lines on test specimen for strain gauge placement. BOTTOM SURFACE OF TEST SPECIMEN Measure for center location and lightly etch reference lines with straight edge and sharp instrument 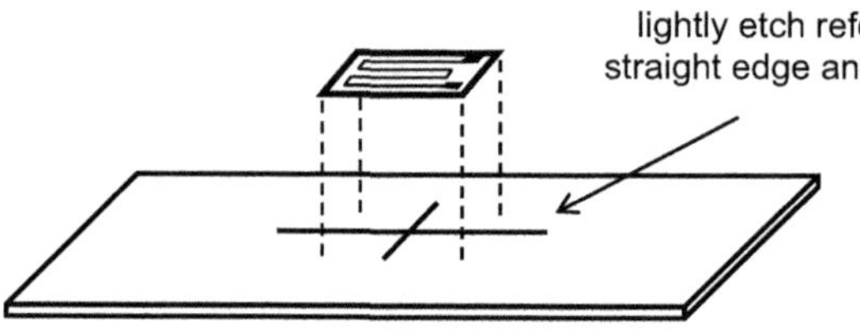
6	CLEAN: Clean the test specimen bonding area again. Repeat till cotton swab is clean.

TABLE L4.D2
Steps to Bond Strain Gauge to Bending Test Specimen

#	ACTIONS
1	PPE: Don a pair of nitrile gloves and safety glasses.
2	STRAIN GAUGE: Use tweezers to grasp and remove a strain gauge from packaging. Note the strain gauge factor (GF) and resistance on the packaging for calculations.
3	PLACE: Place the strain gauge **shiny side up** on the cleaned surface of the test specimen. The dull bonding side of strain gauge should be against test specimen.
4	HANDLING: Use length of clear adhesive tape to roll onto strain gauge for alignment. Tape is then peeled back with strain gauge affixed on tape to apply adhesive. *CAUTION: The strain gauge may "jump" on to the tape due to static electricity. Hold down a corner of the gauge with small utensil to maintain position until the tape is securely in place.*

BOTTOM SURFACE OF TEST SPECIMEN

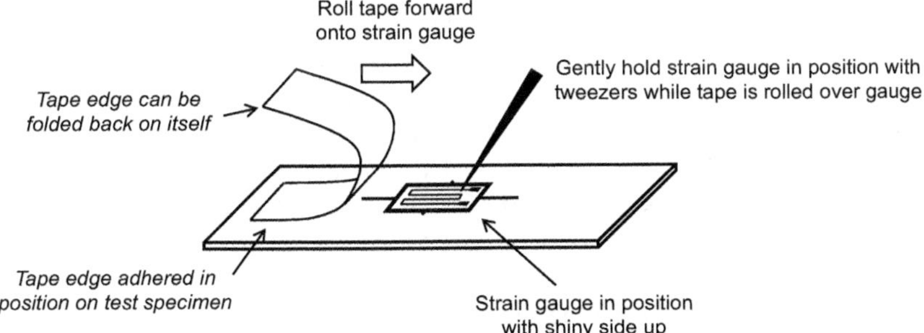

#	ACTIONS
5	PEEL: Peel the tape with strain gauge back to allow application of bonding agents.

BOTTOM SURFACE OF TEST SPECIMEN

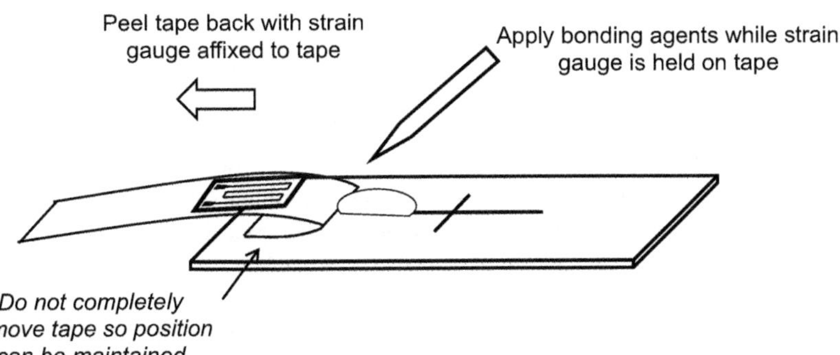

#	ACTIONS
6	CATALYST: Brush a layer of bonding catalyst onto the exposed surface of the strain gauge.
7	ADHESIVE: Dab a small drop of adhesive on test specimen at tape junction. Be sure strain gauge is fully exposed such that adhesive will cover strain gauge in next step.
8	SQUEEGEE: *Squeegee* the adhesive onto the strain gauge as you press the tape forward with strain gauge back into position.
9	SET: Allow the adhesive to set for several minutes.
10	REMOVE: Carefully remove only the tape without debonding the strain gauge

APPENDIX L4.E: COMPUTATIONAL SIMULATION

Several FEA software solutions are available to complete the simulated stress analysis. Specific steps and details vary between solutions. The instructions offered here provide a typical approach associated with professional software packages.

Start the software and select a new part. Establish units according to SI. Begin a new sketch by clicking on the *Sketch* tab. This is the 2D sketch mode to initially draw a 2D shape for subsequently creating a 3D model. Select a sketch plane and create a 2D rectangle on the sketch plane using the *Rectangle* tool. Select the *Dimension* tool and dimension the rectangle according to the length and width dimensions of the bending test specimen. Exit sketch mode.

Select the *Features* tab to turn the 2D sketch into a 3D model. Choose *Extrude Boss* and select *Midplane* from the *Direction* drop-down box. If necessary, ensure the sketch rectangle is selected by clicking on it so that it is highlighted. Extrude the rectangle to the thickness of the test sample. *Commit* the values by clicking the green check mark. The solid model is now complete. Save the file.

Select the *Simulation* tab for a *New Study*. Enable the simulation via add-ins, if necessary. Choose and apply a material to the model based on the test specimen material. Locations to place the load and measure stresses must be created. For the top bending anvils, a *Split Line* must be created on the upper surface of the model to apply the loads. Select the model's upper face and create a new sketch on the face. Draw two lines at the locations of the anvils and exit the sketch. Choose *Split Line* via *Projection* to create the load application lines. Alternatively, two reference planes can be created perpendicular to the model face, and split lines can be created with the *Intersection* method between the face and planes.

To simulate constraints for 4-point bending, the left and right edges of the model must be constrained in two directions. Choose *Roller/Slider* from *Fixtures*. Constrain the model via *Translations* in the vertical and horizontal directions. Ensure the model is still allowed to translate left and right. Apply the constraints to both of the end edges (lines, not faces).

Apply a load to both split lines on the upper surface. Select one of the planned load values from the bending data table (point 4 or 5), and plan to use half the load value for each split line location on the model. Choose *Force* from *External Loads*. Click on a split line on the upper surface of the model so the load arrows are applied only to the line and not the face of the model. The load direction must now be set. Choose a *Direction* and ensure the force points downward. Enter the appropriate load for the data point. If the planned load for point 4 from Table L4.C1 is desired, use half the value for one split line load and half the value for the other split line load.

The model is now ready for the simulation. Select *Run* and allow the simulation to automatically mesh the model and determine stresses. Load the *Results* after the simulation finishes and use a Probe tool to measure the stress on the bottom face of the model, where the strain gauge would be located. Record this stress. Save an image of the stress distribution to include in the lab report.

LAB 5
Fractures

Safety Requirements

Lab coat, safety glasses and nitrile gloves must be worn during all activities
WEAR GLOVES AT ALL TIMES WHEN HANDLING STAINLESS STEEL FOIL
Stainless steel foil must be handled with care – it WILL easily cut skin
Nitrile gloves are the minimum glove requirement to prevent small cuts
Nitrile gloves will not prevent cuts due to careless handling

Equipment Requirements

Computer with data logging software
Load cell test apparatus with tensile test capability and polarizing sheets
Custom load cell fracture test clamps
Steel foil and custom-cut acrylic test specimens to fit clamps
Calipers, scissors, utility razor, straight edge, paper cutter (for steel foil) and file

Procedural Requirements

Do not scratch polarizing sheets and return to plastic sleeves immediately after use
Snug clamp bolts with tools but do not over-tighten

Clean-up Requirements

Discard the stainless steel test samples in the trash
Unplug all lab equipment and return to proper locations

Emergency Actions

MINOR CUTS: Clean affected area and use bandage kit as required.
MAJOR CUTS or INJURY: Stabilize affected area. Call 911.

INTRODUCTION

Ideally, all structures will support loads as designed up to their published material strength. However, several factors can cause a structure to fail at a load well below its material design stress. Flaws in a material, such as cracks due to fatigue, corrosion or damage, can result in material failure well the below design yield stress.

Cracks cause sharp discontinuities in the stress field of a structural component. Stresses concentrate around the tips of a crack at a much higher level than within the overall material. Even though a material may be loaded below yield stress, the localized stresses near a crack tip may push the localized stress past yield stress into failure. As a result, cracks tend to grow in length as high localized stresses cause material failure in an otherwise moderately loaded part.

High localized stresses at a crack tip are quantified with stress intensity factors (SIFs). These SIFs are specific to the geometry of the crack, proximity to other cracks and type of loading, such as tensile, in-plane shear or out-of-plane shear.

Although stress concentrations may be locally high near a crack, some materials can withstand the SIFs better than others. Brittle materials can fail catastrophically, whereas ductile materials may continue to deform without catastrophic failure. Fracture toughness is a material property identifying a material's ability to withstand failure in the presence of crack SIFs. Glass has a low fracture toughness compared to steel. Given similar loadings and cracks, glass will fail with little warning, while more ductile steel may just undergo plastic deformation and slow crack propagation without experiencing brittle fracture.

This lab will consider the impact of a crack on overall material design properties. Similar test specimens of ductile and brittle material with cracks and without cracks are subjected to tensile load to assess fracture toughness. A theoretical stress field is also compared to a photoelastic stress field of the cracked test specimens.

OBJECTIVE

Verify modes of failure between ductile and brittle materials. Verify experimental fracture stress relative to theoretical fracture stress. Qualitatively compare theoretical stress field to photoelastic stress field.

THEORY

A material can fail due to stress exceeding either ultimate stress, σ_{ULT}, or critical fracture stress, σ_C. If a crack is present (or assumed present), both conditions must be checked to determine the failure mode.

Theoretical stress field. The analytic development of the stress field due to an elliptical hole in an infinite plate is lengthy and involves working between both polar and Cartesian coordinate systems. However, the basic geometry in the derivation is

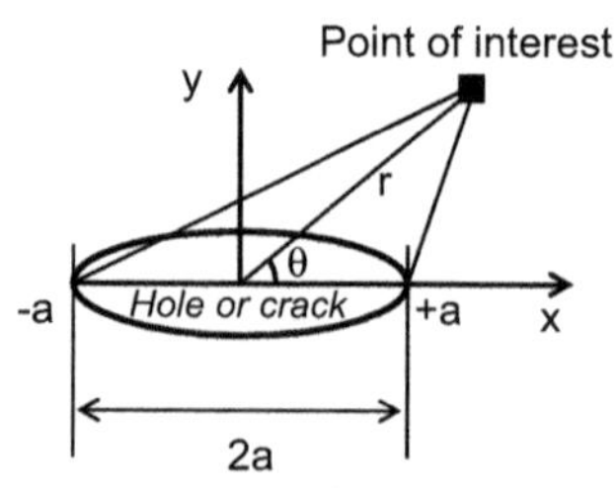

FIGURE L5.1 Geometry of crack associated with derived equations.

given in Figure L5.1 where an arbitrary point of interest is identified in relationship to an elliptical hole. The hole has a length (major axis diameter) of '$2a$' as defined by the points in Figure L5.1. This convention is used such that any crack *within* a material has a crack length of '$2a$'. Cracks on the edge of a material are defined by a crack length of just 'a', which is referred to as the '*half-crack*' length.

Only tensile stress is considered in this development, which is called *Mode I*, or *opening* stress. Also, only y-direction stress in direction of load application is of interest since x-direction stress is generally insignificant in comparison. The generalized analytic solution for stress in the y-direction results in Equation L5.1a and L5.1b. The polar values of 'r' and 'θ' from an x-y coordinate system can be plotted with Equation L5.1 to indicate the entire theoretical stress field associated with a crack. An example of code using Equation L5.1 to plot a stress field is provided in Appendix L5.A.

$$\sigma_{yy} = \frac{\sigma_0 \sqrt{\pi a}}{\sqrt{2\pi r}} f(\theta) \quad \text{(L5.1a)}$$

$$Where: f(\theta) = \cos\left(\frac{\theta}{2}\right)\left[1 + \sin\left(\frac{\theta}{2}\right)\sin\left(\frac{3\theta}{2}\right)\right] \quad \text{(L5.1b)}$$

where: σ_{yy} is stress at any point in the y-direction
σ_0 is the far-field applied stress
a is the 'half-crack' length (crack dimension)
r is the distance from center of crack to point of interest
θ is the angle from x-axis to point of interest

SIFs. The grouping of $\left(\sigma_0 \sqrt{\pi a}\right)$ from Equation L5.1a is a common stress intensity term uniquely identified as the SIF, K, as shown in Equation L5.2. The subscript 'I' in 'K_I' indicates a Mode I SIF (tension). Equation L5.2 is valid for a central crack in an infinite plate based on the geometry of Figure L5.1.

$$K_I = \sigma \sqrt{\pi a} \quad \text{(L5.2)}$$

To correct Equation L5.2 for realistic crack geometries in finite plates, the *function parameter*, Y, is introduced for Equation L5.3

$$K_I = Y\sigma \sqrt{\pi a} \quad \text{(L5.3)}$$

where: K_I is the Mode I SIF
Y is a crack-specific look-up value
σ is the applied stress
a is the 'half-crack' length (crack dimension)

The function parameter, Y, varies based on crack geometries and configurations. As a result, the SIF is a function of crack geometry and configuration. The *Stress Intensity Factors Handbook* by Y. Murakami is considered a definitive reference for SIFs. Therefore, once a crack has been discovered in a material, similar looking crack geometry can be looked up in the reference. A crack embedded in a material can be measured for its length, $2a$ (or *half-length*, a, for edge cracks), and plugged into the appropriate SIF reference look-up formula to calculate the SIF (K) associated with the crack.

When considering *critical* stress intensity associated with fracture, Equation L5.3 should be written with 'C' subscripts to identify the critical computation in Equation L5.4.

$$K_{IC} = Y\sigma_C\sqrt{\pi a_C} \tag{L5.4}$$

The critical SIF 'K_{IC}' in Equation L5.4 is a material property, known as a material's *fracture toughness*, for Mode I loading. As long as the K_I calculated from Equation L5.3 is less than the material property of K_{IC}, the part should not catastrophically fail in brittle fracture. However, a crack may also continue to grow and eventually get to a critical length, a_c (or $2a_c$), where K_I does exceed K_{IC}, and the material fails. Therefore, at fracture, Equation L5.3 is properly written as Equation L5.4, where σ_C is the critical stress at which the part fails due to fracture.

Equation L5.4 can be easily rearranged for its most useful engineering form in Equation L5.5 to determine critical fracture stress for a given crack length, crack geometry and material property.

$$\sigma_C = \frac{K_{IC}}{Y\sqrt{\pi a_C}} \tag{L5.5}$$

where: K_{IC} is the material fracture toughness for Mode I
Y is a crack-specific look-up value
σ_C is the critical fracture stress
a_C is the critical 'half-crack' length at fracture

To determine the maximum critical stress that can be applied to a material, σ_C, look up the material's fracture toughness, K_{IC}. Plug this K_{IC} value into the RHS of Equation L5.5. Use a reference such as Murakami and look up the value for Y (also written as F in Murakami), and plug the Y-value or Y-function into Equation L5.5. Each unique crack geometry will have a different Y, either as a function or simplified as a single value. Measure the crack length specified by the reference, and enter crack length, a, into the RHS denominator radical of Equation L5.5. The planned design stress, σ, must be less than the critical fracture stress, σ_C, calculated from Equation L5.5 to avoid failure due to brittle fracture.

For design purposes, a few common approaches can be used, all which yield the same result since they are variations of the same equation. The first approach is to calculate a critical stress using Equation L5.5 based on material property and a possible crack length, and then ensure design stress is lower than critical. A second approach is to calculate a SIF from Equation L5.2 given a planned design stress or known stress and a crack length, then check if that SIF is less than a material's fracture toughness, K_{IC}. A third approach is to determine the size of crack a material can tolerate with a variation of either Equation L5.2 or L5.4. Enter the published material fracture toughness, K_{IC}, the crack geometry factor, Y, a known applied stress, σ, and solve for critical crack length, a_c. This last approach is useful for parts already in service when a crack is discovered. If the measured crack length is at or close to the critical crack length, the part can be expected to fail. If the crack length is well below the critical value, the part might be placed back into service and the crack is monitored.

For an edge crack in a *semi-infinite plate* under tensile load, the Y-function is approximated as $Y = 1.12$. Equation L5.3 becomes Equation L5.6 for this particular edge-crack geometry. This equation is also used for a plate of *finite* width with acceptable accuracy for the purposes of this lab.

$$K_I = 1.12\ \sigma\sqrt{\pi a} \tag{L5.6}$$

Failure mode. A cracked part of given dimensions, loaded in tension, may fail *either* by exceeding material yield or ultimate stress (σ_{YS} or σ_{ULT}, as appropriate), or it may fail due to the critical stress concentration of the crack, σ_C, according to Equation L5.4. To determine the failure mode, both stresses must be checked. For example, the ultimate stress of 6061-T6 aluminum is about 290 MPa. A part with a small crack will break if loaded to this stress. If the crack is sufficiently long to cause σ_C to be lower than 290 MPa, the part will instead break at the lower critical fracture stress. If critical fracture stress calculates higher than 290 MPa, then the crack does not significantly affect the stress and the part still breaks at 290 MPa.

The fracture toughness of a material is also influenced by the thickness of the part. Thin parts tend to deform more in the lateral direction around a crack than thick parts, so the actual K_{IC} will be higher than the published values for thin parts.

Other failure modes. This analysis has been simplified, where cracked parts will fail *only* due to ultimate stress or brittle fracture. For large test specimens with relatively small cracks, this binary choice may continue to hold true. However, other failure modes and mixed modes are possible as cracks often propagate slowly in ductile materials. As a result, parts may fail at stress higher than the critical fracture stress (no brittle fracture), but still well below ultimate stress. This failure mode is due to a localized region of plastic deformation and plastic zone propagation around the crack tip. A significant amount of energy is absorbed by plastic deformation moving ahead of the crack. This plastic deformation blunts the progress of crack propagation as the material fails due to ultimate stress exceeding a small region around the crack tip. Even though ultimate stress is exceeded locally, the overall stress in the material is less than the material σ_{ULT} when failure occurs.

EXPERIMENTAL METHOD

A commercially available benchtop load cell is typically used for this experiment. Custom clamps are required to hold the test specimens in this experiment. Refer to Appendix L5.B for clamp configuration.

Hand Calculations. Prepare a table similar to Table L5.2 in the results section to record material properties along with theoretical and experimental values. Look up the ultimate stress and fracture toughness values for both 304 stainless steel and acrylic. Enter these values into the appropriate cells of Table L5.2. Use the estimated dimensions provided in Table L5.2 to complete expected results for the lab plan. Use Equation L5.5 to calculate the fracture stress of both materials. The value for Y in Equation L5.5 is given in Equation L5.6 for an edge crack. Compare fracture stress to ultimate stress and indicate whether brittle fracture or ultimate stress failure is expected in the FAILURE MODE column of Table L5.2. Complete all theoretical expected values for the lab plan. Update all parameters during the lab using measured values.

Stress field plot. Generate a plot of the theoretical stress field based on Equation L5.2 for an edge crack. Refer to Appendix L5.A for an example of code to create the plot. This plot will be compared to the acrylic photoelastic image in the lab report. Include the stress plot in the lab plan.

Stress-strain curves. Include two side-by-side sketched stress-strain figures for comparison in the lab plan similar to Figure L5.8. The first sketch should be a typical example of a ductile material stress-strain curve taken to failure. The second sketch should be a typical example of a brittle material stress-strain curve taken to failure. These lab plan sketches will be replaced with stress-strain data curves from the experiment.

Test specimens. Stainless steel and acrylic materials are examined. Both materials are tested with and without a pre-configured crack. The test specimens without the cracks are the control specimens. Test specimen material dimensions are summarized in Table L5.2.

Stainless steel foil test specimens. Handle with care and USE GLOVES AT ALL TIMES when handling stainless steel foil. Use a heavy-duty paper cutter or scissors to cut two rectangular strips of 2-mil stainless steel foil similar to Figure L5.2 with dimensions from Table L5.2. Dimensions are not critical, but cuts should be smooth to avoid unplanned SIFs. Use hand scissors to make a very small but caliper-measurable cut about 1 mm on one edge of one steel test specimen as shown in Figure L5.2.

Acrylic test specimens. Acrylic test specimens are best cut by laser from 1 mm thick sheets. Rectangular specimens can also be cut from sheet acrylic by scoring, breaking and drum sanding to create a narrow test section. The narrow failure zone should be approximately 0.5 cm wide by 1 mm thick to keep ultimate stress within manageable levels. Cut two acrylic test specimens according to dimensions in Table L5.2, as shown in Figure L5.3.

Place one of the acrylic test specimens FLAT on a sharp-cornered table or other sharp-cornered surface with just enough of the test section off the edge of the table to make a uniform vertical notch through the thickness of the test specimen. Use a sharp edge file to create an edge notch in the acrylic test specimen, as shown in

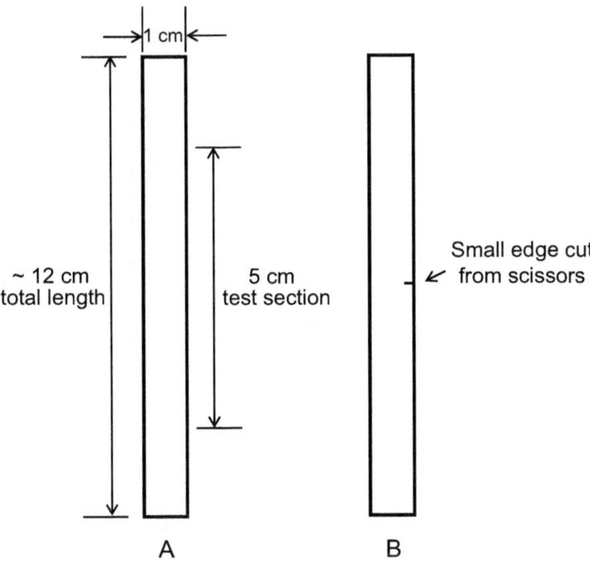

FIGURE L5.2 Steel test specimens, one with no flaw (A) and one with an edge crack (B).

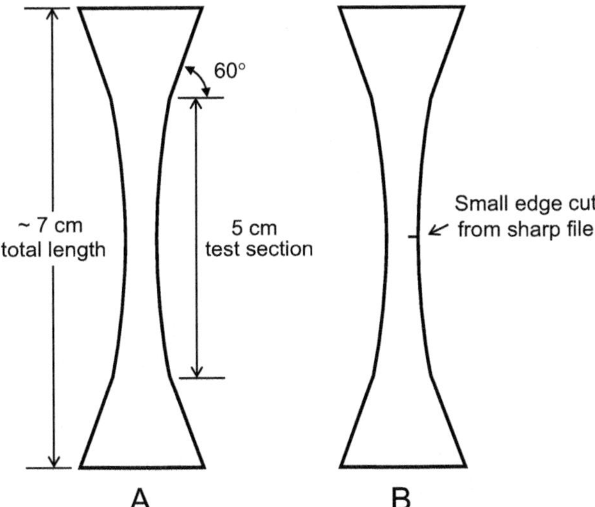

FIGURE L5.3 Acrylic test specimens, one with no flaw (A) and one with an edge cut (B).

Figure L5.4. Be sure to securely hold the acrylic part on the table and avoid any unnecessary stress to avoid breaking the part while filing the notch.

Use calipers to record the cross-sectional dimensions in Table L5.2 for all test specimens at their narrowest locations (two stainless steel and two acrylics) in both width and thickness. Use calipers to carefully measure the *half-lengths*, *a*, of the

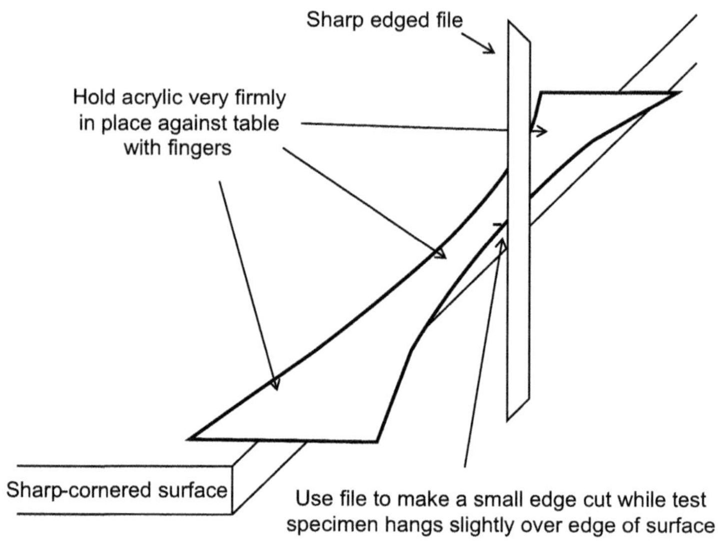

FIGURE L5.4 Creating edge flaw in acrylic test specimen.

cracks in both the steel and acrylic test specimens. Cracks are considered separate from cross-sectional area tensile stress calculations since cracks may or may not exist. Calculate the cross-sectional area for the cracked test specimens exactly as with the non-cracked test specimens. The crack length should not be large enough to significantly change the cross-sectional area or the test specimen should be redone.

Steel Foil Test. Begin the fracture tests with the stainless steel foil specimens. To help avoid slipping during the test, CAREFULLY (to avoid cuts) make a 90 upward bend, approximately 1cm, on both ends of the stainless steel as shown in Figure L5.5 to wrap around the top of the clamp. Use tools such as a straight edge and utility knife to avoid injury while making the bends.

The clamp fixtures should first be secured into the load cell test apparatus. Each clamp fixture consists of three blocks on the top and three blocks on the bottom as seen in Figure L5.6. Loosen the bolts and nuts as necessary to slip the foil sample into place as shown in Figure L5.6. *Lightly* tighten the bolts and nuts just enough to hold the foil in place and ensure the foil specimen is squarely aligned in the load cell for even loading. If not evenly aligned when loaded, one side of the test specimen may be stressed more than the other, yielding less accurate results.

Following the steps in Table L5.1 can facilitate proper alignment of the test specimen in the clamp assemblies.

Set up the data recording software to record loads for the experimental run. Perform a calibration of the equipment as necessary. Create a data column for stress where STRESS = [Force]/(cross-sectional area). Use the cross-sectional area of the test specimen. Include a column for strain where STRAIN = [Displacement]/(original length). For original length, use the test specimen's nominal length between the clamps subject to strain.

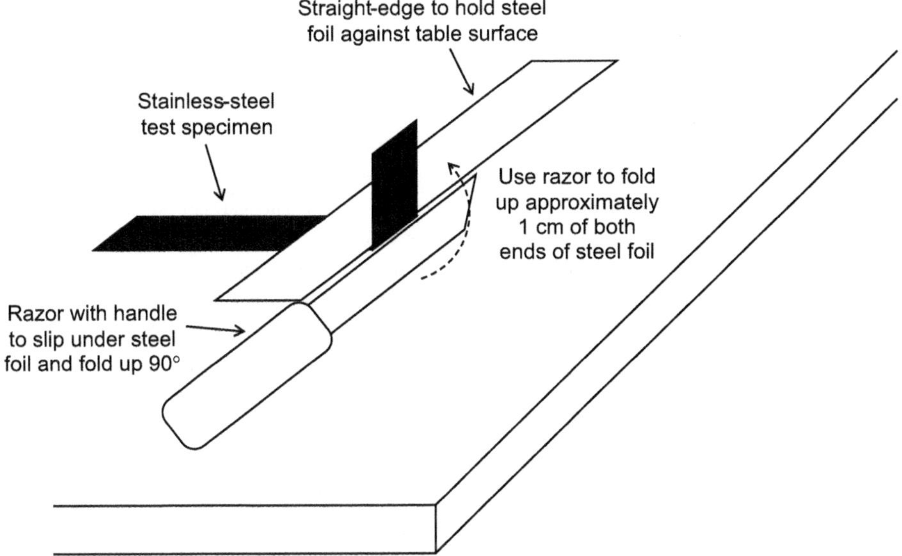

FIGURE L5.5 Fold up both ends of steel foil with tools to facilitate placement into clamps.

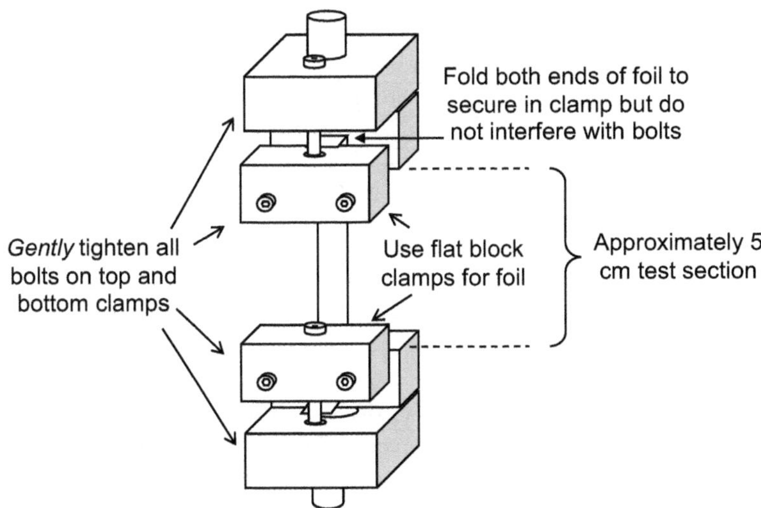

FIGURE L5.6 Securing steel foil test specimen into clamps.

When properly configured for recording data, begin the test run by loading the non-cracked steel test specimen all the way to failure where the specimen breaks into two pieces. Record the peak failure stress along with both stress and strain to create a stress-strain curve. After part failure, remove the foil test sample and repeat the process with the cracked foil test specimen.

TABLE L5.1
Loading Test Specimen in Test Apparatus

1. Lightly secure test specimen in top clamp.
2. Adjust load cell height until bottom of foil specimen is at correct position to fit into bottom clamp.
3. Slip test specimen into opened bottom clamp then lightly tighten bottom clamp bolts.
4. Check test specimen is evenly aligned between top and bottom clamps. If not, adjust position.
5. Once evenly aligned between top and bottom clamps with no kinks or bends, moderately tighten both upper and lower clamp nuts and bolts to secure the test sample in place.
6. The foil test section should have about 5 cm between top and bottom clamps.
7. Apply a slight pre-load to the specimen and confirm evenly taut.

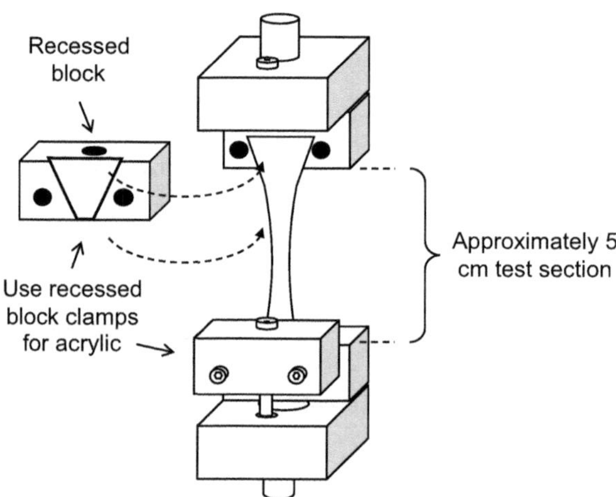

FIGURE L5.7 Securing acrylic test specimens into clamps.

Acrylic test with photoelastic analysis. Reconfigure the load cell clamps by replacing one set of flat blocks with the recessed blocks to hold the acrylic specimen in place as shown in Figure L5.7. Follow the same basic process in Table L5.1 to properly secure the non-cracked acrylic test specimen in the clamps. Do not twist or stress the acrylic test specimen when securing it into the clamps or it will fracture. If desired, place polarizing sheets on either side of the non-cracked acrylic test specimen with a backlight for viewing. The non-cracked test specimen will have even loading, so the photoelastic analysis is somewhat trivial in this case.

Begin loading the non-cracked acrylic specimen in the load cell and record stress and strain data. Apply the load SLOWLY. If the load is applied too quickly, the part may unexpectedly fail without sufficient data points to accurately determine peak failure stress. Load the acrylic specimen to failure and record the peak failure stress.

After the non-cracked sample is loaded to failure, remove the parts and replace with the cracked test specimen, taking care not to twist or bind the part. Attach polarizing sheets in front of the test specimen and behind the test specimen. Use a backlight to illuminate the acrylic part. Begin recording data and SLOWLY begin to load the part. Note when the visible stress concentration patterns develop through the polarizing sheets and pause the load application. Take a picture of the photoelastic stress concentrations before loading the part to failure. Slowly continue loading the test specimen to failure while recording data. Record the peak failure stress.

RESULTS

Material properties, theoretical calculations and experimental results are summarized in Table L5.2. Clearly identify the failure mode for each material (ultimate stress or fracture stress) in Table L5.2. Determine the error between theoretical and

TABLE L5.2
Theoretical Fracture Analysis

Property	304 SS Foil	Acrylic
σ_{ULT}	*(Look-up)*	*(Look-up)*
K_{IC}	*(Look-up)*	*(Look-up)*
Width	*(1.0 cm)*	*(0.5 cm)*
Thickness	*(0.0508 mm)*	*(1.0 mm)*
A_{XC}		
A	*(0.5 mm)*	*(0.5 mm)*
THEORETICAL TENSILE FAILURE		
σ_{ULT}	*(Look-up)*	*(Look-up)*
P_{ULT}	$P_{ULT} = (\sigma_{ULT})*(A_{XC})$	$P_{ULT} = (\sigma_{ULT})*(A_{XC})$
THEORETICAL FRACTURE FAILURE		
σ_C	*(Equations L5.5 and L5.6)*	*(Equations L5.5 and L5.6)*
P_C	$P_C = (\sigma_C)*(A_{XC})$	$P_C = (\sigma_C)*(A_{XC})$
EXPERIMENTAL FAILURE VALUES		
Non-cracked P_{FAIL}		
Non-cracked σ_{FAIL}	$(\sigma_{FAIL}) = (P_{FAIL})/(A_{XC})$	$(\sigma_{FAIL}) = (P_{FAIL})/(A_{XC})$
Cracked P_{FAIL}		
Cracked σ_{FAIL}	$(\sigma_{FAIL}) = (P_{FAIL})/(A_{XC})$	$(\sigma_{FAIL}) = (P_{FAIL})/(A_{XC})$
EXPERIMENTAL FAILURE MODE (Indicate σ_{ULT} or σ_C)		
% ERROR FOR FAILURE MODE (Experimental relative to theory)		
% STRENGTH REDUCTION FOR NON-CRACKED VS CRACKED ACRYLIC		

Note that stainless steel foil has a thickness of 2 mils, which is 0.002 inches, not 2 mm.

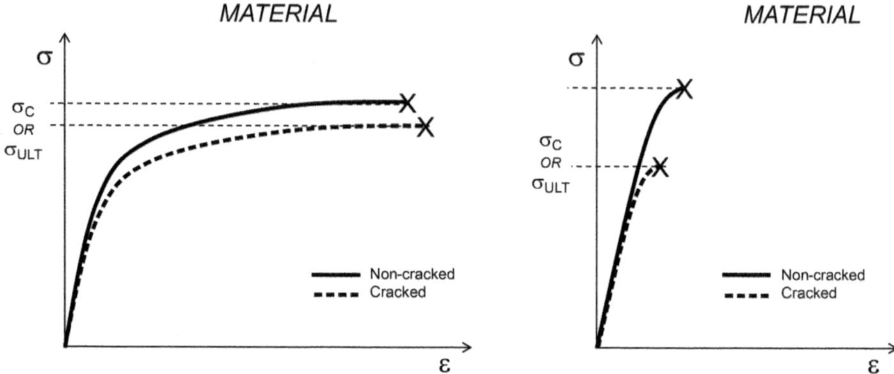

FIGURE L5.8 Stress-strain curves for ductile and brittle materials.

experimental results only for the failure mode. For the acrylic brittle fracture case, also provide a percentage difference between experimental failure stresses of the cracked versus non-cracked test specimens, relative to the non-cracked value. This will indicate the reduction of strength in the material due to the presence of a crack.

The two stress-strain curves in Figure L5.8 depict the fracture behavior of the ductile and brittle material for both the cracked and non-cracked cases. Identify the material associated with each curve. The lab plan stress-strain plots should be sketches of expected behavior which are replaced with stress-strain curves based on lab data for the report. For the lab report, clearly indicate the value of failure stress on the curves and whether it was due to ultimate stress or critical stress.

DISCUSSION

This lab considers only two discrete failure modes: ultimate stress failure and critical fracture failure. Table L5.2 should indicate the *most appropriate* failure mode for both materials. If the cracked steel foil failed at a stress above the critical fracture stress but well below ultimate stress, discuss how the crack might have caused localized crack tip stresses to exceed ultimate stress while localized plastic deformation absorbed energy to blunt rapid crack growth. Also, there is a distinction between failure stress and ultimate stress for ductile materials. If the stress-strain curve peaks at ultimate stress, then loses strength prior to failure, for the purposes of this lab, the failure can still be considered due to ultimate stress.

Discuss how ductile failure compared to brittle failure. Clearly state the reduction in strength for each material. Discuss which scenario might provide an indication of impending failure.

Compare the computational stress field plot from the example code in Appendix L5.A using Equation L5.1 with a picture from the photoelastic test. Note the similarities and differences.

Discuss possible sources of error for the results. Do not make references to human error.

CONCLUSIONS

Based on the objective, cite the percentage errors between experimental results and theory and assess whether experimental data appears valid compared to theory. Make an assessment about the impact of a crack on the design strength of a material. Discuss what considerations must be made during the design, manufacturing and service of a part to account for SIFs. Discuss where any improvements could be made in *your* experimental setup. Do not make any references to what was good or bad about the lab. Do not make any references to whether it was enjoyable or not. Limit the discussion to the details of the lab and results.

APPENDIX L5.A: SAMPLE CODE FOR EDGE CRACK BY TERRY W. ARMSTRONG, PHD, COPYRIGHT 2015

```
x=0:0.05:2;       % Set x-dimension & resolution
y=-4:0.05:4;      % Set y-dimension & resolution
[X,Y]=meshgrid(x,y);     % Define the plot mesh
[theta,rho] = cart2pol(X,Y); % theta & rho need to be a matrix of X & Y (not x & y)
sigmaNot=100;     % Set a nominal far-field applied stress
a=0.5;            % Set a nominal edge crack length
% Calculate entire theoretical stress field in x- & y-directions
sigmaX = (sigmaNot .* (pi .* a).^0.5 ./ (2 .* pi * rho).^0.5) .* (cos(theta ./ 2)
.* (1 - sin(theta ./ 2) .* sin(3 .* theta ./ 2)));
sigmaY = (sigmaNot .* (pi .* a).^0.5 ./ (2 .* pi * rho).^0.5) .* (cos(theta ./ 2)
.* (1 + sin(theta ./ 2) .* sin(3 .* theta ./ 2)));
Z = sigmaY;       % Set Z of surface plot to y-direction stress of interest
(x-direction insignificant)
surf(X,Y,Z)       % Generate a surface plot
pbaspect([1 2 1]) % Set rectangular aspect based on x & y dimensions above
shading interp    % Interpolate colors for shading
view(2)           % View 2D-only plot
```

APPENDIX L5.B: CUSTOM FRACTURE TEST SPECIMEN CLAMPS

Several commercially available clamps have been used in this experiment with poor results. Both the steel foil and acrylic specimens slip from typical jaw-style clamps. Attempts to fold or loop foil around a pin and using cyanoacrylate glue to provide grip to the acrylic have all equally failed. Screw-down clamps with sharp grips result in stress concentration failure at the grip point. The custom-machined clamps modeled in Figures L5.B1–L5.B3 fit the threaded rod of the test machine in use, and they have been thoroughly effective. These clamps provide a large clamp-up surface for the steel foil and a v-grove with clamp-up for the acrylic.

Each clamp assembly consists of three blocks of 6061T6 aluminum. There are two clamp assemblies, one for the top and one for the bottom. The top and bottom clamps are reflectively symmetric about a horizontal plane separating the top from bottom clamps. The uppermost and lowermost caps have a large central through-hole for the load cell test machine threaded rod. The large concentric recess contains the nut holding the cap and assembly onto the load cell threaded rod. Overall dimensions are SAE with all bolt assembly holes sized to 7 mm to accept 6 mm cap bolts.

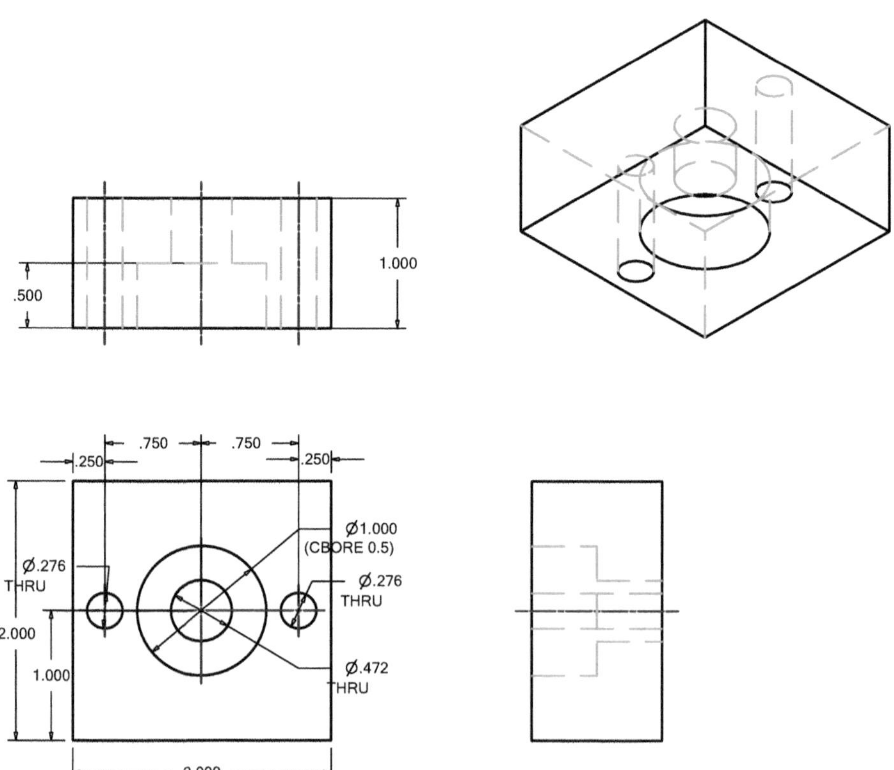

FIGURE L5.B1 Top and bottom clamp caps. Two caps required. All dimensions given in SAE (inches). Bolt through-holes are loose tolerance 7 mm holes for 6 mm cap bolts.

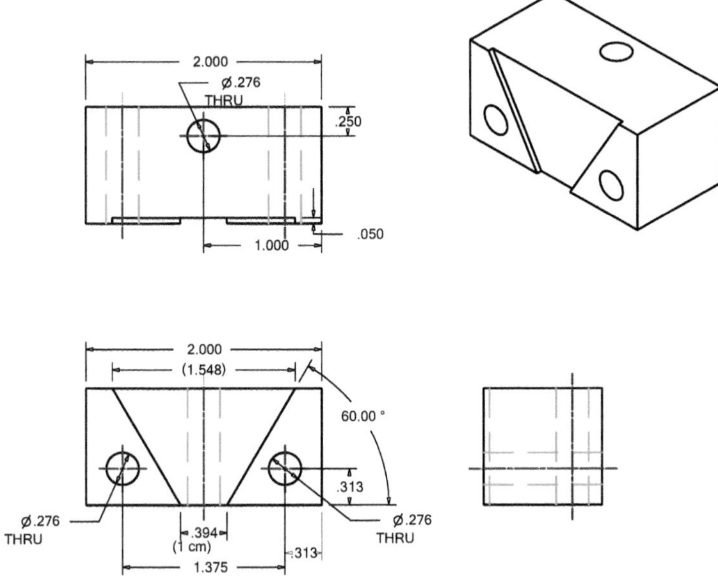

FIGURE L5.B2 Top and bottom acrylic clamp blocks. Two acrylic clamps required. All dimensions given in SAE (inches). Dimension at top of v-recess is a trivial result of the 60° angle. The bottom of the recess should accept the 1cm wide acrylic specimen.

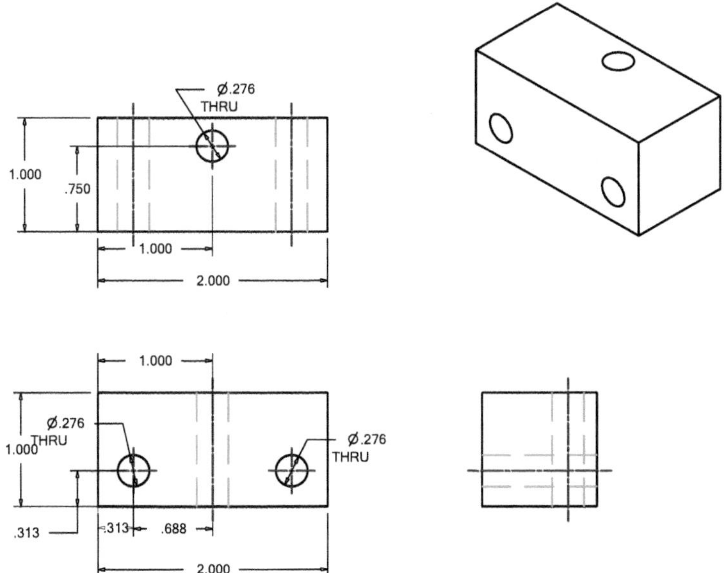

FIGURE L5.B3 Top and bottom steel foil clamp blocks and backing block clamps for both acrylic and steel. Four total blocks required; two blocks always remain in place as backing support and two blocks are swapped out for the acrylic clamp blocks.

APPENDIX L5.C: ACRYLIC TEST SPECIMEN DIMENSIONS

Figure L5.C1 depicts the acrylic test specimen suitable for the custom clamps in Appendix L5.B.

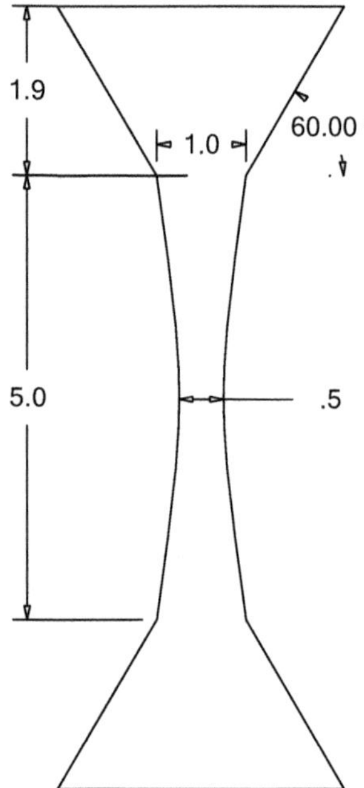

FIGURE L5.C1 Acrylic test specimen. Test specimen is sized to fit in custom clamps. Failure zone dimensions are 0.5 cm by 1 mm thick to limit failure loads to a manageable level for desktop tensile test machines. Dimensions are not critical. Specimens are best cut via laser.

LAB 6

Non-Destructive Inspection

Safety Requirements

Lab coats and safety glasses must be worn
Take care not to drop heavy test specimens

Equipment Requirements

Metallic test specimens of the same or different materials, a few to several cm thick
  One test specimen is defect free
  One test specimen has holes drilled to different depths on the bottom surface
Oscilloscope
Calipers
Ultrasonic pulse generator
Ultrasonic transducers
Coupling agent (lotion is the "couplant")

Procedural Requirements

Do not drop or damage expensive ultrasonic transducers

Clean-up Requirements

Wipe lotion off all components
Unplug and return all equipment to the proper locations

Emergency Actions

No unusual hazards

INTRODUCTION

Once a material is put into service within a structure, it is subjected to the ravages of the environment. It bears the stress of the structure along with other environmental loads due to wind, impacts, over-use, friction, heat or cold. Other conditions such as corrosion due to water, chemicals and galvanic interaction affect the material. All of these actions can quietly alter the condition of the material with cracks or distortion as well as change it chemically through processes such as oxidation. If these changes go completely undetected, they can eventually lead to the failure of the material within a part and structure as a whole.

A multitude of tools and techniques are available to detect material flaws or degradation. Dye penetrants can be sprayed onto metallic surfaces to visually detect small cracks. The dye penetrates into the crack to make the crack readily stand out during the visual inspection. Metal parts can also be examined for flaws by inducing an electrical current through electromagnetic interaction with a probe. The *eddy currents* are measured by the device and material anomalies appear as signal changes on the display screen. These inspection techniques are just a couple of techniques that do not require tearing into the structure to determine whether a problem exists. Any inspection process that does not damage the serviceability of the part is considered a non-destructive inspection technique. The terms non-destructive inspection (NDI), non-destructive examination (NDE) or non-destructive testing (NDT) are all considered equivalently interchangeable terms.

This lab investigates ultrasonic NDI as a common industry technique where high-frequency sound waves are propagated through a material to check for the depth of hidden defects. The velocity of wave propagation is calculated based on theory and then checked experimentally with a material of known thickness using an oscilloscope. Once verified, an ultrasonic wave is used to determine the depth of a flaw within a material.

OBJECTIVE

Experimentally verify the theoretical wave velocity of ultrasound with a known material thickness. Use verified ultrasound properties to experimentally measure the depth of a flaw within a material. Compare experimental results relative to truth values.

THEORY

Sound waves with a frequency greater than 20 kHz are above the human threshold of hearing and are classified as ultrasonic or ultrasound. With very short wavelengths, ultrasonic waves offer higher imaging resolutions compared to longer wavelengths of audible sound. A tightly-packed group of short-length waves will *ping* clearly with higher resolution of a structure rather than a long undulating wave that is already echoing before the sound pulse has even finished emitting.

A sound wave emitted from an ultrasonic transducer will propagate through a homogeneous material until it reaches a surface as shown in Figure L6.1. The surface will reflect the sound wave back to the transducer, just as sound waves traveling in the air will bounce back from a distant surface to create an audible echo.

The time it takes between sound wave emission and return can be used to calculate distance traveled, D, based on the speed of sound, v, through the material. This basic formula is given in Equation L6.1 where the measured time-of-flight, TOF, must be divided by two to account for the total distance the sound wave travels back and forth across the material to the transducer.

$$D = v\left(\frac{TOF}{2}\right) \tag{L6.1}$$

The velocity of sound, v, within the material is dependent on both material density, ρ, and bulk modulus, B, as given in Equation L6.2.

$$v = \sqrt{\frac{B}{\rho}} \tag{L6.2}$$

The speed of sound in air is in the vicinity of 350 m/s while sound travels much faster through aluminum at around 6300 m/s. A stiffer material with a higher modulus will propagate sound faster than a less stiff material at similar density.

Bulk modulus, B, is related to the modulus of elasticity, E, and Equation L6.2 can be rewritten as Equation L6.3.

$$v = \sqrt{\frac{E(1-\upsilon)}{\rho(1+\upsilon)(1-2\upsilon)}} \tag{L6.3}$$

where: v is the speed of sound within a material
E is the Young's (elastic) modulus of material
ρ is the material density
υ (nu) is Poisson's ratio

The complete derivation of Equation L6.3 is rather lengthy as it begins with stress-strain relationships, equations of motion and the partial differential wave equation. However, a brief synopsis of the derivation follows.

For one dimension, Hooke's law relates stress to strain via Young's modulus as $\sigma = E\varepsilon$. For the three dimensions, Poisson's ratio also enters the equation. Normal stress, say σ_{xx} in the x-direction, can be expressed in terms of Young's modulus, Poisson's ratio and strain in the normal directions. By applying Newton's second law, mass (density), enters the equation along with the second derivative of displacement (acceleration). A second-order PDE is formed, relating normal stresses to density and acceleration. Using Hooke's law and strain displacement relationships, a simplified one-dimensional wave equation is developed based on velocity and the material properties of elastic modulus, Poisson's ratio and density as given in Equation L6.3.

EXPERIMENTAL METHOD

Controlling variables and error. For the lab plan, think through the experimental method and consider potential sources of error. Identify at least one potential source of error and provide a mitigating strategy (control, repetition, replication and randomization) to minimize the effect.

For example, the ultrasound signal must pass from the transducer through the couplant and into the test sample. This is a noisy interface where the signal tends to bounce around, causing multiple peaks around the desired signal peak. Locating and measuring from the correct peak is now a matter of observational interpretation. This is a *random type* of error since any one person or other team members might measure it slightly different each time. The error *source* is noise from the interface due to an acoustic impedance mismatch. This source of error enters the results as variations in the measured time signal and ultimately affects the experimentally determined depth of the flaw. This source of error will show up in the error calculation.

A potential strategy to control the error is repetition and randomization. Take several measurements of the signal and use the mean value. Since the error is a type random, the Central Limit Theorem states the measurements should be normally distributed about the likely truth value.

The measurements can also be randomized by using more than one person to take several measurements. Although the error source itself is random, the person taking the measurements may have a bias in how they read the results. Randomly assigning various people to take measurements will both randomize any potential bias (systematic error) and further normalize the results.

A well-conceived lab plan considers all opportunities for error or failure in the experiment and provides strategies to minimize the impact. The results or conclusion section should then be used to discuss the effectiveness of the strategy and whether it can be improved.

Identify at least one potential source of error for the lab plan (and report). Do not use the same error just discussed; provide a different possibility. Discuss the *type* of error (random or systematic) and the *source* of the error. Provide a controlling or mitigating strategy.

Test specimens. Two cylindrical blocks with the dimensions and properties given in Table L6.1 are used for the ultrasonic test. One block is homogeneous, while the other block has *defect* holes drilled into the bottom side at various locations and depths. Confirm all materials and dimensions in the lab and update Table L6.1 accordingly.

Hand calculations. Create a table similar to Table L6.1 with columns to record each material's published speed of sound velocity (celerity), theoretical velocity, experimental velocity and error between theoretical and experimental velocities. Look up the material properties for the test specimens and record the properties in Table L6.1. Use Equation L6.2 or L6.3 to determine the theoretical velocity of the ultrasonic pressure wave within the materials identified in Table L6.1. Theoretical calculations must be completed in the lab plan using the best available information.

Create a second table similar to Table L6.2 for the test specimen with embedded defects. Include cells for defect depths measured by calipers and cells for experimentally measured defect depths via ultrasound. Include entries for ultrasonic measurement errors relative to physical measurements. The caliper measurements of the defect hole depths can be considered truth values as known quantities.

Setup. Connect a function generator, an oscilloscope, and a piezoelectric transducer as shown in Figure L6.1. Use a small dab of *coupling agent* such as common hand lotion to better transmit and receive signals to the test specimen as shown in Figure L6.1. The coupling agent serves as an intermediate between the transducer and test specimen to help reduce the acoustic mismatch from the air gap interface. The ultrasonic pulse generator has two frequency selectors. The highest frequency should be in the MHz range. Set this frequency to the rated frequency of the transducer.

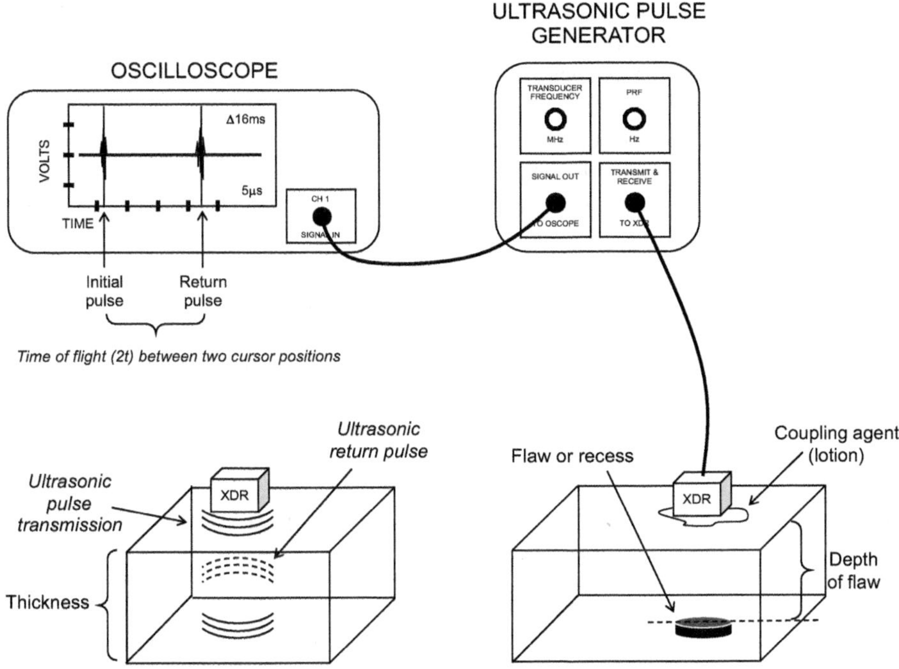

FIGURE L6.1 Experimental setup.

The lowest frequency should be in the tens to hundreds of Hertz range. Set this low-frequency range to allow enough TOF between signal pulses. An upper limit low-frequency can be computed, but any low-frequency setting should be adequate. Refer to Appendix L3.C for oscilloscope setup and operation.

Speed of wave propagation. Measure the thickness of both test specimens. Update Table L6.1 with actual thicknesses. Beginning with the test specimen without flaws (no holes drilled into the bottom surface), place a dab of coupling agent on the surface of the test specimen and place the ultrasonic transducer onto the couplant, pressing down to ensure firm contact between the transducer and test specimen. Adjust the oscilloscope as necessary to see two clear signal peaks. Measure and record the TOF between the signal peaks on the oscilloscope. Use Equation L6.1 to compute the velocity of sound in the test specimen based on the caliper-measured thickness, D, and oscilloscope time of flight, TOF. Record this experimentally determined velocity in Table L6.1. This will be the verified velocity used for defect measurements.

If the defect test specimen is a different material, attempt to perform the same velocity determination as with the non-defect test specimen by ultrasonic measurement over a clear part of the specimen. Once a satisfactory velocity has been obtained, ensure the defect holes are at the bottom and begin scanning the smooth top surface of the test specimen with the transducer. Look for shorter TOF peaks on the oscilloscope compared to the thickness return signal and record the TOF in Table L6.2. Take note of the defect location associated with the TOF. Since the test specimen is covered with couplant, use a sheet of paper and a reference point on the specimen to annotate the defect locations.

When the defect holes have been adequately identified, turn the test specimen over to measure the hole depths with a caliper. Use the protruding end of the caliper, opposite the clamp ends, to measure into the holes. Record these values as truth values in Table L6.2.

RESULTS

Material properties, theoretical calculations and experimental results are summarized in Table L6.2. Compare the theoretical and experimentally determined velocities in the test specimens and indicate the percentage error relative to the theory.

Complete all entries in Table L6.2. Use Equation L6.1 with the experimentally verified velocity value and experimental TOF values to calculate the depth (distance), *D*, of the defect within the test specimen. Compute the errors relative to the caliper-measured truth values for each defect hole depth.

TABLE L6.1
Test Specimen Properties

Material	*(AL6061T6)*	*(1018 STEEL)*
Bulk or elastic modulus	*(Look up)*	*(Look up)*
Poisson's ratio	*(Look up)*	*(Look up)*
Density	*(Look up)*	*(Look up)*
Published velocity	*(Look up)*	*(Look up)*
Theoretical velocity	*(Equation L6.2 or L6.3)*	*(Equation L6.2 or L6.3)*
Experimental velocity		
% Error		
Measured thickness (Truth value)	*(7 cm)*	*(5 cm)*
Experimental thickness		
TOF (2 × Thickness)	*(O'scope)*	*(O'scope)*

TABLE L6.2
Test Specimen Defect Measurements

Material	*(AL6061T6)*
Measured defect depth 1 (truth value)	
Ultrasonic measured defect depth 1	
% Error	
Measured defect depth 2 (truth value)	
Ultrasonic measured defect depth 2	
% Error	

DISCUSSION

Describe the results. Discuss the ability to accurately determine the depth and location of defects within a material using ultrasonic techniques. Discuss whether the controlling strategy for the error identified in the lab plan was effective.

CONCLUSIONS

Based on the objective, cite the percentage error and assess whether experimental results agreed with truth values. Assess whether the measurement techniques were effective for determining celerity and defect depth. Offer a suggestion how on to better control error for future experiments.

LAB 7
Monte-Carlo Simulation

Safety Requirements

None

Equipment Requirements

Computer with programmable code or spreadsheet

Procedural Requirements

No lab plan required
Code development
- Individually develop code and physically type code at an individual computer
- Do *not* copy code from online sources or classmates
- Do *not* email code to classmates
- Collaborate with lab teammates *only* to understand code concepts
- Full code must be included in the lab report
- Final code must be fully commented to explain the functionality

Written report
- Reports must be *a 100% individual effort* with no collaboration allowed
- No length requirements; only length sufficient to cover the material
- Report must be uniquely and originally written
- Include all report requirements: cover page, abstract, sections and references

Clean-up Requirements

None

INTRODUCTION

Most real-world processes are subject to variability, which makes it difficult to predict an exact outcome. Probabilistic events include weather patterns, manufacturing processes and life expectancies. Modeling these processes based on available information, input variations and relationship between parameters helps provide insight into the most likely outcome.

The Monte-Carlo simulation is a *stochastic* algorithm that models these uncertain processes by using randomized values for the independent variables of a governing equation. The simulation harnesses the power of the computer to run this simulation many times as variables are randomized each run to return a statistical distribution, which can be plotted as a histogram and the mean as the most likely result.

Tensile load failure for a material due to ultimate stress is typically cited as a single number based on a material's ultimate stress and cross-sectional area. However, measurements are inexact numbers and measurement instruments return a value with a given level of uncertainty. Likewise, values for ultimate stress for any particular material can vary based on the published source. Manufacturing processes, different alloy batches and testing techniques introduce a certain level of variability in the ultimate stress for a material.

This lab is a computational-only exercise to predict the most likely failure load of a 3D printed polylactic acid (PLA) part. A Monte-Carlo simulation is run based on the variability of part dimensions and the ultimate stress of PLA.

OBJECTIVE

Perform a Monte-Carlo simulation to *computationally predict the most likely tensile failure load of a 3D printed PLA part based on varying cross-sections and material properties.*

THEORY

A part will fail due to tensile load based on its material property of yield stress, σ_{YS} or ultimate stress, σ_{ULT}, (as appropriate) and cross-sectional area according to Equation L7.1.

$$P_{FAIL} = \sigma_{FAIL} A_{XC} \tag{L7.1}$$

where: P_{FAIL} is the failure load
σ_{FAIL} is the material failure stress (yield or ultimate stress, as appropriate)
A_{XC} is the part cross-sectional area

Equation L7.1 is generally sufficient for a nominal value based on a single failure stress value and a single cross-sectional area value. However, for all manufacturing processes, both the failure stress and cross-sectional areas can be expected to vary.

3D printed PLA parts often vary by ± 0.5 mm from the design dimensions. PLA plastics also vary in tensile strength based on the specific formulation and manufacturer. These variations for the independent variables in Equation L7.1 mean the exact failure load for a 3D printed part is not a single value but an uncertain range of values based on the variability of the two input parameters.

The Monte-Carlo algorithm in Appendix L7.A or Appendix L7.B can be used to model the variability in Equation L7.1 and provide a distribution of failure loads. The mean of this distribution is the most likely failure load.

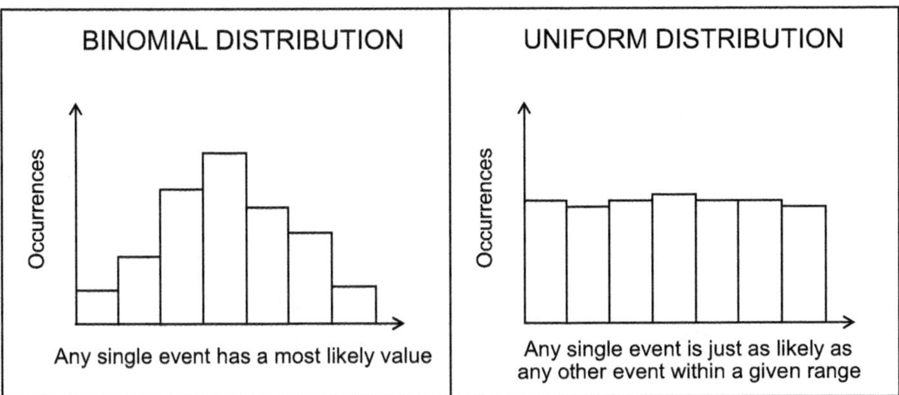

FIGURE L7.1 Common discrete distributions are the binomial and rectangular distributions.

Different random distributions of the input variables are possible. The two most basic distributions are the binomial distribution and the uniform (rectangular) distributions shown in Figure L7.1. Other distributions are possible, to include functions based on empirically obtained data.

Computer programs and spreadsheet programs can usually generate either a uniformly distributed random number or a normally distributed random number with a single function call. The results of how these random numbers combine according to Equation L7.1 (as the simplest of examples) become less certain, which is where the Monte-Carlo simulation provides clarity.

Alternative implementations of the Monte-Carlo are possible. The basic recommendation in this lab merely identifies the likely failure load from the resultant distribution mean. An alternative is to identify how many 3D parts will be rejected based on unacceptable results. If a part is designed at specific dimensions with a minimum accepted failure load, the Monte-Carlo simulation can be used to determine how many parts must be discarded. For example, a design failure load might be 100 N with a minimum acceptable failure load of 98 N. The Monte-Carlo simulation might return a range of failure loads between 95 N and 105 N based on the input variations. If ten parts are printed, the probability distribution from the Monte-Carlo simulation can provide insight as to how many of those ten parts will not meet the minimum acceptable failure load of 98 N.

COMPUTATIONAL METHOD

Prepare a table similar to Table L7.1 in the results section. Choose any desired 3D print fused deposition modeling (FDM) filament and look up various resources for typical ultimate strength values. Enter these values into Table L7.1. Decide upon the most likely distribution (normal, uniform or other) of ultimate strength values and enter the planned distribution in Table L7.1.

Choose a test specimen cross-sectional profile such as circular or rectangular. Include cell entries in Table L7.1 to record any desired radius or diameter for circular cross-sections or width and thickness for rectangular cross-sections. Look up typical precision tolerances for FDM printing. Calculate the minimum and maximum cross-sectional area of a 3D printed part based on chosen cross-sectional dimensions and printer tolerances. Enter these values in Table L7.1. Decide upon suitable distribution for the printer tolerances and enter this choice in Table L7.1.

Use Equation L7.1 to calculate an expected theoretical failure load for the planned test specimen using a nominal ultimate stress value and the design cross-section. Perform this calculation as normal for most engineering calculations and enter the value in Table L7.1 for the expected failure load.

Refer to Appendix L7.A or L7.B to write and run a Monte-Carlo simulation of failure load based on the parameters in Table L7.1. Write the algorithm code completely and do not use any *canned* Monte-Carlo commands if available in the programming language. The program can be written in any desired programming language. The code can also be successfully written using only functions from a spreadsheet program.

Plot a histogram of possible failure loads based on the simulation. Identify the most likely failure load based on the histogram mean. Enter the likely failure load in Table L7.1. Alternatively, identify a minimum acceptable failure load and indicate the percentage of parts that fail to meet the criterion. Include the simulation histogram in the results section.

RESULTS

The Monte-Carlo test specimen properties and results are summarized in Table L7.1.

Include a histogram of the simulation in this section. Label all appropriate features of the histogram chart. Depending on the choice of random number distributions, and the relative impact of either variable in Equation L7.1, the histogram may appear as a neat normal distribution, a very rectangular distribution without a clear central tendency, or a skewed distribution with a flat peak. Identify the mean on the histogram. Do not calculate an error percentage between the expected failure load and the simulated failure load.

TABLE L7.1
3D FDM Filament Specimen Properties

MATERIAL			
Cross-sectional profile and dimensions			
	MINIMUM	DISTRIBUTION	MAXIMUM
Ultimate tensile strength range			
Cross-sectional area range			
Expected failure load			
Simulated failure load			

DISCUSSION

Describe what the histogram means. Discuss your choice of input distributions and the likely impact on resultant distribution. Compare the simulation result to the expected result. Discuss how real-world manufacturing processes might be modeled by your results.

CONCLUSIONS

Based on the objective, discuss the ramifications of looking up a single failure stress for design calculations compared to using the Monte-Carlo simulation mean. Assess how the simulation provides greater insight into the failure load. Briefly provide another example of how the Monte-Carlo simulation can be used for an engineering design problem.

APPENDIX L7.A: EXAMPLE MONTE-CARLO PROGRAMMING CODE ALGORITHM

% Dimension All Arrays to Number of Runs

```
N = "a lot" (such as 10^4, 10^5 or 10^6)        % Plan many iterations
InputValue_1 ... MEANofRESULTS                   % Dimension all variable arrays
For i = 1 to N                                   % Run simulation many times
% OBTAIN THE RANDOM VARIABLE(S)                  % Randomize each iteration
InputValue_1 = Random(Low to High)               % Randomize first independent variable
InputValue_2 = Random(Low to High)               % Randomize second independent variable
%COMPUTE DESIRED FUNCTION                        % Run all necessary calculations
EndResult = (formula)                            % Evaluate appropriate formula(s)
SUMofRESULTS = sum(EndResult)                    % Sum results for a mean calculation
MEANofRESULTS(i) = SUMofRESULTS/i                % Calculate mean and enter into array
Next i                                           % A single pass is now complete
Histogram(MEANofRESULTS)                         % Plot histogram of all engineering results
```

APPENDIX L7.B: EXAMPLE MONTE-CARLO SPREADSHEET ALGORITHM

*To create a histogram chart: Plot **FREQ** (vert) vs **BINS** (horz).*

STRESS_ULT:		**AREA_XC:**	**LOAD_FAIL:**	**BINS:**	**FREQ:**
ULT_low:	DIA_low:	19.5E–3	FAIL_low:	Number:	(COUNTIF)
40.0E+06	DIA_hi:	20.5E-3	11.9E+3	10	
ULT_hi:	AREA_low	2.99E–04	FAIL_hi:	Range	
60.0E+06	AREA_hi	3.30E–04	19.8E+3	786	
RANDOM:		RANDOM:	RANDOM:	DIVISIONS	OCCURRENCES
5.57E+07		3.03E–04	1.69E+04	11.9E+3	3
4.22E+07		3.07E–04	1.30E+04	12.7E+3	9
4.18E+07		3.25E–04	1.36E+04	13.5E+3	6
4.09E+07		3.22E–04	1.32E+04	14.3E+3	2
4.19E+07		3.16E–04	1.33E+04	15.1E+3	5
5.12E+07		3.12E–04	1.60E+04	15.9E+3	2
5.27E+07		3.30E–04	1.74E+04	16.7E+3	7
5.27E+07		3.22E–04	1.70E+04	17.4E+3	3
5.46E+07		3.20E–04	1.75E+04	18.2E+3	1
5.29E+07		3.01E–04	1.59E+04	19.0E+3	1
4.09E+07		3.03E–04	1.24E+04	19.8E+3	0
4.94E+07		3.12E–04	1.54E+04	20.6E+3	
5.09E+07		3.06E–04	1.56E+04		
5.03E+07		3.10E–04	1.56E+04		
4.19E+07		3.13E–04	1.31E+04		
5.41E+07		3.17E–04	1.72E+04		
5.16E+07		3.06E–04	1.58E+04		

Example Cell Formulas

STRESS	=RANDBETWEEN($B$6,$B$7)
AREA	=RAND()*($E$7-$E$6)+$E$6
LOAD	=I11+$I$7
BINS	=I11+$I$7
FREQ (COUNTIF)	=COUNTIFS($G$11:$G$49," >"&$I11,$G$11:$G$49," <="&$I12)

LAB 8
Thermocouples

Safety Requirements

Use caution with hot plate and boiling water
Secure thermocouple wires to avoid tipping beaker and spilling hot water
Lab coat, safety glasses and insulated gloves are required

Equipment Requirements

Thermocouple wire, wire cutter, wire stripper and needle-nose pliers
Millivoltmeter or bench multimeter
Thermometer, test stands and beaker or pan
Ice bath
Hot plate to boil water
Insulated gloves and beaker tongs

Procedural Requirements

None

Clean-up Requirements

Dispose of water down drains
Dispose of wire remnants in trash
Unplug all equipment and return to proper locations

Emergency Actions

BURNS or SCALDS: Run cold water on affected area. Call 911 as appropriate.

INTRODUCTION

The average kinetic energy of molecules or atoms vibrating within a substance is indicative of the substance's internal energy. As one of the more commonly measured variables in engineering, temperature is a measure of this internal energy. Differences in temperature indicate differences in internal energy and potentials for heat transfer, expansion or contraction of a substance, the potential of thermal failure of a material and a multitude of other engineering considerations.

The zeroth law of thermodynamics states that if two bodies are in thermal equilibrium with a third body, they are in thermal equilibrium with each other. This law forms the basis of temperature measurement and generally implies a thermometer as the third body used to determine whether objects are in thermal equilibrium with each other.

Several devices are available for temperature measurements. One of the oldest reliable temperature measurement devices uses the volumetric expansion of a substance within a glass tube, and glass thermometers are still in use today. Bi-metallic strips of metal expand or contract differently relative to each other to provide an indication of temperature. Electrical devices to measure temperature include resistance temperature detectors (RTDs) and thermistors, both of which vary electrical resistance based on temperature. Thermocouples measure temperature by creating an electrical potential between two different conductive materials.

This lab demonstrates how thermocouples are used for temperature measurements. Two thermocouples are constructed and connected for measuring a voltage potential. In the first experiment, one thermocouple junction is submerged in ice water as a reference, while the other thermocouple junction is used to measure the temperature of heated water. Voltages generated by the thermocouple circuit are recorded and compared to published data. The second experiment uses tepid water as a temperature reference instead of ice water. This non-standard tepid temperature reference is adjusted by formula and the results are compared to published data.

OBJECTIVE

Experimentally verify published thermocouple voltages for both an ice point reference and a non-standard reference. Compare experimental values to published values.

THEORY

Thermocouples rely on the Seebeck effect (discovered by Thomas Seebeck) where an electromotive force (voltage potential, or EMF) is produced when two dissimilar metals are joined together to form a circuit in the presence of a temperature differential. This voltage potential varies linearly with temperature and can be measured with a voltmeter (millivoltmeter).

If the entire thermocouple circuit is at the same temperature, there is no voltage potential. Also, even with a temperature differential in the circuit but no fixed temperature reference, only temperature difference information is available and not the actual temperature. The solution is to create two thermocouples within one circuit and use one thermocouple junction as a known temperature reference. The circuit will then return a standardized EMF potential correlated to a temperature scale. This configuration is shown in Figure L8.1.

A standard, repeatable and known temperature reference is water's ice point. The ice point of water is 0°C, which is easily reproduced and accurate. Although the boiling point of water is also an option, the standard-day, one-atmosphere temperature of 100°C varies with atmospheric pressure. Water's boiling point is reduced by approximately 1°C per thousand feet of pressure altitude, such that water boils at approximately 95°C at 5000' elevation. In electronic equipment, the ice point reference is usually created with a thermistor circuit. Linearity, robustness, and other

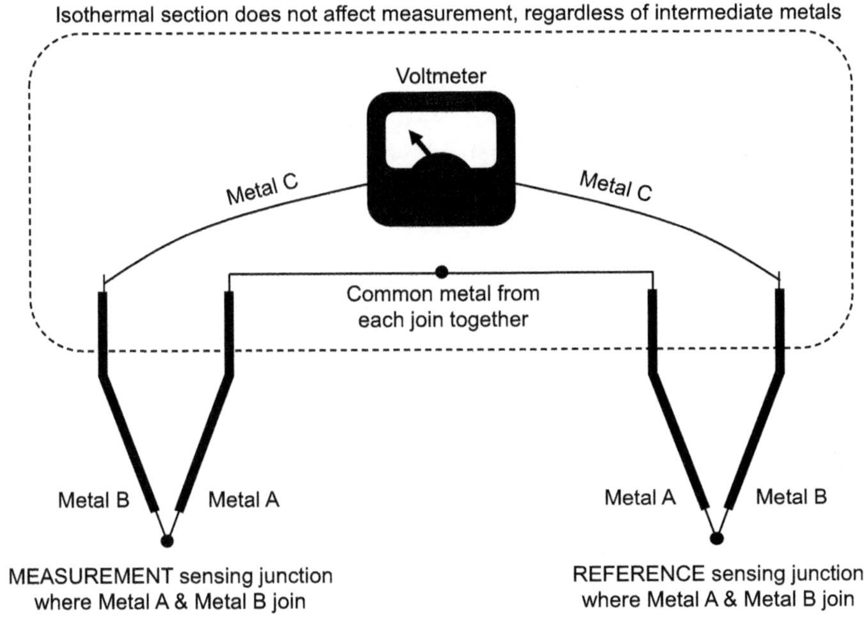

FIGURE L8.1 Thermocouple circuit with measurement and reference junctions.

issues make thermocouples more suited for temperature measurements and thermistors more suited for specific calibration circuits.

Different thermocouple metal combinations offer a different set of characteristics such as high sensitivity, corrosion resistance or high-temperature capabilities. Some of these attributes are presented in Table L8.1.

Thermocouples come in paired wires according to the metal combinations in Table L8.1. One wire is positive and the other is negative to indicate the EMF potential direction. The inner insulation on the individual wires is color-coded to indicate the thermocouple type and polarity. The red wire is always negative (*and the correct one to cut first, at least according to the movies*).

If the 0°C ice point is not used as a reference temperature, a correction must be performed to correlate thermocouple voltages to published values. The Law of Intermediate Temperatures for thermocouples essentially states that voltages generated by thermocouples must add up. If two thermocouples of a specific type are joined together with the reference junction operating at 0°C and the other junction at 100°C, they will generate a given voltage. If these thermocouples now operate at 20°C and 100°C, the measured voltage must now be less. It is less by the amount of voltage that would otherwise be generated between 0°C and 20°C. In other words, the voltages generated between 0°C to 20°C and 20°C to 100°C must still add up to the original voltage between 0°C and 100°C. This law is expressed in Equation L8.1a where Equation L8.1b is a more descriptive form of the equation.

TABLE L8.1
Thermocouple Metal Combinations and Unique Attribute (Based on Data by Figliola and Beasley)

T/C	Metal 1 (+)	Metal 2 (−)	Attributes
Type E	Chromel (PURPLE)	Constantan (RED)	High sensitivity, accurate
Type K	Chromel (YELLOW)	Alumel (RED)	General purpose, high-temperature range, rugged
Type J	Iron (WHITE)	Constantan (RED)	General purpose, modest temperature range, can lose effectiveness if the iron metal oxidizes
Type T	Copper (BLUE)	Constantan (RED)	Low temperatures, resistant to oxidation

Source: Figliola, R. & Beasley, D. (2015). *Theory and design for mechanical measurements*. Jon Wiley & Sons. p.348.

$$V_{0\to T1} + V_{T1\to T2} = V_{0\to T2} \qquad \text{(L8.1a)}$$

$$V_{0\to REF(TABLE)} + V_{REF\to MEASURED} = V_{0\to MEASURED\ (TABLE)} \qquad \text{(L8.1b)}$$

EXPERIMENTAL METHOD

Prepare two separate tables for manually recording spreadsheet data, similar to Tables L8.2 and L8.3 in the results section. Create multiple rows to record 10°C increments from 0°C to 100°C according to the "water temperature" column.

The first table is for the ice point referenced data. Identify the thermocouple type and create columns to record temperatures, millivolt values and error, as shown in the table. Use any available reference to look up standard thermocouple millivolt values for a Type K or Type J thermocouple. Fill in the published millivolt values column for each of the 10°C temperature increments. This column must be completed for the lab plan.

The second table is for the non-ice point reference data. This table is similar to the first table but includes a "reference-temperature corrected" column based on the non-standard reference temperature. For the lab plan, calculate these expected values for the reference-temperature corrected column. Estimate a tepid water reference temperature and identify that temperature in each entry of the "reference temperature" column. Use Equation L8.1 to calculate the expected millivolt values by solving Equation L8.1 for $V_{T1\to T2}$. This is the expected measurement voltage based on the non-zero reference temperature. The other two terms in Equation L8.1 are taken from the published data.

Both tables with all calculations should be entered into a spreadsheet and ready for use during the lab. During the lab, confirm the thermocouple type and update the tables and spreadsheets, if necessary.

Thermocouple preparation. Two separate lengths of thermocouples wire are required, both approximately 12–18″ (30–46 cm) long. Thermocouple wire consists of two dissimilar metal wires, individually insulated, surrounded by an overall insulating layer. Within the overall insulating layer, there *may* be a third wire without insulation (bare) or aluminum shielding which will not be used and can be cut away, if present.

For both ends on both thermocouple wires, use a wire stripper to carefully remove approximately ½″ of the outer insulation layer to expose the individually insulated wires. Use correctly sized wire strippers to strip away about ¼″ of insulation from each of the two inner wires on both ends as shown in Figure L8.2.

For only one end on both thermocouple wires, use a pair of needle-nose pliers to firmly hold the wires together in place and another pair of pliers to tightly twist the bare wires together as shown in Figure L8.2. Do not overly twist the wires such that they break. The other end of both thermocouples should be stripped but not twisted, shaken or stirred.

Setup. Configure the equipment and thermocouple wires as shown in Figure L8.3. Securely twist together the red wires (negative) from the free ends of both thermocouple wires. Connect the yellow wire (Type K thermocouples) free ends to the multimeter (or millivoltmeter) with the cold reference ice point thermocouple wire connected to the negative terminal. If millivolt readings are increasingly negative with higher temperatures, the connections have been reversed. Use lab test stands as necessary to help hold the thermocouple wires and thermometers in a stable position to prevent tipping over beakers.

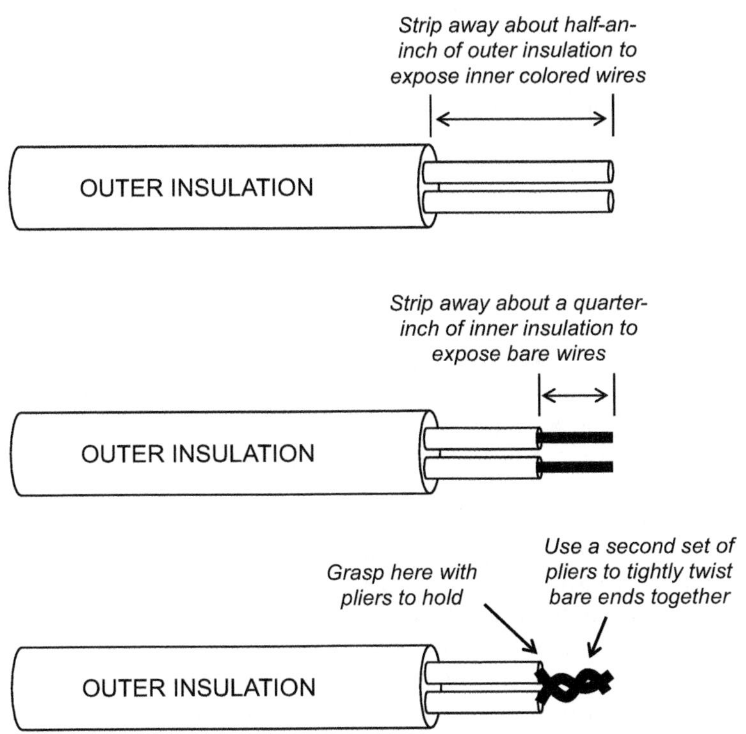

FIGURE L8.2 Thermocouple wire, properly prepared and twisted together.

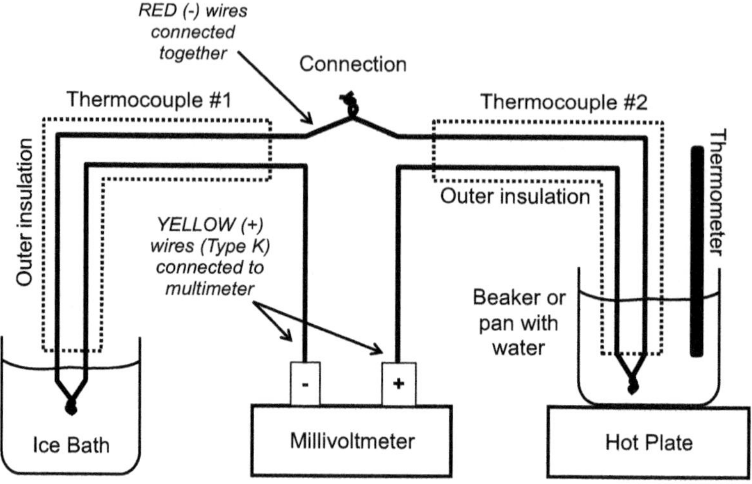

FIGURE L8.3 Electrical configuration. A test stand can be used to help steady the thermocouple wires in place. Do not allow the sensing junction of the thermocouples to touch the edges or bottoms of the beakers.

Fill a beaker with ice and water for the reference junction. Fill another beaker or pan with cool tap water to sit on the still turned-off hot plate. To ensure a more complete data range, use cool to cold tap water rather than warm water. Use a glass or electronic thermometer as a known temperature measuring device placed into the beaker of water on the hot plate. Ensure the thermometer and the thermocouples do not touch each other nor touch the sides or bottom of the beakers. Keep the thermocouples and thermometers suspended in the water or the results will have errors. Turn on the hot plate and begin recording data. As temperature of the hot plate water increases, record the millivolt values at the appropriate increments in Table L8.2. Cease recording when the water reaches a full rolling boil.

Repeat the test run with the tepid water reference temperature. Empty the ice bath beaker, fill with tepid tap water and place it back into the experimental setup with the thermocouple in the water. Use the thermometer to measure the water temperature and record this reference temperature in Table L8.3. Carefully empty the hot water from the hot plate using insulated gloves or tongs. Refill this beaker with cool tap water and replace back into the experimental setup with the thermocouple and thermometer in place in the water.

Realize if the *reference* temperature water is greater than the water sensed on the hot plate, the millivolt values will be negative until crossing the reference junction temperature threshold. For example, if the reference temperature water is 30°C and the measured temperature in the hot plate beaker is initially at 20°C, the millivolt values will be negative. When both temperatures are at 30°C, the millivolt values will be zero. *A temperature differential is necessary to drive the circuit.* This is also why the published thermocouple voltage tables indicate zero volts at a temperature of 0°C since the reference is set at 0°C. At temperatures below 0°C, the published values are negative. The thermocouple equation, Equations L8.1a and L8.1b provides an *offset* from the published table data for the non-zero reference temperature. Be sure to record all values, including any negative values.

Controlling Variables and Error. For the lab plan, think through the experimental method and consider potential sources of error. Identify at least one potential *source* of error and indicate the *type* of error (random or systematic). Provide a planned mitigation strategy (control, repetition, replication, randomization) to minimize the effect.

A well-conceived lab plan considers all opportunities for failure or error in the experiment and provides strategies to minimize the impact. For the final report, the results and conclusion section should then be used to discuss the effectiveness of the strategy and whether it can be improved.

RESULTS

Table L8.2 summarizes experimentally measured thermocouple values relative to published values for the standard 0°C ice point reference. For the lab report, Table L8.2 should be replaced with a chart to plot both published millivolt values and experimental millivolt values. Table L8.2 should be moved to an appendix as supporting data for the chart.

Table L8.3 summarizes experimentally measured thermocouple values relative to published values for the non-standard reference temperature. For the lab report,

TABLE L8.2
Thermocouple Data Using Ice Point Reference

		Thermocouple Type:		
Ref Temp	**Water Temp**	**Published Millivolt Value**	**Experimental Millivolt Value**	**% Error**
0°C	0°C	*(Look up)*	*(Lab Multimeter)*	*(Calculate)*
0°C	10°C	↓	↓	↓
...	...			
0°C	100°C			

TABLE L8.3
Thermocouple Data Using Non-Ice Point Reference

Thermocouple Type:

Ref Temp	**Water Temp**	**Published Millivolt Value**	**Ref Temp Corrected Millivolt**	**Experimental Millivolt Value**	**% Error**
	0°C	*(Look up)*	*(Equation L8.1)*	*(Lab Multimeter)*	*(Calculate)*
	10°C	↓	↓	↓	↓
	...				
	100°C				

Table L8.2 should also be replaced with a chart to plot both published millivolt values and experimental millivolt values as corrected by Equation L8.1. Table L8.3 should also be moved to an appendix as supporting data for the chart.

Consider the published millivolt values as truth values and compute the percentage error for each data point as shown in Tables L8.2 and L8.3. For the final report, include an overall average error with each chart plot.

DISCUSSION

Describe what is occurring with the data. Review sources and potential sources of error to include any mitigation strategies. Identify any observed data anomalies as possible errors and discuss possible reasons for anomalies. Discuss whether the controlling strategy identified in the lab plan was effective. Briefly discuss how a non-zero reference temperature can still be used for temperature measurements with Equation L8.1.

CONCLUSIONS

Based on the objective, cite the average percentage error for both cases and assess whether experimental results validate published values for both the standard and non-standard reference temperature. Perform some quick research and discuss how the ice point reference is created in electronic circuits.

LAB 9

Thermal Loading Simulation

Safety Requirements

None

Equipment Requirements

Lab computer station with computational fluid dynamics (CFD) heat transfer software capabilities

Procedural Requirements

A lab coat is required and safety glasses are optional at computer stations
Use lab session time to allow all team members to practice at the computer station
Each team member must create a solid model from start to finish
Only one team-completed full CFD analysis is required for lab session
Physically measure the diameter of Nichrome wire used in the next lab
The simulation should be run using correct gauge of Nichrome wire for the next lab
Without direct lab access, run the simulation using 32-gauge Nichrome wire
If the simulation does not match the actual wire gauge, it should be re-run

Clean-up Requirements

None

Emergency Actions

No unusual hazards (computer simulation only)

INTRODUCTION

This lab is an introduction to computational fluid dynamics (CFD) and heat transfer in direct support of the thermal loading lab in the next session. A length of Nichrome is virtually modeled for the thermal properties used in the next lab. Both labs will resolve the power-carrying capacity of Nichrome wire for free-air and forced-air convection.

The theoretical power capacity of wire is a two-variable and one-equation problem since the electrical current, *I*, and convective heat transfer coefficient, *h*, are unknowns. Although many CFD packages have the ability to perform a parametric study to automatically solve this two-variable problem, this lab requires a manually iterative process to resolve the simulation parameters.

For the iterative process, an initial estimate of power is entered into the simulation. The CFD simulation is then run, and the resultant temperature of the wire and convective heat transfer coefficient are obtained. If the temperature is not close to the melting point of Nichrome wire, a different value of power (higher or lower) must be entered into the simulation to achieve a closer result. This process is repeated until the approximate melting temperature of Nichrome wire has been simulated.

A more complete theoretical development of the thermal management problem is presented in the subsequent lab since only the CFD simulation is explored in this lab.

OBJECTIVE

Numerically determine the following values necessary to melt a section of Nichrome wire: power, free-air convective heat transfer coefficient and forced-air convective heat transfer coefficient.

THEORY

Equation L9.1, developed in the next lab, describes the energy balance of Nichrome wire subjected to Joule heating as current travels through the wire.

$$I^2R = mc\frac{dT}{dt} + \sigma\varepsilon A_S\left(T_{WIRE}{}^4 - T_\infty{}^4\right) + hA_S\left(T_{WIRE} - T_\infty\right) \tag{L9.1}$$

where: I = current

R = resistance of the wire

I^2R as a collective term = power

m = mass of the wire = (density)*(volume) of the wire = ρV

c = specific heat capacity of the wire material

$\frac{dT}{dt}$ = the time-rate-of-change temperature of the wire

σ = Stefan-Boltzmann constant = 5.67×10^{-8} W/m^{2}K^4

ε = emissivity of the wire

A_s = surface area of the wire

T_{WIRE} = surface temperature of the wire (in K)

T_∞ = room temperature = 20°C (convert to K)

h = convective heat transfer coefficient

h = 0.5–1000 W/m^{2}K for dry gasses, free and forced

For steady-state conditions, temperature no longer changes with time, so the term, $mc\frac{dT}{dt}$, of Equation L9.1 goes to zero. The LHS of Equation L9.1 is *power*, a function of current and resistance.

For clarity, the steady-state version of Equation L9.1 can be simplified into conceptual terms in Equation L9.2.

$$(\textit{Total power dissipated}) = (\textit{radiation heat transfer}) + (\textit{convective heat transfer}) \quad \text{(L9.2)}$$

Equation L9.2 states that, for steady-state conditions, the total power applied to the wire as a result of a current running through a resistance element (Joule's law for power, combined with Ohm's law, I^2R) will be dissipated by a combination of radiation and convection. As power is applied, the wire heats up, and the only two means modeled to dissipate the input power as equilibrium is achieved are radiation and convection. With even more applied power (electrical current), the wire reaches its melting point since radiation and convection heat transfer become insufficient to dissipate enough power and stabilize the wire at a lower temperature. However, this energy balance can be adjusted in the last term by using forced-air convection instead of free-air convection, thereby changing 'h'. A higher convective heat transfer coefficient is obtained by moving higher velocity air over the wire. This allows the wire to dissipate more heat in the convective heat transfer term of the equation, and compared to free-air convection, more power can be added on the left-hand side before the wire melts. Both scenarios are explored.

The convective heat transfer coefficient, h, has a very broad published range of 0.5–1000 W/m^{2}K. Free-air convection is often published in the range of 5–10 W/m^{2}K, but can also vary significantly above these values, depending on the temperature differentials. Forced-air convection values are at the higher end of the range.

For a given applied current, most of the parameters in Equation L9.1 are essentially known or fixed, except for the very broad-ranging value of the convective heat transfer coefficient, h. A CFD simulation is used to determine the convective heat transfer coefficient for free-air and forced-air convection and power at the 1400°C melting point of Nichrome wire.

COMPUTATIONAL METHOD

Create a table to record all information similar to Table L9.1 in the results section. Since both free-air and forced-air convection are considered, ensure the table includes parameters for both conditions as shown. Use a spreadsheet program to have *all formula calculations ready prior to the lab.*

For expected results in the lab plan, use 32-gauge Nichrome wire and look up the diameter of this gauge of wire. Plan a wire length of 5.0 cm. Enter these values into Table L9.1 and calculate the surface area of the wire, A_s, as a function of the wire length and diameter. Total wire resistance can be estimated from a value of 23 Ω/m. Nichrome wire has a published melting temperature of 1400°C. The emissivity of Nichrome wire is generally given in the range of 0.65 to 0.79. However, emissivity can increase with temperature, especially as the surface begins to visibly glow. Use an emissivity, ε, of approximately 0.95. Clearly indicate all values in Table L9.1 in the expected results section of the lab plan.

Calculations. Follow the notations associated with Table L9.1 as a guide to assist the table calculations. The goal of these calculations is to determine an upper and lower bound for total power dissipation, I^2R, to begin the CFD simulation. Begin with the free-air convection table and repeat the following process for the forced-air table.

The heat transfer coefficient, h, should be estimated for the *low* and *high bounds* of both free-air and forced-air convection conditions in Table L9.1. These nominal values are discussed in the theory section of the lab manual. It is also highly recommended to perform some additional research to determine a typical high-end range for free convection associated with high temperatures since the Nichrome wire will reach 1400°C and may exceed the anticipated bounds. Since the heat transfer coefficient has such a broad range, the concept is to establish a nominally bound range to calculate power. Enter these *expected* heat transfer coefficient upper and lower bound values into the table.

Use the steady-state version (previously discussed) of Equation L9.1. Calculate the lower and upper bound values for power, I^2R, in Table L9.1. Enter the associated value for current, I. Complete the calculations for both the free-air and forced-air convection tables. The forced-air values should be higher.

CFD simulation. Begin the CFD lab session by physically verifying the properties and dimensions of the Nichrome wire. If necessary, update all values in Table L9.1 with the Nichrome wire values obtained in the lab. Nichrome wire resistance is sometimes directly written on the spool of wire in units of Ω/L. If so, multiply this value by the wire length to determine the total resistance, R. Correct resistance values may also be available from the manufacturer's online datasheet. A benchtop multimeter or millivolt meter can also be used to measure resistance over a 5.0 cm length.

Run the CFD simulation for the free-air convection case, then adjust the computational volume and run the simulation again for the forced-air convection case. Refer to Appendix L9.A for the detailed CFD configuration and analysis process.

Create a Nichrome wire solid model in the software by extruding an appropriately sized circle to a length of 5.0 cm. Once the solid model is complete, set up a sufficiently sized computational domain as necessary, depending on the software requirements. Ensure sufficient volume in the vertical dimension for a numerical solution in the free-air case and likewise in the horizontal direction for the forced-air case. Refer to Figure L9.1 for suitable domain configurations. For mesh size, limit cells to around 50K to save computational effort. The simulation should be configured to complete within a few minutes to complete the entire experiment in a timely manner.

Free-air convection. Use *Air* as the simulation fluid with an initial velocity of zero. Begin the heat flow simulation by entering the *dissipated power* into the simulation with units of watts. Since it is uncertain how much power the Nichrome wire can dissipate at just the melting point, choose a value within the upper and lower bounds computed in the I^2R column (power) of Table L9.1. Given the high temperatures of the Nichrome wire, if the upper bound of 'h' was not properly considered, this initial guess for power will likely be under-estimated. To save time in the simulation, consider a power estimate (I^2R) using $I = 3A$.

When the first simulation run is complete, check if the Nichrome wire is near its melting point of 1400°C (1673 K). If the simulation returns a final temperature that is more than 50°C different from the melting temperature, choose a new value for current, I, and re-run the simulation until the final temperature is within 50°C.

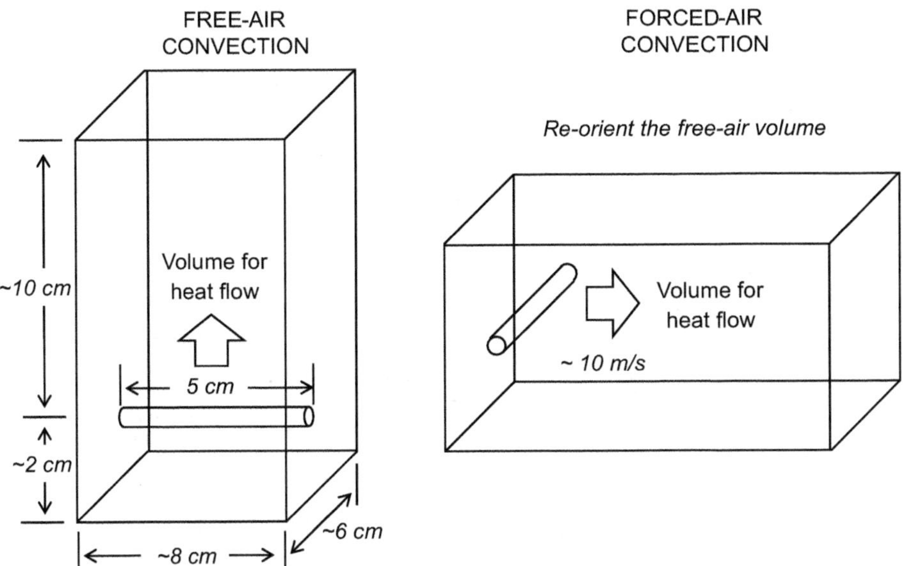

FIGURE L9.1 Example of CFD computational domains for both free-air and forced-air convection simulations.

Once the simulation returns a final maximum temperature close to the melting point of Nichrome, record the *average* heat transfer coefficient from the simulation result into Table L9.1 in the COMP (CFD) row. Since the model simulates exposed free ends of the wire rather than clamped ends in a physical experiment, the computational model has a higher *maximum* heat transfer coefficient due to air flowing around the ends of the wire. Enter the corresponding values for current, I, and power, I^2R, for the successful simulation run into Table L9.1 in the COMP (CFD) row.

Forced-air convection. Once the free-air convection simulation is complete, reconfigure the computational domain according to Figure L9.1 and re-mesh the domain. Change the input data to a horizontal velocity (x-direction) of 10 m/s. Use the higher dissipated power term within the upper and lower bounds from Table L9.1 for forced air. Run the simulation again as before and iterate until the maximum temperature is within 50°C of the Nichrome melt temperature. Record the average heat transfer coefficient from the simulation into the COMP (CFD) row of Table L9.1 along with the current (amperage), I, and power, I^2R.

RESULTS

Table L9.1 provides a summary of Nichrome wire parameters and the theoretical power capacity of the Nichrome wire. The computational results for power capacity of the Nichrome wire are also summarized in the table with the computationally determined heat transfer coefficients.

For the lab report, update Table L9.1 with the physically measured parameters of the Nichrome wire. Retain the same low and high values estimated for the heat

TABLE L9.1
Summarized Parameters, Heat Transfer Coefficients and Power Computations for Free-Air and Force-Air Cases

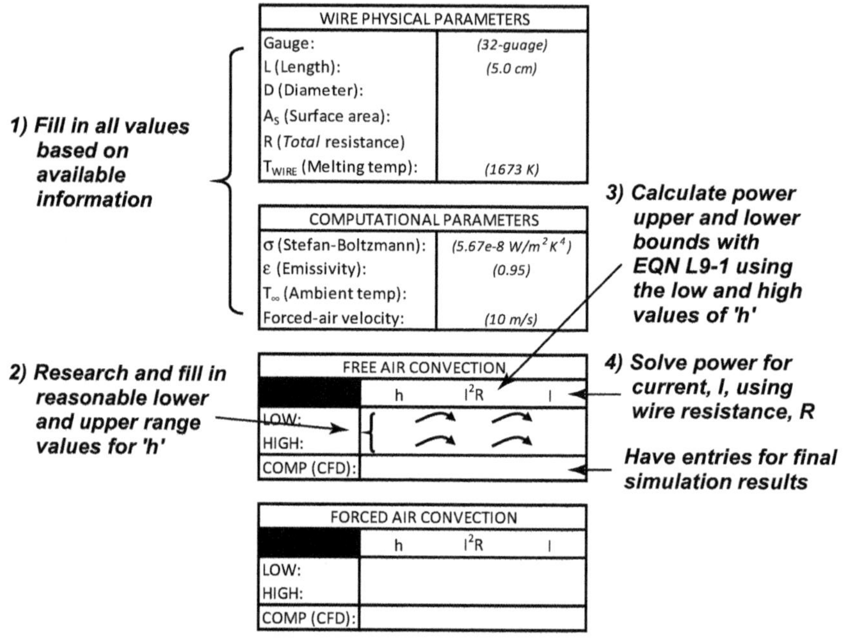

WIRE PHYSICAL PARAMETERS	
Gauge:	*(32-guage)*
L (Length):	*(5.0 cm)*
D (Diameter):	
A_S (Surface area):	
R (*Total* resistance)	
T_{WIRE} (Melting temp):	*(1673 K)*

COMPUTATIONAL PARAMETERS	
σ (Stefan-Boltzmann):	*(5.67e-8 W/m^2K^4)*
ε (Emissivity):	*(0.95)*
T_∞ (Ambient temp):	
Forced-air velocity:	*(10 m/s)*

FREE AIR CONVECTION			
	h	I^2R	I
LOW:			
HIGH:			
COMP (CFD):			

FORCED AIR CONVECTION			
	h	I^2R	I
LOW:			
HIGH:			
COMP (CFD):			

Notations are included to assist table completion.

transfer coefficient. Record the computationally determined heat transfer values in the table for both free-air and forced-air convection. Do not attempt to compute an error between the computational and theoretically estimated values.

Include at least one image plot for each of the temperature flow fields for both free-air and forced-air convections as figures in the lab report. Additional figures can be included to support any discussions of the CFD analysis in the appendix. Do not include more charts, tables or figures than necessary to succinctly present the results.

Clearly indicate in Table L9.1 the simulated power capacity of Nichrome wire for both free-air and forced-air convection along with the corresponding heat transfer coefficients. Clearly state these values in the discussion text as the objective of the experiment.

DISCUSSION

Discuss how many iterations were necessary to determine the maximum current capacity of the wire and the final temperature of the Nichrome wire. Identify whether the computed heat transfer coefficients fell within the expected bounds or outside the expected bounds. If the values fell outside the expected bounds, offer an explanation as to why, based on the parameters of the experiment. Consider the Nichrome wire's high temperature at 1400°C and small diameter and how these parameters might have been a cause.

There are a few likely sources of error in this experiment. The emissivity was estimated. The actual resistance at high temperatures was not considered. The boundary conditions of the wire were not properly simulated. Identify likely sources of error in the simulation and how they might impact results.

CONCLUSIONS

Based on the objective, clearly state the simulated power capacity of Nichrome wire for both free-air and forced-air convection. Offer a conclusion as to whether the computationally determined heat transfer values are reasonable based on published values. Assess whether these values are suitable for use in engineering calculations.

APPENDIX L9.A: CFD SIMULATION

Several CFD software solutions are available to complete the simulation. The software must offer a heat transfer module in conjunction with the basic CFD analysis. Specific steps and details vary between solutions. The instructions offered here provide a typical approach associated with professional software packages.

Solid model. Begin the simulation by creating a *New Part* for the solid model. Set *Units* to SI or MKS (meters-kilograms-seconds). Begin a new sketch by clicking on the *Sketch* tab to enter the 2D sketch mode for drawing 2D shapes. Click on the *Front Plane* to select a 2D drawing plane. Choose the circle tool at the top of the toolbar, click on the plane origin point and draw a 2D circle on the front plane.

Select the dimension tool and dimension the circle to the diameter of Nichrome wire by entering the value on the left panel. Use the diameter of Nichrome wire verified by measurement in the lab. Do not accidentally dimension the circle with a radius value (this is a common mistake). Commit the dimensions and exit the sketch mode by clicking *Exit Sketch*.

Select the *Features* tab to turn the 2D sketch into a 3D model. Click on the sketch circle and choose *Extrude Boss*. Choose *Midplane* for the direction option. Extrude the circle to the total length of the Nichrome wire. Save the solid model and set up the CFD simulation.

CFD configuration. Ensure the CFD module is available by selecting any *Add-ons*. Click on the CFD simulation tab to begin. If a *Wizard* is available for the setup, use the wizard to step through the configuration. Ensure units are SI. This analysis is *External* to the solid model, not *Internal*, so select the appropriate external option. Heat conduction is simulated for both *Heat conduction in solids* and *Radiation*, so these options should be selected. Do not select an entity calling for *Heat conduction in solids only*. If an option for gravity is included, ensure that gravity is enabled for the appropriate direction.

Choose *Air* as the fluid for both *Laminar* and *Turbulent* flow fields. Choose the solid model material as a *Nichrome* alloy. Set up the *Initial* and *Ambient* conditions.

The initial solid model temperature should be room temperature, approximately 295K. Leave velocities at zero for the initial free convection simulation. In the subsequent simulation, velocity must be adjusted for forced-air convection.

Complete the wizard as appropriate and set up the *Computational Domain* and *Edit Definitions* to size the domain to the flow area of interest. For a hot wire, a smaller zone on the bottom and sides is sufficient, compared to a larger volume above the wire where the heat will be rising. Too large of a computational domain will require more computational effort, while too small of a domain may result in inaccurate results due to insufficient model space. Size the domain according to Figure L9.1. Ensure the wire's solid model is *completely enclosed* in the computational domain by reorienting the view as necessary.

Select the simulation *Goals* so the simulation will know when to cease computational effort on the parameters of interest. Choose *Insert global goals* and choose the following: *velocity* (max), *heat transfer coefficient* (avg and max), *heat transfer rate, heat transfer rate* (convective) and *temperature of solid* (max). For example, by selecting the max temperature of the solid, the computational effort will be satisfied when the maximum temperature is no longer changing. This parameter is of interest since as soon as any portion of the wire reaches the highest temperature of 1400°C, it will melt and the goal is achieved.

Mesh. The computational domain needs to be meshed. A global mesh is usually sufficient to fill the entire domain. However, a local mesh can be added to refine detail around the wire model. As appropriate, right-click on *Global mesh* to *Edit Definitions*. Depending on the computational power of the computer, the mesh should be limited to around 50K cells to permit solutions within a few minutes. A *Manually* configured global mesh of the following divisions will ensure a reasonably sized mesh: $x = 50$, $y = 80$ and $z = 12$. Check the mesh configuration before attempting to run the complete simulation. Deselect the *Solve* option and *Run* the mesh to ensure the cells are not above 100K. If necessary, choose different x, y and z mesh parameters or resize the computational domain.

Power dissipation. Before running the computational simulation, the *power dissipated* in watts must be entered as *Input Data*. Choose the *Heat Source* property and select *Surface Source*. Ensure the outer surface of the wire solid model is selected as the heat source. An initial estimate for power should be used from the range calculated in Table L9.1. An initial estimate with $I = 3A$ is usually appropriate for the first computational run. Power dissipated is given in Equation L9.1 as I^2R.

Solve. Always ensure a *New Calculation* for each simulation and select *Run*. When the computation is complete, note the temperature of the Nichrome wire and the convective heat transfer coefficient. If the temperature is significantly below or above 1400°C (1673 K), increase or decrease the current, I, and enter a new value for dissipated power. Re-run the simulation until finding a current (amperage) resulting in a reasonably close melting temperature of Nichrome wire. Ensure the resulting temperature is within 50°C. Record the power dissipation term used for the successful simulation run into Table L9.1 on the CFD row. Enter the result, I, (amperage) used in the power calculation into Table L9.1 on the CFD row. Record the free-air convective heat transfer coefficient from the simulation into Table L9.1 on the CFD row. These values will be used in the next lab. Save a cut plot image of the simulation to include in the lab report.

For forced-air convection, go back into the *Input Data* option for *General Settings*. Select *Initial and Ambient Conditions and* change the velocity to approximately 10 m/s in the horizontal direction (x-direction). Ensure the computational domain is reconfigured as shown in Figure L9.1. Once the computational domain is reconfigured, the domain must be re-meshed. Delete any local meshes and reconfigure the global mesh to: $x = 80$, $y = 50$ and $z = 12$. Re-run the mesh. Select a higher power dissipation term from the force-air bounds in Table L9.1 and run the simulation iterations as before. Once the simulation solves to a maximum temperature within 50°C of the Nichrome melting point as before, record the final heat transfer coefficient value from the simulation. Enter this value into Table L9.1 on the CFD row. Enter the corresponding current, I, and power, I^2R, into Table L9.1 for the successful CFD run. Save an image of the simulation for the lab report.

LAB 10
Thermal Loading

Safety Requirements

Ensure the power supply is *off* before touching any uninsulated wires or components
Verbally confirm "*power supply off*" before touching bare wires or terminals
Use only insulated tools or gloves to remove hot wires after the experiment has run
Use a heavy insulated glove to actuate and hold the toggle switch on during the experiment
Turn air flow away from hands and personnel during forced-air convection test runs
Keep all personnel, items and loose papers away from the test fixture while running experiment
Safety glasses, gloves and lab coats must be worn at all times during the lab

Equipment Requirements

Custom *bridgewire* test fixture, high-amperage power supply and benchtop multimeter
Duct fan with optional nozzle constriction to increase air velocity or carpet fan
Nichrome wire, wire cutters, calipers and a handheld anemometer

Procedural Requirements

None

Clean-up Requirements

Ensure all melt wires have cooled to near ambient conditions (2–3 min) before discarding
Discard all cooled melt wire fragments
Unplug power supplies and return all unused wire and tools to their proper locations

Emergency Actions

BURNS: Run cold water on the affected area. Call 911.
FIRES: Locate the nearest fire extinguisher. If necessary, activate the fire alarm.

INTRODUCTION

Thermal management is controlling energy, and therefore temperature, of a system by an appropriate method of heat transfer. Systems generally have the ability to both store and transfer heat to some degree. Heat transfer into or out of a system occurs by one or more of the basic mechanisms of conduction, convection or radiation. When energy is transferred into a system or energy is created within the system, the energy level of the system increases. If the energy within the system exceeds its functional capacity to manage heat, thermal failure can occur. Thermal failure modes include excessive oxidation, fracture, melting or fire.

Electrical wire can melt if excess heat builds up due to internal Joule heating while conducting current. Conductive wires have an internal resistance to current based on the resistivity of the material, the cross-sectional area and the length of the wire. Power is required to overcome this electric resistance and drive a circuit as designed. If the wire cannot manage the applied power, it can thermally fail.

An engineer can manage thermal capacity of a wire by increasing its cross-sectional area, using a lower resistivity material or increasing rate of heat transfer. However, thermal management options may be limited if a system is already in place. In this experiment, the thermal loading of a Nichrome wire is managed by increasing the convective heat transfer coefficient. Numerically determined heat transfer coefficients are used for theoretical calculations in this lab. The free-air convection case is the control condition, and the forced-air convection case is the experimental test parameter case.

OBJECTIVE

Compare the power-carrying capability of a Nichrome wire under forced-air conditions versus free-air conditions both experimentally and theoretically. Assess the merits of thermal management by forced-air convection compared to free-air convection.

THEORY

Ohm's law describes the relationship between voltage, V, resistance, R, and current, I, within an electrical circuit or component. For a simplified, one-dimensional case, this relationship is given in Equation L10.1.

$$V = IR \tag{L10.1}$$

All conductive wires have internal resistance to conducting electricity and generally follow Ohm's law. A wire's resistance is a function of cross-sectional area, A, length, L, and the material's inherent property of resistivity, ρ, as given by Equation L10.2.

$$R = \frac{\rho L}{A} \tag{L10.2}$$

The resistivity of Nichrome is published in the range of $\rho = 100e{-}8\ \Omega m$ to $150e{-}8\ \Omega m$. If manufacturer data is not available for Nichrome wire, Equation L10.2 can be used to estimate the total resistance of a length of wire for a given gauge size (diameter).

Energy is required to overcome a wire's internal resistance to conduct electricity. To continuously conduct electricity, energy must be supplied at a constant rate. The rate of energy supply is power, P. For an electrically conductive component, power is a function of both current and resistance, given in Equation L10.3.

$$P = I^2 R \tag{L10.3}$$

Equation L10.3 can be understood in a few different ways. First, Equation L10.3 states how much power is required to drive a current for a given resistance. It also states how much power the wire or device must be able to manage for a given current and resistance. Finally, Equation L10.3 is also a measure of heat generated by the wire since power is an energy rate and heat is a form of energy. Equation L10.3 is also referred to as Joule's law for power, and heat generated by electrical current is referred to as Joule heating.

Substituting Ohm's law from Equation L10.1 into Equation L10.3 provides another useful power formula as a function of voltage and current in Equation L10.4.

$$P = IV \tag{L10.4}$$

Since all real-world systems operate over a period of time, heat energy is generated and dissipated as a rate, quantified as energy per unit time. Heat as a mathematical quantity, Q, is therefore managed in rate form, $\dot{Q}$ (heat, as a form of energy, per time) and power, P (energy per time).

Heat transfer model. To create a heat transfer and therefore a heat management model, the first law of thermodynamics requires accounting for all forms of energy flowing into and out of a system. Most models are simplifications of real-world problems, so the model may only consider the most significant forms of energy accounting relevant to the problem. Less significant forms of energy transfer are essentially mixed in with the more relevant terms without any undue effect. If a non-considered form of energy appears to have an effect, the model is inadequate, and that form of energy must be broken out into its own term.

For any given length of wire, L, with diameter, D, as shown in Figure L10.1, current flowing within the wire generates heat at a rate of $\dot{Q}_{GEN}$, according to Equation L10.3. This generated heat builds up within the wire at a rate of $\dot{Q}_{STORED}$ as the wire gets hotter. Heat also both convects away into the air as $\dot{Q}_{CONVECT}$ and begins to significantly radiate into space as $\dot{Q}_{RADIATE}$ once the wire begins to glow red. These are the most significant energy input, storage and transfer terms modeled in Figure L10.1.

The energy model in Figure L10.1 is mathematically balanced in Equation L10.5. The energy generation term is a direct function of external power put into the system, so this term appears on the LHS of Equation L10.5. All other terms on the RHS of Equation L10.5 signify where all of this input energy goes in its various forms.

$$\dot{Q}_{GEN} = \dot{Q}_{STORED} + \dot{Q}_{CONVECT} + \dot{Q}_{RADIATE} \tag{L10.5}$$

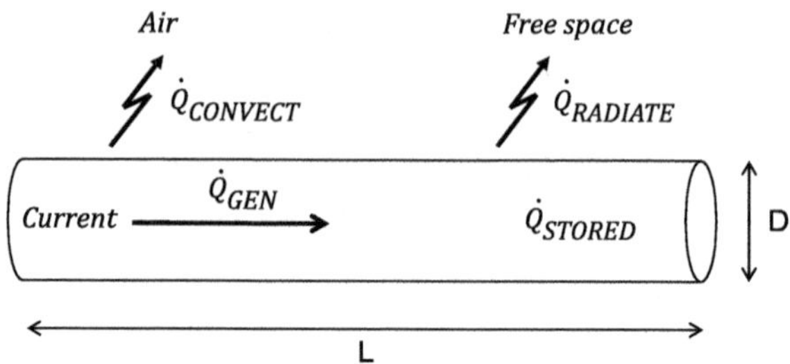

FIGURE L10.1 Energy conservation model of Joule heating in a conductive wire.

where: $\dot{Q}_{GEN}$ is the total energy input (rate) within the wire
$\dot{Q}_{STORED}$ is the energy (rate) retained by the wire causing it to get hot
$\dot{Q}_{CONVECT}$ is the energy (rate) lost by warming the surrounding air
$\dot{Q}_{RADIATE}$ is the energy (rate) lost by electromagnetic radiation into space

In reality, no wire is just floating in space, radiating and convecting. The wire is also attached to posts or other components and therefore *conducts* heat into these attachments. For long, thin wires, heat transfer via conduction is considered negligent compared to the other heat transfer terms in Equation L10.5. If the wire was short and relatively thick with rather conductive attachments, the heat transfer model should be revisited to include the conduction term.

To manage the terms in Equation L10.5 at a more fundamental level, each term must be explored in more detail with available mathematical relationships.

$\dot{Q}_{GEN}$. Heat generated within the wire is exactly the power put into the wire as previously discussed with Equation L10.3. If no power is applied to the wire, no energy is created and no heat transfer occurs. Equation L10.6 is therefore a re-statement of Equation L10.3 in terms of heat generation rate.

$$\dot{Q}_{GEN} = I^2 R \tag{L10.6}$$

$\dot{Q}_{STORED}$. Any material substance can absorb and store heat energy to a certain extent based on the material's specific heat capacity constant, *c*. Water has a relatively high specific heat constant of about 4.2 J/g°C, so it can retain a fair amount of heat. Metals have a lower specific heat capacity than water, and Nichrome has only about one-tenth the capacity of water. The ability to store heat within a substance is also a function of the total mass, *m*, of the material, since more material can store more heat.

Both specific heat, *c*, and mass, *m*, are static internal energy storage capacity terms. As more heat energy is stored within a substance, the temperature, *T*, increases. Therefore, internal heat energy is proportional to temperature. In rate form, change in internal energy storage rate, $\dot{Q}_{STORED}$, is proportional to change in temperature per time, $\frac{dT}{dt}$.

These internal energy terms fully describe the energy storage of a substance and combine into the basic thermodynamic relationship in Equation L10.7.

$$\dot{Q}_{STORED} = mc\frac{dT}{dt} \tag{L10.7}$$

$\dot{Q}_{CONVECT}$. Convective heat transfer follows Newton's law of cooling where the convective heat transfer rate is a function of temperature difference between a substance, (T_{WIRE}, in this case) and its surroundings (T_∞). A hot object placed in an equally hot environment will not transfer heat energy. A hot object placed in a cold environment will transfer heat energy. A hot object placed in cold water will transfer heat quicker than in equally cold air since the convective heat transfer coefficient for water is higher than for air. A larger surface area, A, will also facilitate greater heat conduction. All of these terms combine into Newton's law of cooling for a Nichrome wire in Equation L10.8.

$$\dot{Q}_{CONVECT} = h\ A\left(T_{wire} - T_\infty\right) \tag{L10.8}$$

The convective heat transfer coefficient, h, depends on the complexity of interactions between the heated surface and the surrounding fluid (liquid or gas). The heat transfer coefficient values published for air generally range from 0.5 to 1000 W/(m²K).

$\dot{Q}_{RADIATE}$. All objects above absolute zero emit radiant energy, most often in the form of infrared heat wavelengths. This radiant heat transfer rate is proportional to the fourth power of an object's surface temperature, T^4, where the constant of proportionality is the Stefan-Boltzmann constant ($\sigma = 5.67e{-8}$ W/m²K⁴). The ideal standard for a radiant heat emitter is a perfect *blackbody*. Emissivity, ε, is the correction factor for real objects that are less like a blackbody emitter. Shiny metallic surfaces are more reflective than emissive and have emissivities closer to zero. Objects with larger surface areas, A_S, also have a greater ability to transfer radiant heat. These terms combine to form the radiant heat transfer term in Equation L10.9.

$$\dot{Q}_{RADIATE} = \varepsilon\ \sigma\ A\left(T_{wire}^4 - T_\infty^4\right) \tag{L10.9}$$

Emissivity of Nichrome wire is generally given in the range of $\varepsilon = 0.65$ to $\varepsilon = 0.85$. Since emissivity changes with temperature and the surface of the Nichrome is visibly glowing prior to melting, an emissivity of $\varepsilon = 0.95$ is more suitable.

Energy balance. Each of these individual heat transfer terms can now be entered into the energy balance using Equation L10.5. The result is Equation L10.10, which fully models the energy balance for a Nichrome wire with sufficient detail to manage specific aspects of the heat transfer.

$$I^2R = mc\frac{dT}{dt} + \varepsilon\sigma A_S\left(T_{WIRE}{}^4 - T_\infty{}^4\right) + hA_S\left(T_{WIRE} - T_\infty\right) \tag{L10.10}$$

where: I = current

R = resistance of the wire

m = mass of the wire = (density)*(volume) of the wire = ρV

c = specific heat capacity of the wire material
$\frac{dT}{dt}$ = the time-rate-of-change temperature of the wire
ε = emissivity of the wire
σ = Stefan-Boltzmann constant = 5.67×10^{-8} W/m^{2}K^4
A_s = surface area of the wire
T_{WIRE} = surface temperature of the wire (in K)
T_∞ = room temperature = 20°C (convert to K)
h = convective heat transfer coefficient
h = 0.5–1000 W/m^{2}K for dry gasses, free and forced

For steady-state conditions, temperature does not change with time, so the $\frac{dT}{dt}$ term in Equation L10.10 goes to zero. For this experiment, power is applied slowly near the melting temperature of the Nichrome wire, such that the final temperature change can be considered insignificant and relatively steady state.

EXPERIMENTAL METHOD

A custom bridgewire test fixture is required for this experiment. Details of this simple fixture to melt Nichrome wire are included in Appendix L10.B. The duct fan used in this experiment can be improved by adding a nozzle to increase air velocity as described in Appendix L10.C. Alternatively, a squirrel cage fan such as a carpet fan provides a higher velocity airflow.

Create a table or set of tables similar to Table L10.1 in the results section to summarize each of the parameters in Equation L10.10, theoretical calculations and results of the experiment. Use a spreadsheet program with these tables for all formula calculations entered and ready prior to the lab. Complete all theoretical calculations prior to entering the lab. Although not required, a table similar to Table L10.A2 in Appendix L10.A should be created to record voltage and amperage values during each run. This table will facilitate monitoring values and changing Nichrome wire resistance with temperature.

Begin by filling in all available parameters in Table L10.1. Use Equation L10.2 to calculate wire resistance or use a manufacturer value. For the theory sub-table, enter the *numerically determined* (CFD simulated) heat transfer coefficients, h, for both free-air and forced-air convection from the *previous simulation lab.* If these values are not available, use the suggested values in parentheses. Use Equation L10.10 to solve for the theoretical power-carrying capacity of the Nichrome wire, I^2R. Ensure Equation L10.10 is modified to reflect steady-state conditions. Once power is calculated, use the wire's resistance to solve power for current, I. The theory sub-table should now be completely filled in. Transfer these values into the appropriate cells of the free-air and forced-air convection error sub-tables in Table L10.1 for later comparison.

Create two more tables with a spreadsheet similar to the example in Appendix L10.A. Label these tables as "Table L10.A1: Free-air convection data" and "Table L10.A2: Forced-air convection data". (If also using Table L10.A2 for data points during each run, rename this Table L10.A3.) Tables L10.A1 for free-air and L10.A2 for forced-air are for recording run data and creating histograms of the data. Since the experiment will consist of several data runs, only the average values from these appendix tables will be entered into the experimental value cells in Table L10.1.

Setup and procedures. Configure the test setup as shown in Figure L10.2. Arrange all equipment and materials so there is nothing downwind of or close to the Nichrome wire test section. Connect the bridgewire test fixture wires to a high-amperage power supply capable of supplying at least 10 amps. Do not use a low-amperage power supply. The polarity of the wires does not matter. The power supply should be easily accessible and away from the bridgewire test section to adjust amperage during the experimental run.

Cut 20 lengths of Nichrome wire, each approximately 8–10 cm long. This length should be just long enough to fit into the 5 cm bridgewire test section, as shown in Figure L10.3. Ten wires are for free-air convection and ten wires for forced-air convection. Verify the resistance of any section of Nichrome wire with a benchtop multimeter by placing the probes on the wire spaced 5 cm apart. Update the Table L10.1 resistance value as necessary.

Prior to every test run. Ensure the power supply is turned off before placing the Nichrome wires in the test fixture. Ensure the power supply amperage (current) knob is **fully turned down** (counter-clockwise) to prevent burning up the wire upon initial application of power. Safety glasses and lab coats are an absolute requirement for all personnel. Nichrome wire can melt and fly off in an unpredictable direction.

After every test run. Ensure the power supply is turned off before attempting to remove Nichrome wire fragments. Use insulated gloves and allow Nichrome fragments to cool for about a minute prior to removing them from the test fixture.

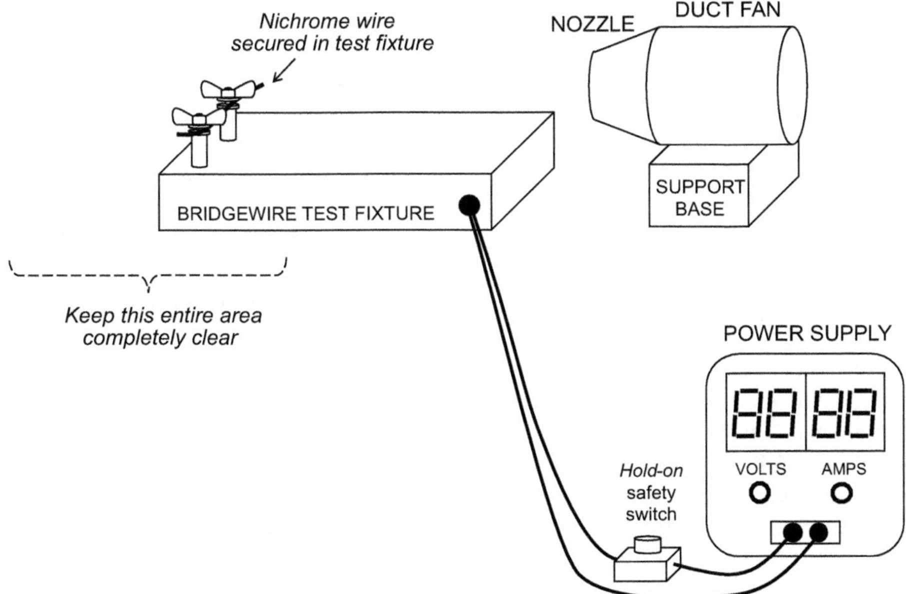

FIGURE L10.2 Experimental test setup. Keep the area near and downwind of the Nichrome wire clear. Keep the power supply accessible to make adjustments during the experiment.

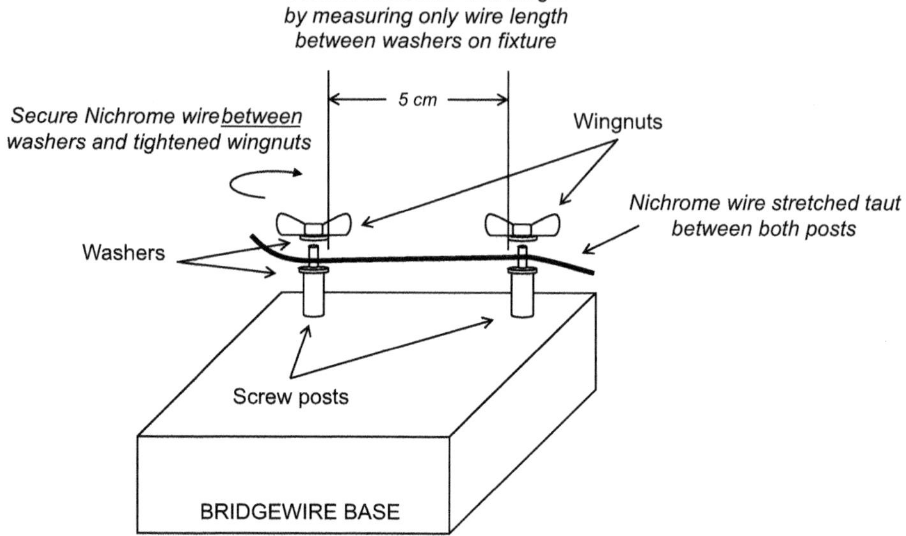

FIGURE L10.3 Securing Nichrome wire into bridgewire fixture.

With the power supply off, secure the first Nichrome wire test specimen into the test fixture as shown in Figure L10.3. Straighten the wire between the two posts and ensure the wire is *between* the washers as shown in the figure. Tighten one of the post wingnuts to hold the wire in place. Pull the wire taut to the other post and tighten the wire into place. Use calipers to measure the length of Nichrome wire only between the washers in the test fixture. Update Table L10.1 accordingly.

Free-air convection. Keep the fan turned off for this series of runs. With the Nichrome test specimen secured in the fixture, ensure the current adjustment knob is turned fully down and turn on the power supply. Have one teammate press and hold the hold-on safety switch to complete the electrical circuit to the test fixture while the test is in progress. Have another teammate ready to record data into Table L10.A1 (and Table L10.A2, if desired) in Appendix L10.A. *Slowly* begin to turn up the amperage. As amperage increases, the Nichrome wire will begin to glow hot. Continue slowly increasing the amperage while paying close attention to the supplied current and voltage. When the melting point is reached, the wire will fail in an instant and the voltage and current will immediately drop back to zero. Note the final peak current and voltage values in Table L10.A1 just as the wire melts. Turn off the power supply and follow the procedures for another run.

Sometimes the Nichrome wire will melt unexpectedly without actually increasing the amperage as resistance increases due to temperature. If amperage is not correctly recorded or otherwise missed, consider this as a failed data point and discard that test run. The failed run can be re-accomplished if time and supplies are available. Otherwise, complete the experiment with a reduced data set.

Forced-air convection. For this set of test runs, the fan is turned on and air velocity is recorded with a handheld anemometer. Follow procedures to load a Nichrome wire into the test section. Turn on the duct fan and hold the anemometer *at the same location* as the Nichrome wire to record air velocity at the Nichrome wire. Adjust the fan power or position to register about 10 m/s on the anemometer. This air velocity should be approximately the same as the velocity simulated in the previous simulation lab.

Keep the fan running and remove the anemometer. Ensure the area is clear and begin the test run as with the free-air convection. Slowly increase the amperage until the wire melts and record amperage and voltage at the peak value just prior to melting. Follow procedures and complete the test runs for each test specimen.

RESULTS

Table L10.1 summarizes the Nichrome wire parameters, theoretical calculations and experimental results.

Analyze the data for both free-air and forced-air convection runs according to the example in Appendix L10.A. Use Equation L10.4 to calculate power from recorded voltage and amperage. Calculate resistance values using Equation L10.1. Determine the mean value and standard deviation of the power capacity of the wire. Create a histogram plot for only the power data. Use any desired program to create the histogram. A spreadsheet example is provided in Appendix L10.A using a *smoothed curves* plot of data.

Use the mean current values and mean power values from Tables L10.A1 and L10.A2 to fill in values for the experimental sub-table in Table L10.1. Since every parameter in Equation L10.10 is explicitly known, use Equation L10.10 to calculate the actual heat transfer coefficient, h, and enter this value into the experimental sub-table in Table L10.1.

When all of the experimental and theoretical values have been entered into Table L10.1, compute the percentage change and percentage error as indicated by the table.

DISCUSSION

Identify any anomalous data points in the data sets. Discuss any potential sources of error encountered during the experiment. Discuss any challenges with recording values. Discuss the distribution of data in the histogram and whether the data appears normally distributed. Discuss whether additional data points should be considered based on the histogram. Discuss the change in heat transfer coefficients between the two scenarios as the specific mechanism for managing thermal loads in this case. Discuss how close the experimentally determined heat transfer coefficient compares with the simulated heat transfer coefficient.

CONCLUSIONS

Based on the objective, clearly state the *power capacity* percentage improvement with forced-air convection compared to free-air convection. Assess whether forced-air convection is an effective strategy to help manage thermal loads. Review Equation L10.10 and discuss other potential means for managing thermal loads.

TABLE L10.1
Summarized Parameters, Heat Transfer Coefficients and Power Computations for Free-Air and Force-Air Cases

WIRE PHYSICAL PARAMETERS	
Gauge:	*(32-guage)*
L (Length):	*(5.0 cm)*
D (Diameter):	
A_S (Surface area):	
R (*Total* resistance):	
T_{WIRE} (Melting temp):	*(1673 K)*

COMPUTATIONAL PARAMETERS	
σ (Stefan-Boltzmann):	*(5.67e-8 W/m^2K^4)*
ε (Emissivity):	*(0.95)*
T_∞ (Ambient temp):	
Forced-air velocity:	*(10 m/s)*

THEORY	h	I^2R	I
Free-air convection:	*(200)*		
Forced-air convection:	*(800)*		
% Change:			

EXPERIMENTAL	I	I^2R	h
Free-air convection:			
Forced-air convection:			
% Change:			

FREE-AIR CONVECTION ERROR	h	I^2R	I
Experimental:			
Theory:			
% Error:			

FORCED-AIR CONVECTION ERROR	h	I^2R	I
Experimental:			
Theory:			
% Error:			

APPENDIX L10.A: DATA COLLECTION AND ANALYSIS TABLES

Table L10.A1 is an example for analyzing the data runs into a histogram. Table L10.A2 is recommended to record voltage and amperage values as voltage is increased during each run. While the table is not required for the lab, it can facilitate monitoring values and change in resistance as the wire heats up. The number of data points is not specified. The table indicates five points for each run, but the spreadsheet should accommodate ten or more entry rows per run for convenience.

TABLE L10.A1
Example Data Collection and Analysis Table. A Table Is Necessary for Both Free-Air and Forced-Air Convection Cases

DATA			CALCULATE		
RUN	VOLTS	AMPS	RESISTANCE	POWER	
1		5.5			
2		6.2			
3		5.7			
4		5.9			
5		6.3			
6		5.9			
7		5.8			
8		5.7			
9		6			
10		6.1			
MEAN		5.91			=AVERAGE(E2:E11)
STD DEV		0.25			=STDEV(E2:E11)

POWER DATA		
BINS	FREQ	
5	0	=COUNTIFS($C$3:$C$12,">"&$H2,$C$3:$C$12,"<="&$H3)
5.1	0	
5.2	0	
5.3	0	
5.4	0	
5.5	1	
5.6	0	
5.7	2	
5.8	1	
5.9	2	
6	1	
6.1	1	
6.2	1	
6.3	1	
6.4	0	
6.5	0	
6.6	0	
6.7	0	
6.8	0	
6.9	0	
7	0	

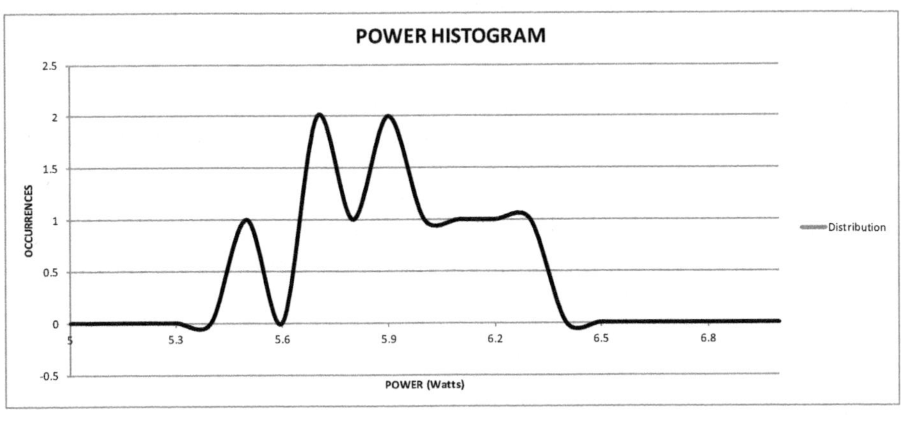

TABLE L10.A2
Suggested Data Point Collection Table

	Free-Air Convection				Forced-Air Convection			
	V	I	$R=V/I$	$P=IV$	V	I	$R=V/I$	$P=IV$
RUN #1:								
RUN #2:								
↓								
RUN #10:								

APPENDIX L10.B: BRIDGEWIRE TEST FIXTURE

The bridgewire test fixture is simply constructed from any wood, metal or plastic box, drawer or container turned upside down. Dimensions are not critical since any size is suitable. Refer to Figure L10.B1 for the basic configuration.

Space two drill holes for bolts approximately 5 cm apart as the innermost dimension according to the figures. Drill the bolt holes for the Nichrome wire posts. Any standard ¼″ or 3/8″ bolts about 2″ long are suitable for the posts to provide working clearance from the base. An aluminum tube extension collar spacer should be cut to length as shown to raise the working area off the base. Use a locking nut such as a nylon stop nut to secure the bolts and extension tubes in place. Use four large fender washers for each post assembly, as shown in the figure. The stop nut secures the assembly, while the final wing nuts hold the Nichrome wires for testing.

Prior to securing the post bolt assemblies in place, cut a minimum of 3″ of standard 16/2 wire for the electrical connections. Strip the insulation off all ends of the wire. Wrap the bare wires around the bolt heads prior to the washers, as shown in the figure. Assemble the posts into place and securely tighten the top nut. The wing nuts should remain only be lightly tightened in place. Drill a hole in the side of the base to run the wires out to the power supply.

Cut one of one of the 16/2 wires near the free end, close to the power supply attachment point. Insert an SPST momentary-on button rated for 120V. Use a small enclosure, shrink tubing or hot glue to insulate the connections. This safety switch could also be mounted directly on the bridgewire test fixture base, but this has proven diffficult due to the proximity to the hot Nichrome wire and the need to use two hands to both steady the fixture and hold the switch.

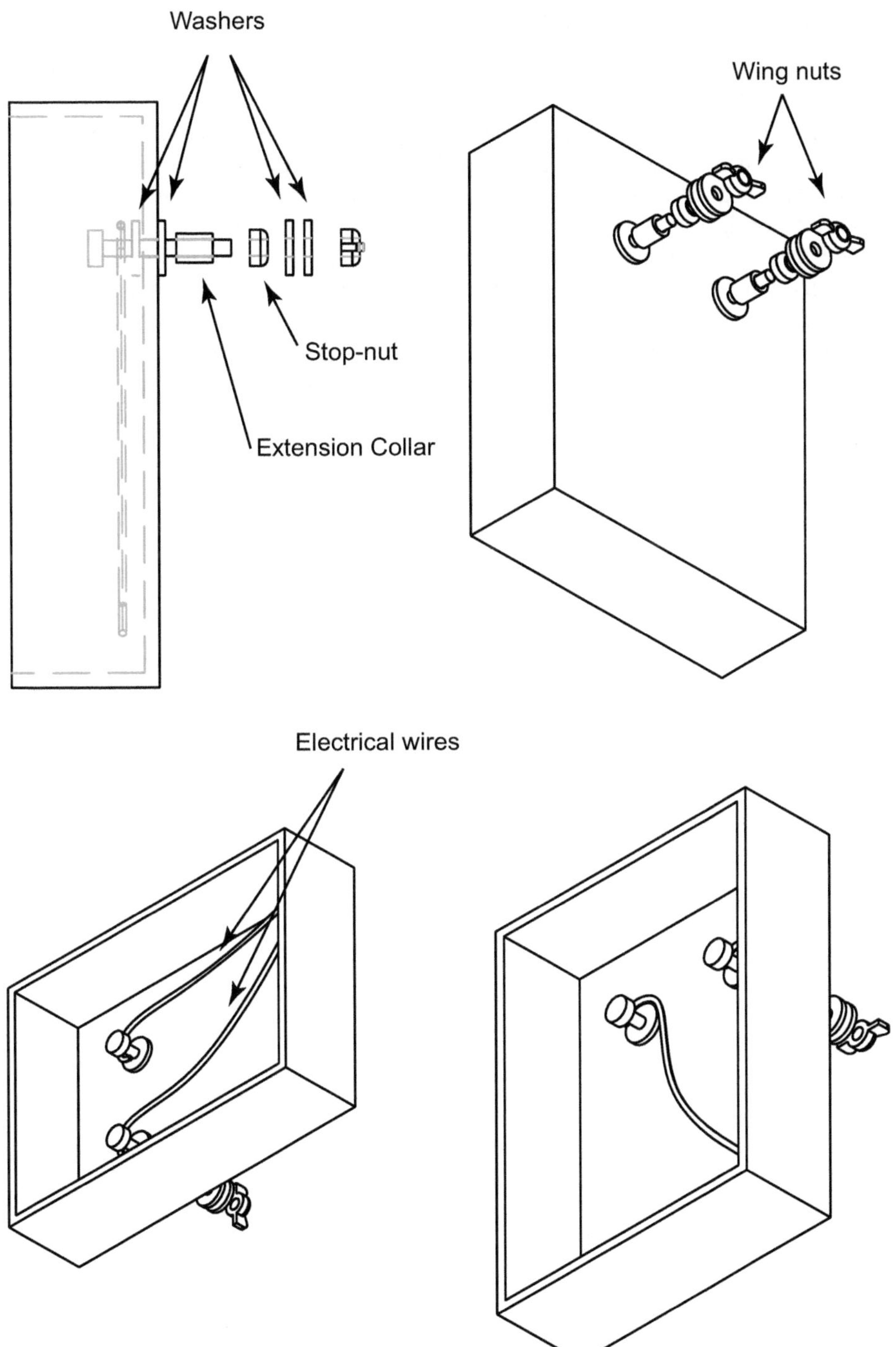

FIGURE L10.B1 Bridgewire test fixture design features.

APPENDIX L10.C: ADDITIONAL NOTES

A large duct fan with variable speed control is most suitable to provide a forced air source. A slightly constricting nozzle is required to increase air velocity. This is easily achieved with any thin-gauge metal duct pipe round reducer. Any reducer section sized for the duct fan can be taped to the duct fan with HVAC tape. A squirrel cage "carpet fan" is also suitable if the fan can be situated to create an unobstructed flow toward the bridgewire.

Voltages and currents associated with this lab are well within safe levels. The Nichrome wire is a burn hazard but easily managed. Nichrome wire should be limited to a maximum of 30 gauge. Thicker gauges do not melt completely and tend to fragment away from the test fixture in unpredictable directions. Gauge sizes of 32 AWG to 36 AWG are best for this experiment, providing safe and predictable behavior.

LAB 11

Fins

Safety Requirements

Do not touch the heating elements or surfaces of conductive metals while hot
Do not allow wires or objects to touch hot metal surfaces
If necessary to handle hot surfaces, use insulated gloves or an appropriate tool
Terminate heating at least 10 minutes prior to the end of the lab to allow for cooling
Do not run cables that obstruct work areas, causing hazard to people or equipment

Equipment Requirements

Fin experiment apparatus, power supplies (2) and handheld thermocouple-based thermometer
Thermocouples, data acquisition system and computer

Procedural Requirements

Do not over-tighten the aluminum screws that secure thermocouples to fins because they can easily shear off in the fixture
Do not completely remove screws; they are easy to lose
Do not exceed ~12 V on the power supply

Clean-up Requirements

Unplug all lab equipment and power supplies
Disconnect and stow DAQ system as appropriate

Emergency Actions

MINOR BURNS: Run cold water on the affected area. Seek medical attention if required

INTRODUCTION

Thermal *conduction* is the movement of internal energy within an object or between two objects in physical contact with each other. The Second Law of Thermodynamics states that heat energy moves from a hotter region of an object to a colder region. This flow of heat (energy) via conduction can be measured as a temperature gradient within the object.

Thermal *convection* is the transfer of energy between an object and a surrounding fluid, such as water or air. Whereas conduction transfers heat through an object, convection transfers heat through the movement of a fluid adjacent to the object. This convective heat transfer can be increased by either increasing the temperature difference between the object and its environment or by increasing the surface area of the object.

Fins are structures attached to the surface of an object to both pull heat away from the object via conduction and then dissipate the heat into the surrounding atmosphere via convection. This lab compares the performance of two different fin configurations, both with the same approximate volume.

The first fin has a circular cross section with a larger volume-to-surface ratio. This fin is more *efficient* since it tends to retain heat along its length and increases convection to the surrounding air by an improved temperature differential. The second fin has a rectangular cross section with a greater surface-to-volume ratio. This fin is more *effective* since it tends to increase convection in the surrounding air due to its increased surface area. Both fins are compared, where all parameters are the same except surface area. The cylindrical fin is the control case where perimeter, and therefore surface area, is increased with the rectangular fin.

OBJECTIVE

Use numerical simulation to determine a suitable heat transfer coefficient for fins. Verify fin temperature distributions relative to theory. Assess the cooling capability of a rectangular fin relative to a cylindrical fin.

THEORY

Power. The primary purpose of a fin is to move heat away from a heat-generating device, such as an electronic component or internal combustion engine. Power is a measure of heat energy the fin must continually manage. For electrical devices, the power consumed and dissipated by a device can be calculated from Joule's power equation and Ohm's law, with two useful forms in Equations L11.1a and L11.1b.

$$P = I^2 R \tag{L11.1a}$$

$$P = IV \tag{L11.1b}$$

where: P is the power input, used, produced or dissipated
I is the electrical current in amps
R is the electrical resistance
V is the electrical voltage

Heat distribution within a fin. A fin dissipates power by pulling heat energy from a source and conducting it down the length of the fin. Along the length of the fin, this energy is also convected into the surrounding air. The temperature distribution along the length of the fin changes since energy is continuously lost along its length. Knowing this temperature distribution is helpful for correctly sizing a fin. Excess fin length that no longer contributes to heat dissipation is wasted material and unnecessary bulk.

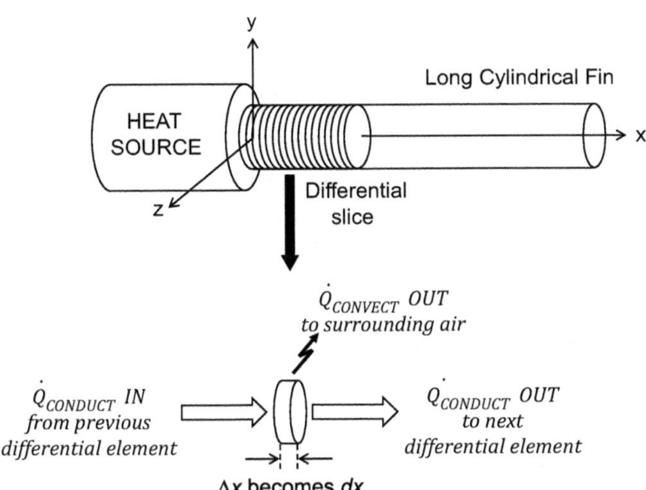

FIGURE L11.1 Energy balance model of a fin.

To model a fin's heat distribution, an energy balance is performed on a differential slice of the fin, as shown in Figure L11.1.

Heat energy conducts into the differential slice from the previous hotter slice and conducts out of the slice to the next cooler slice. The differential slice also convects some energy to the surrounding fluid. Radiant heat transfer is considered negligible at the relatively low fin operating temperature. The energy balance equation for Figure L11.1 is written in Equation L11.2. A rate-form energy balance is used since the process is ongoing.

$$\dot{Q}_X^{CONDUCT} = \dot{Q}^{CONVECT} + \dot{Q}_{X+\Delta X}^{CONDUCT} \tag{L11.2}$$

where: $\dot{Q}_X{}^{CONDUCT}$ is the total heat energy (rate) entering the slice
$\dot{Q}^{CONVECT}$ is the energy (rate) lost from slice to surrounding air
$\dot{Q}_{X+\Delta X}^{CONDUCT}$ is the energy (rate) lost from slice into next cooler slice

For convenience in the solution process, move all terms in Equation L11.2 to the LHS, change the signs and group the two conduction terms for the form in Equation L11.3. Perform a term-wise analysis of Equation L11.3 for more complete insight into the problem.

$$\dot{Q}^{CONVECT} + \left(\dot{Q}_{X+\Delta X}^{CONDUCT} - \dot{Q}_X^{CONDUCT}\right) = 0 \tag{L11.3}$$

$\dot{Q}^{CONVECT}$. Newton's law of cooling states that convective heat loss is a function of the temperature *differential* between an object and its surrounding environment. The actual temperatures do not matter, since it is only the difference that drives convection. Convection is also a function of an object's surface area, since a greater surface area exposes more of the heat energy for transfer into the surroundings. Finally, convection is also driven by the heat transfer coefficient, which describes how active the fluid is at removing heat from the object. Fast-moving air over a hot surface carries heat away more effectively than stagnant air. These terms are captured by Newton's law of cooling in Equation L11.4.

$$\dot{Q}^{CONVECT} = h\ A_{SURF}\ (T - T_{\infty}) \quad \text{(L11.4)}$$

where: h is the convective heat transfer coefficient
A_{SURF} is surface area exposed to surrounding air
T is temperature of the fin
T_{∞} is ambient temperature of the surrounding air

To apply Equation L11.4 to the differential slice in Figure L11.1, a few more adjustments are necessary. For each differential slice, the exposed surface area is the perimeter of the fin, P, times the slice element length, Δx. Also, since heat is continually lost along the length of the fin, the temperature of any particular differential slice is a *function* of position along the fin, x. With these substitutions, the convective heat transfer term from Equation L11.4 is revised into Equation L11.5.

$$\dot{Q}^{CONVECT} = h(P\Delta x)[T(x) - T_{\infty}] \quad \text{(L11.5)}$$

Substitute Equation L11.5 into Equation L11.3 to arrive at Equation L11.6.

$$h(P\Delta x)[T(x) - T_{\infty}] + \left(\dot{Q}_{X+\Delta X}^{CONDUCT} - \dot{Q}_{X}^{CONDUCT}\right) = 0 \quad \text{(L11.6)}$$

Divide Equation L11.6 through by Δx for Equation L11.7.

$$hP[T(x) - T_{\infty}] + \frac{\left(\dot{Q}_{X+\Delta X}^{CONDUCT} - \dot{Q}_{X}^{CONDUCT}\right)}{\Delta x} = 0 \quad \text{(L11.7)}$$

$\dot{Q}^{CONDUCT}$. Equation L11.7 is now ready to take the limit as Δx approaches zero to set up for integration. However, it is first important to note the difference in conduction in the numerator, which becomes differential heat conduction through the differential slice. The entire conduction term in Equation L11.7 collapses into a differential heat rate term relative to position along the x-axis per Equation L11.8.

$$hP[T(x) - T_{\infty}] + \left(\frac{d\dot{Q}}{dx}\right)^{CONDUCT} = 0 \quad \text{(L11.8)}$$

The differential heat conduction term can now be refined with Fourier's law of heat conduction. Fourier's law states the rate of heat conduction is a function of a material's thermal conductivity, cross-sectional area transferring the heat and the temperature differential driving the heat transfer. These terms are summarized with Fourier's law in Equation L11.9.

$$\dot{Q}^{CONDUCT} = -kA_{XC}\frac{dT}{dx} \tag{L11.9}$$

where: k is the material thermal conductivity coefficient
A_{XC} is the cross-sectional area engaged in heat transfer
$\frac{dT}{dx}$ is the temperature difference across a differential distance

Fourier's law in Equation L11.9 has a negative sign to account for the negative correlation between changing temperature, dT, and changing position, dx. As distance, x, increases along the fin, temperature, T, decreases. Temperature in Equation L11.9 is a function of position, x. For clarity and consistency between terms, this notation is included in Equation L11.10.

Just the heat rate term, $\dot{Q}$, within the differential in Equation L11.8 can now be substituted into Equation L11.9 to arrive at the second-order differential term in Equation L11.10.

$$hP\left[T(x)-T_{\infty}\right]+\frac{d}{dx}\left(-kA_{XC}\frac{dT(x)}{dx}\right)=0 \tag{L11.10}$$

For constant cross-sectional area fins, the constants can be pulled from the differential and collected into a single term with Equation L11.11.

$$-\frac{hP}{kA_{XC}}\left[T(x)-T_{\infty}\right]+\frac{d^2T(x)}{dx^2}=0 \tag{L11.11}$$

Temperature distribution solution. Equation L11.11 can be re-arranged to simplify the integration. A helpful step is to collect the several constants into a single constant. With some insight into the second-order solution process, this constant is squared to simplify the solution roots. The temperature function notation, $T(x)$, is also simplified for clarity. The result is a second-order, non-homogeneous, ordinary differential equation with constant coefficients in Equation L11.12.

$$T'' - c^2T = -c^2T_{\infty} \tag{L11.12}$$

where: T is the temperature function of x-position along the fin
T_{∞} is the ambient surrounding temperature of the fin
$c^2 = \frac{hP}{kA_{XC}}$, a collection of all constants

Equation L11.12 has a readily available solution. The homogeneous solution is provided in Equation L11.13a. Using the method of undetermined coefficients to solve an assumed polynomial, the particular solution resolves to ambient temperature in Equation L11.13b. The general solution is the combination of solutions in L11.13c.

$$T_h = Ae^{cx} + Be^{-cx} \tag{L11.13a}$$

$$T_P = T_\infty \tag{L11.13b}$$

$$T = Ae^{cx} + Be^{-cx} + T_\infty \tag{L11.13c}$$

Boundary conditions are checked to resolve coefficients in L11.13c. At the fin base, $x = 0$, the fin has the base temperature of the heat source, $T(0) = T_b$. Assuming an appropriately sized fin, the fin tip, $x = L$, reaches ambient temperature, $T(L) = T_\infty$. With these boundary conditions and a lot of algebra, Equation L11.13c resolves into the final fin temperature distribution with Equation L11.14.

$$T(x) = (T_b - T_\infty)e^{-cx} + T_\infty \tag{L11.14}$$

where: $T(x)$ is the temperature distribution along the fin
$(T_b - T_\infty)$ is a constant value
$c = \sqrt{\frac{hP}{kA_{XC}}}$, fin properties and heat transfer coefficient

The preceding development made use of several assumptions as follows:

1. Steady-state heat transfer process
2. Constant material properties (independent of temperature)
3. No internal heat generation
4. One-dimensional conduction
5. Uniform cross-sectional area
6. Uniform convection around the surface area
7. Infinitely long fin such that the fin tip achieves ambient temperature

Two terms are considered for fin design: efficiency and effectiveness. Fin efficiency describes a fin's ability to readily *conduct* heat away from the hot object. This theoretically ideal case creates an entire fin at the hot source temperature and maximizes the temperature differential for convection to the environment. Fin effectiveness describes a fin's ability to *convect* heat into the surrounding environment by maximizing fin surface area. This creates a more rapid temperature drop along the fin, making it less efficient but more effective.

Efficiency (η). Efficiency compares the heat transfer of an actual fin with a temperature gradient to that of a theoretical fin completely at the hot source temperature throughout. Without derivation, efficiency can be calculated with Equation L11.15.[1]

$$\eta = \frac{\dot{Q}_{fin,\ actual}}{\dot{Q}_{fin,\max temp}} = \frac{1}{L}\sqrt{kA_{XC}/hp} \tag{L11.15}$$

The implication of Equation L11.15 is that a *thicker* fin with less surface area will stay hotter and maximize the temperature differential parameter in Newton's law of cooling from Equation L11.4.

Effectiveness (ε). Effectiveness compares heat transfer from a heat source with and without a fin. Clearly, a fin adds significant surface area to a heat source compared to not having a fin, which maximizes the surface area parameter in Newton's law of cooling in Equation L11.4. The effectiveness equation is offered without derivation in Equation L11.16.[2]

$$\varepsilon_{long} = \frac{\dot{Q}_{fin}}{\dot{Q}_{no\ fin}} = \sqrt{kp/hA_{XC}} \tag{L11.16}$$

The implication of Equation L11.16 is that a more effective fin will be cooler than a more efficient fin since heat is readily transferred with the greater surface area.

Time constant (τ). Equation L11.14 has the mathematical form of Equation L11.17a, where temperature decays exponentially along the fin as a function of *position, x*. The exponential constant, $-c$, dictates how steeply the function decreases.

$$y(x) = e^{-cx} \tag{L11.17a}$$

Temperature also decays exponentially as a function of *time, t*, according to the same mathematical form in Equation L11.17b. The exponential constant, $-k$, dictates how steeply the function decreases.

$$y(t) = e^{-kt} \tag{L11.17b}$$

Since the exponent in Equation L11.17b must be mathematically unitless, the constant, k, must have units of s^{-1}, to cancel with units of seconds in time, t. Therefore, the inverse of k, $1/k$, has units of seconds and is called the time constant, τ. The time constant is also a direct measure of how quickly the exponential function decays, measured in units of time. For any time-temperature pair in an exponential function, the time constant is calculated using Equation L11.18.

$$\tau = -\frac{\Delta t}{\ln\left[\dfrac{T_C - T_\infty}{T_H - T_\infty}\right]} \tag{L11.18}$$

where: τ = time constant, tau

Δt = elapsed time between any two temperature points

T_∞ = ambient, equilibrium steady-state temperature

T_H = any hotter temperature point to define Δt

T_C = any cooler temperature point to define Δt

$k = 1/\tau$ (exponential decay constant)

COMPUTATIONAL METHOD

This lab requires waiting time for temperatures to heat up and cool down. To utilize time efficiently, the fin experiment should first be configured and heating commenced to steady state. The computational simulation can start once the experiment is underway, so activities can be completed simultaneously.

The theoretical temperature distribution in Equation L11.14 requires knowledge of a reasonable heat transfer coefficient, h. The convective heat transfer coefficient, *h*, for free convection in air is usually published toward the lower end of a broad range of 0.5–1000 W/m^{2}K. Higher temperatures (relative to ambient) result in higher heat transfer coefficients and vice versa for lower temperature differences. To determine a suitable heat transfer coefficient for theoretical calculations, model only one fin, either the cylindrical or rectangular fin for computational analysis. Refer to Table L11.A1 for fin parameters. Confirm all parameters in the lab.

Start the simulation software and create a new part for one of the fins. Refer to Appendix L9.A for detailed instructions on how to create a model and set up a surface heat source. For this simulation, the heat source must be applied to an *END face* of the fin. Set up a computational domain sufficient to enclose the fin dimensions with more space in the vertical to allow for a complete solution.

Refer to Table L11.A3 for expected voltage and amperage. Use Equation L11.1 to calculate power. Enter this power into the simulation as power dissipated by the fin. Complete the computational simulation and record the average heat transfer coefficient value in Table L11.A1. Use this value for theoretical computations in the final lab report. Include one lengthwise cut plot image of the simulation temperature distribution along the length of the fin in the lab report.

NOTE: Be efficient with the computational analysis part of the lab. Both the computational and experimental methods require intensive computer resources that need to be accomplished separately. Do not use more than half of the lab session for computational efforts since at least half of the time should be dedicated to the experimental efforts.

EXPERIMENTAL METHOD

Data collection tables. Prepare data collection tables similar to the tables in Appendix L11.A. Use Table L11.A1 to record parameters for theoretical computations. Confirm all values in the lab. Use Table L11.A2 in Appendix L11.A to record theoretical and experimental temperatures for both rectangular and cylindrical fins at each of the temperature measurement stations along the fin's length. Use Equation L11.14 to calculate the theoretical temperature distribution data in Table L11.A2 where the base temperature occurs at $x = 0$. Complete all theoretical calculations prior to the lab and update values as necessary during the lab.

Plan another table similar to Table L11.A3 in Appendix L11.A for fin cool-down data. Table L11.A3 can be copied from data acquisition device (DAQ) measurements or hand measurements taken at specific intervals. The DAQ can be run at any desired speed, from one sample per second per channel down to about one sample per minute per channel. The data in Table L11.A2 should be limited to only enough points to support the charts in Figure L11.3.

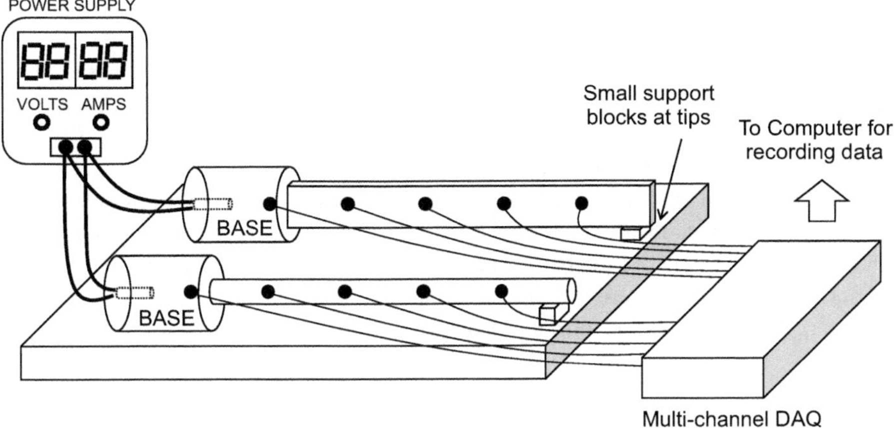

FIGURE L11.2 The experimental setup.

Prepare a quad-chart similar to Figure L11.3 in the results section to plot data from Table L11.A2. All theoretical data should be plotted in Figure L11.3 prior to the lab. Prepare another chart similar to Figure L11.4 in the results section to plot cool-down data from Table L11.A3 in Appendix L11.A1. This chart should be fully configured for the lab plan but will not include actual data for the lab plan. However, include a hand sketch of the anticipated cool down curves in Figure L11.3 as expected results. Sketch the anticipated exponential curves in their relative positions for the cylindrical fin versus the rectangular fin. These hand sketches will be replaced with actual data for the report.

Setup. Configure the cylindrical and rectangular fin experiments as shown in Figure L11.2. Connect either one power supply to both cartridge heater leads in parallel as shown or connect each fin to an individual power supply. Polarity does not matter. Connect thermocouple leads to DAQ. Ensure all wires are neatly routed and do not contact hot surfaces.

Configure the DAQ to record temperature data for the correct thermocouple type. Turn on the power supply to the recommended voltage range in Table L11.A3. Turn on the DAQ monitoring to read temperatures, but do not record temperatures until reaching steady state. Steady-state temperatures should be in the range of 100°C–150°C. Do not exceed 150°C. If separate power supplies are used, small adjustments can be made near steady state to ensure similar base temperatures. Monitor the setup and temperatures for steady-state conditions.

Commence the computational simulation while waiting for the fins to reach steady-state temperature.

When temperatures have stabilized (approximately 10–20 minutes), record a single snapshot of the temperature values at each station. If the DAQ has a bad or suspect measurement input, check for unintentional electrical shorting of the thermocouple leads at the DAQ input. Bare thermocouple wires should not touch each other except where joined together at the temperature-sensing junction on the fins. A handheld temperature measurement device can also be used to manually record temperatures at any station with suspect readings. In this case, compare any manual temperature values to DAQ values and use the temperature source determined to be the most reliable. Do not use an infrared style temperature measurement gun since the emissivity of bare aluminum will not provide an accurate reading.

Use the power supplies to record the power dissipated by each fin. If the power supply indicates the watts, record that value as dissipated power. If watts are not indicated on the power supply, record both the voltage and amperage and use Equation L11.1 to calculate the power dissipated.

Cool down data. After the temperatures for both fins at each measurement station have been recorded, ensure the DAQ is still recording data. Turn off the power supply and allow the fins to cool the heat source. If possible, allow the fin base temperature to reach ambient temperature. However, this may take too long, so record the available cool-down data and terminate the experiment when necessary.

RESULTS

Use the raw temperature data recorded in Table L11.A2, and normalize all values for the table. To normalize the values, each data set must be divided by the maximum base temperature for that data set. The base temperature will always have a value of 1.0, and every station temperature will be less than one. The temperatures do not decay to zero since the normalized ambient temperature at the tip is a small value greater than zero. Use normalized values to plot Figure L11.3.

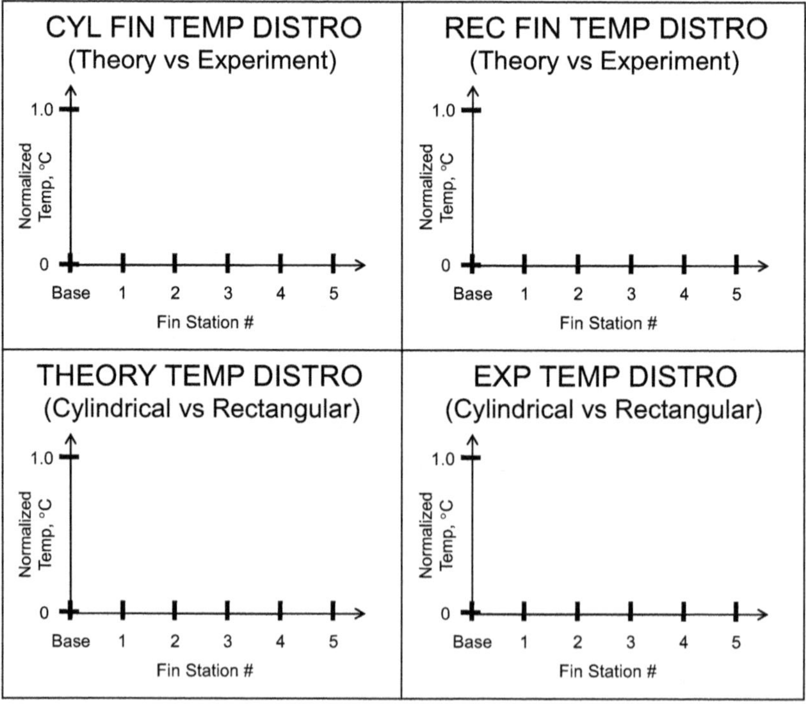

FIGURE L11.3 Fin temperature distribution comparisons.

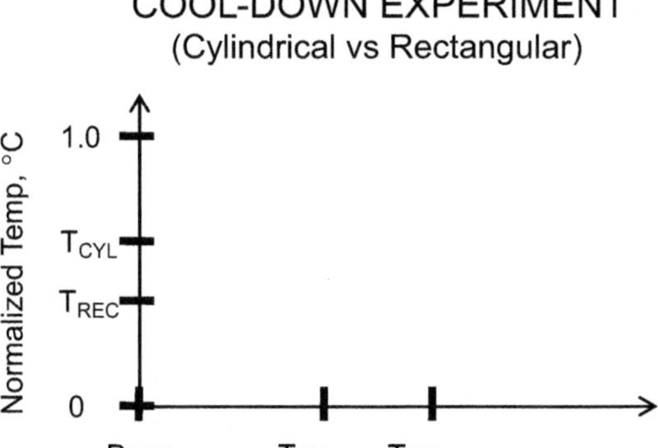

FIGURE L11.4 Fin cooling performance comparison.

Plot the cool-down data in Figure L11.4. Use Equation L11.18 to compute the time constant for both the cylindrical and rectangular fins. Record the time constants in Table L11.A3. If sufficient data was recorded, mark the time constant locations on the chart in Figure L11.4. Compute the increase (or decrease) in fin performance using the time constant values. Use the percent error formula for the difference between time constants, with the cylinder fin time constant in the denominator of the error formula, to indicate the percentage in Table L11.A3. Depending on the fin configuration, the percentage performance difference between the rectangular and cylindrical fins may be negligible.

DISCUSSION

Describe the data and how temperature decays for each fin. Based on the data, discuss which fin provides improved cool-down performance and relate this to the concept of effectiveness. Identify any anomalies in the data and discuss any potential sources of error. Discuss how the free-air convective heat transfer coefficient is at the lower end of the published range compared to a much hotter glowing Nichrome wire.

CONCLUSIONS

Based on the objective of the experiment, assess whether the mathematical models for fin temperature distributions were accurate compared to the experimental data. Clearly state if the rectangular fin was theoretically more effective than the cylindrical fin. Clearly state the percentage improvement of the rectangular fin compared to the cylindrical fin based on cool-down data. Assess whether experimental performance data supports theoretical calculations that the rectangular fin is more effective at cooling than the cylindrical fin. Make an assessment as to whether an extremely thin rectangular fin would become less effective and why.

APPENDIX L11.A: TEMPERATURE DATA TABLES

TABLE L11.A1
Fin Temperature Distribution Parameters

Common Parameters for Aluminum 6061T6 Fins			
Ambient temp, T_∞ (K):	*(24°C+273°C)*	Fin material conductivity coeff, k:	*(247 W/mK)*
Source temp, T_{BASE}:	*(150°C+273°C)*	Estimated HTC, h:	*(11 W/m²K)*
		Computational HTC, h:	
CYLINDRICAL FIN		**RECTANGULAR FIN**	
Diameter (m):	*(6.35e−3 m)*	Thickness (m):	*(1.59e−4 m)*
Perimeter, P (m):		Width (m):	*(1.91e−2 m)*
Cross-sectional area, A_{XC} (m):		Perimeter, P (m):	
		Cross-sectional area, A_{XC} (m):	
Distribution decay constant, c:		Distribution decay constant, c:	

TABLE L11.A2
Steady-State Fin Temperature Distribution Data

Temp Distribution:	T_{BASE}	T_{STA_1}	T_{STA_2}	T_{STA_3}	T_{STA_4}	T_{STA_5}	T_{STA_6}
$RECT_{THEORY}$ (Raw)							
$RECT_{THEORY}$ (Norm)	1.0						
$RECT_{EXPER}$ (Raw)							
$RECT_{EXPER}$ (Norm)	1.0						
CYL_{THEORY} (Raw)							
CYL_{THEORY} (Norm)	1.0						
CYL_{EXPER} (Raw)							
CYL_{EXPER} (Norm)	1.0						

TABLE L11.A3
Cool-Down Temperature Data Summary

	Rectangular	**Cylindrical**
VOLTAGE:	*(10–13 V)*	*(10–13 V)*
AMPERAGE:		
POWER:	*(Equation L11.1b)*	*(Equation L11.1b)*
Time (s)	**BASE TEMP**	**BASE TEMP**
(DAQ)	*(DAQ)*	*(DAQ)*
$t=0$	↓	↓
$t=t_1$		
$t=t_2$		
↓		
$t=t_n$		
TIME CONSTANT, τ:		

% Improvement of rectangular time constant relative to cylindrical fin:

APPENDIX L11.B: FIN DESIGN

Fin experiment designs have been successful with large, 1-m-long fins using commonly available hardware store aluminum rods and bars. Experiments have also been successful with smaller desktop size setups of about 30 cm in length. Overall dimensions of the configuration are non-critical, but fin lengths should be long enough to reach near ambient temperatures. The fin dimensions should be chosen such that the cross-sectional area is very similar so the only experimental test parameter is the ratio of perimeter to cross section. This can be determined by equating the two cross-sectional areas, $(t^*w)_{RECT}=(\pi r^2)_{CYL}$ and solving for one or the other based on available materials. A readily available combination is a $^3\!/_4 \times {}^1\!/_{16}$ inch aluminum bar and $^1\!/_4$ inch diameter aluminum rod.

Small cartridge heaters approximately 6 mm × 20 mm should be used for heating. Voltages around 10 V should provide a relatively safe operating temperature of around 100°C. Temperatures should be limited to 150°C, which is around 13 V.

Although the cartridge heaters can be attached directly to the fins, a small thermal mass should be used for the base to properly simulate the fin cooling performance of the source. A small 1″ diameter aluminum bar, cut to about 2″ in length, is adequate to countersink the cartridge heaters and fins.

The fins should be drilled with approximately 5 cm or greater spacing to record a minimum of four temperature stations in addition to the base temperature measurement station. Eight measurement stations will provide greater fidelity to the exponential function. Ensure the same spacing between the cylindrical and rectangular fins. Drill and tap each of the measurement stations for small aluminum screws to

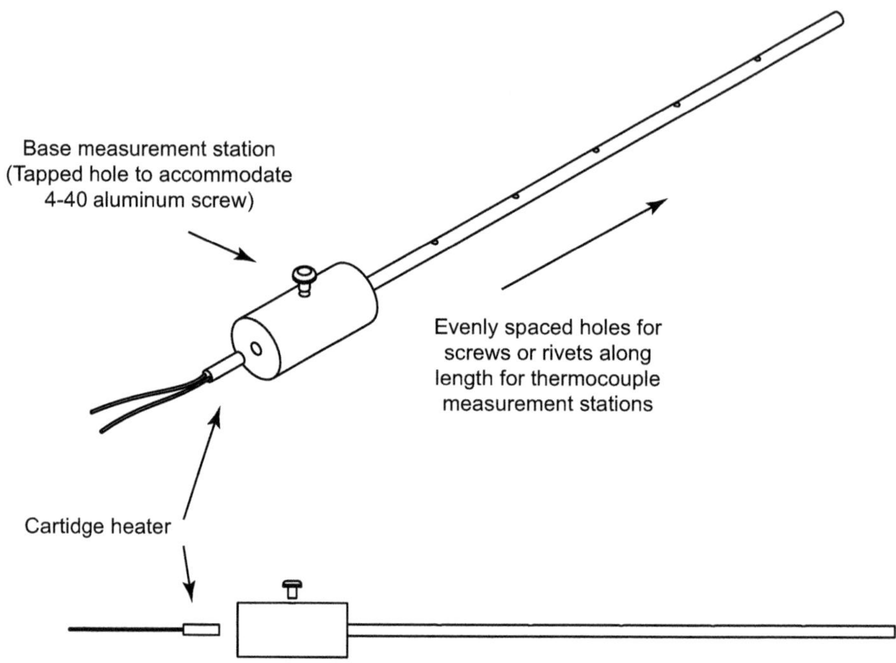

FIGURE L11.B1 Fin design configuration. Only the cylindrical fin is depicted. Dimensions are non-critical. Bore holes to insert cartridge heater and rod. Use a slot for the rectangular fin. Use thermally conductive grease to improve heat transfer before permanently installing components. Pot the entire base assembly in RTV silicone.

secure the thermocouples. Small 4-40 aluminum screws are readily available. To save effort, an alternative to tapping the holes is permanent thermocouple installation with aluminum pop rivets.

The base thermocouple can be located at any convenient location on the thermal mass. Once complete, the thermal mass base should be potted in high-temperature RTV silicone. Only the fin should protrude from the silicone. If the base is not sufficiently insulated, the cool-down data will be obscured by the overall cool-down of the thermal mass base. The base cool-down should be driven by the fins.

The experiment should be operated horizontally to ensure even heat convection across the fin surfaces. Small support blocks can be used to stabilize the setup. The overall fin configuration is depicted in Figure L11.B1.

NOTES

1 Cengel, Y. (2003). *Heat transfer, a practical approach*. McGraw-Hill Higher Education. p.161.
2 Cengel, Y. (2003). *Heat transfer, a practical approach*. McGraw-Hill Higher Education. p.164.

LAB 12

Heat Pipes

Safety Requirements

Never point a torch, lit or unlit, toward yourself or another person
Use a *fire blanket* under vise and on the surfaces prior to using a torch
After using a torch, assume that all surfaces are hot until they have been verified cooled
Propane cylinders must be stored upright with the torch removed
Do not heat a sealed heat pipe directly with an open flame source
Do not point the awl at the body when applying force

Equipment Requirements

Eye protection, lab coat, insulated gloves and absolutely no shorts or open-toed shoes are permitted in the lab
Bench vise, fire blanket, propane torch, solder, flux, round wire brush, sandpaper and awl
¼″ copper tubing (approximately 30 cm length), copper tube cut-off tool and pinching tool
Vacuum pump, vacuum gauge, Teflon tape, vinyl tubing and syringe
Modified refrigerant valve core tool, ¼″ ID access valve and high-pressure Shrader valve
Thermocouple wire (approximately 15″), wire cutters, wire strippers and DAQ system
Water, hot plate, beaker, lab stands, tongs and wood or PVC cap for tamping tube

Procedural Requirements

Order of steps is important; prepare a checklist of construction steps

Clean-up Requirements

Return all equipment to proper locations and clean up any water spills
Desolder and salvage all valves IAW instructions
Wash hands with soap

Emergency Actions

MINOR BURNS: Run cold water on the affected area. Seek follow-up medical treatment.
MAJOR BURNS: Run cold water on the affected area. Call 911.
GAS LEAK: Turn off open flame sources. Do not activate electrical equipment. Place source in fume hood if able. Evacuate. Call 911 from a safe location.
MINOR FIRE: Alert others. Use fire extinguisher. Activate the fire alarm if necessary.
MAJOR FIRE: Alert others. Evacuate area. Close doors. Activate the fire alarm. Call 911.

INTRODUCTION

Heat pipes provide a means of heat transfer several hundred times greater than conduction. Heat pipes are hollow, sealed tubes containing a working fluid that undergoes a phase change from liquid to vapor and vice versa. The latent heat of vaporization removes heat at one end of the heat pipe as the working fluid boils and transfers it to the other end of the heat pipe, where the heat is given off as the working fluid condenses. This two-phase transition process creates a significant heat transfer rate.

A thermosyphon is similar to a heat pipe, with the exception that a heat pipe includes an internal wick. The wick within a heat pipe facilitates the movement of condensed working fluid via capillary action. A thermosyphon has no mechanism for the capillary movement of fluid. As a result, thermosyphons rely on gravity for working fluid condensate to flow back down the insides of the tube to the bottom to begin the cycle again. A heat pipe can be oriented in any direction since capillary action can move condensed working fluid against the force of gravity. For the purposes of this lab, the term *heat pipe* is considered synonymous to *thermosyphon* since the operating principles are generally the same.

The selection of a working fluid within a heat pipe is based on the operating temperatures desired from the pipe. The temperature range must occur between the fluid's triple point and critical point to assure two-phase operation. Other considerations for the choice of working fluid include compatibility between the working fluid and the pipe to avoid corrosion or other issues such as the formation of contaminating gasses within the heat pipe.

The pressure within the heat pipe also dictates the operating temperatures of the heat pipe. Water boils at 100°C at 1 atm, so a heat pipe with water as the working fluid at 1 atm will not become effective until at or above 100°C when both liquid and gas phases can exist. However, if the pressure within the heat pipe is reduced to approximately one-third of an atmosphere, water will begin to boil at approximately 70°C and the heat pipe will function.

For this lab, a heat pipe (thermosyphon) is built with water as the working fluid and compared to a control heat pipe without a working fluid. This lab takes place over two lab sessions. The first session is dedicated to constructing and testing the heat pipe and control pipe. The second session is the test session for working temperature and rise time assessments.

OBJECTIVE

Assess the heat transfer capability of a heat pipe to a copper control pipe of similar dimensions without a working fluid. Theoretically predict the working temperature of the heat pipe. Compare the experimental working temperature relative to the theoretical working temperature.

THEORY

Heat pipes follow a thermodynamic cycle with no moving parts. The working fluid is heated in the evaporator (boiler) section where it flows adiabatically to the condenser section and gives off heat. This functional cycle is depicted in Figure L12.1.

The thermodynamic cycle in Figure L12.1 transpires physically as shown in Figure L12.2 where a small amount of fluid boils at the bottom, heat is given off at the top and the condensate travels back down along the pipe walls.

The thermodynamic cycle of a heat pipe can be depicted on a temperature-entropy (T-s) diagram, as shown in Figure L12.3. The working fluid begins the cycle as a compressed liquid at State 1. As heat is added, the liquid moves into the mixture phase and then exits the evaporator section as a saturated vapor at State 2 or a superheated vapor at State 2′. The working fluid travels quickly and therefore (ideally) adiabatically into the condenser section at the top of the vertically oriented tube as

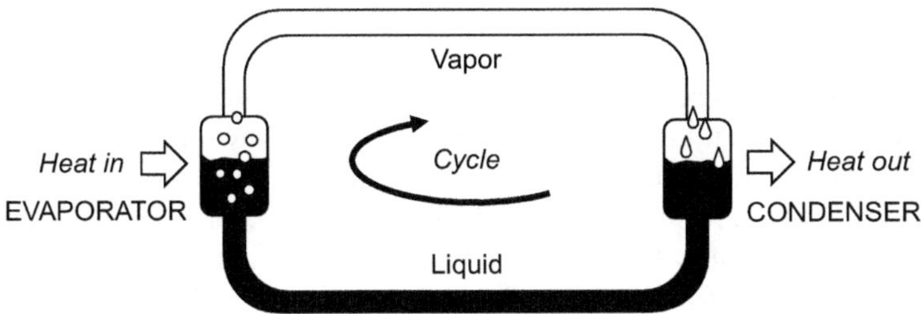

FIGURE L12.1 Functional thermodynamic cycle of a heat pipe.

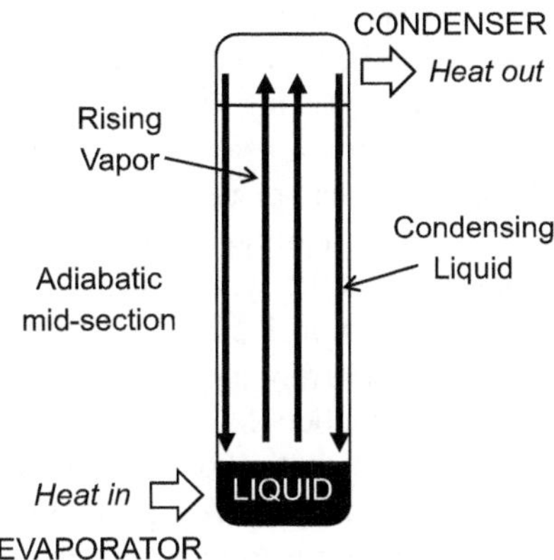

FIGURE L12.2 Thermodynamic cycle within a heat pipe.

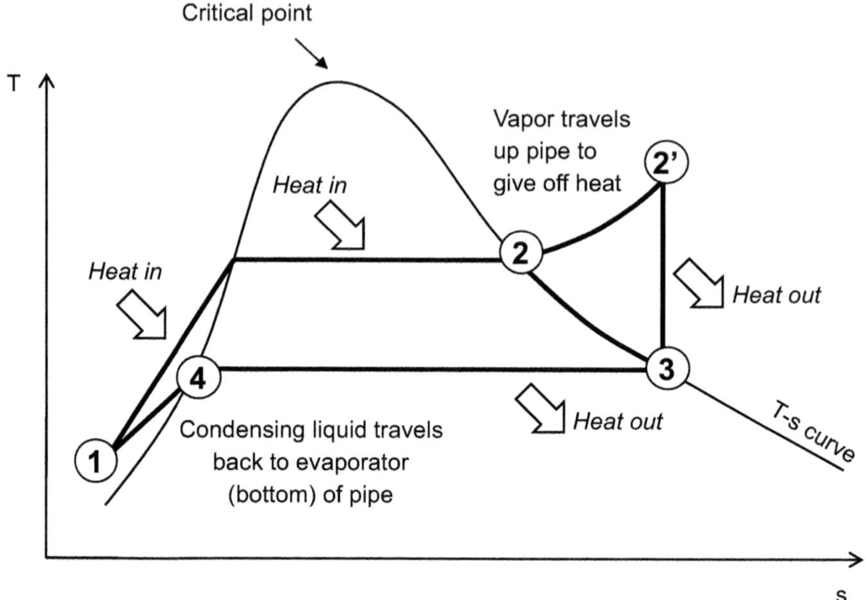

FIGURE L12.3 *T-s* cycle diagram of a heat pipe.

a superheated vapor, where it then gives off heat, condenses and travels back down the walls of the tube to begin the process over. The temperature-entropy cycle of a heat pipe is shown in Figure L12.3. The work performed by the heat pipe is the area within the *T-s* cycle, which both drives the cycle and moves the heat.

Only pure substances should be considered for the working fluid of a heat pipe. Differing vapor pressures and other physical properties will generally cause only one of a mixture of fluids to undergo the thermodynamic cycle. The other fluid will remain in a single state and take up volume within the heat pipe while not contributing to the operation. Working fluids such as alcohol are hygroscopic and can contain soluble water, which can degrade performance of the heat pipe unless the working fluid is purified. For the same reason, ambient air should be evacuated from a heat pipe. As air heats with the action of the heat pipe, it expands, increasing internal pressure and inhibiting the action of the working fluid.

The volume of working fluid within a heat pipe must be adequate to sustain the liquid-vapor cycle. Insufficient liquid can result in *dry out* where there is not enough fluid to continue the cycle. Too much liquid will force liquid into the condenser section, blocking the condensing action of the cycle. Since steam is about 1600 times less dense than water, only a small amount of liquid water is required to completely saturate the heat pipe. Referring to the relatively small evaporator section in Figure L12.2, the recommended fill volume is only about half of the evaporator section.

Lower viscosity fluids are generally better for heat pipe working fluids. The vapor pressure difference between the condenser and evaporator sections creates the pumping action to drive the cycle. If the fluid is too viscous, the pressure differential may be insufficient to sustain the thermodynamic cycle.

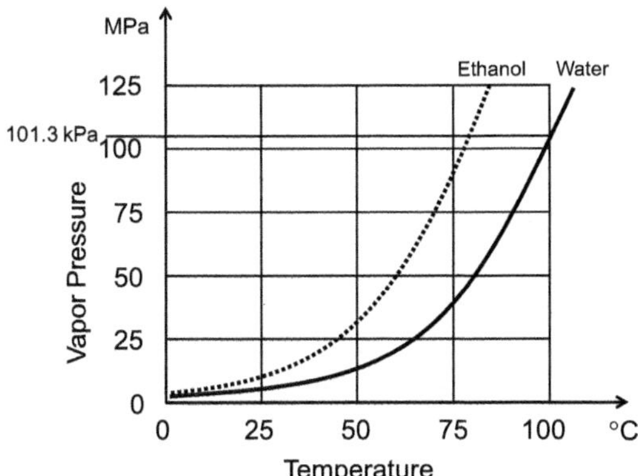

FIGURE L12.4 Vapor pressure chart of water and ethanol.

The heat transfer capability of a heat pipe, $\dot{Q}$, is a function of several parameters. Some of the more significant parameters directly associated with the working fluid include density, ρ, surface tension, σ, latent heat of vaporization, L_V, and dynamic viscosity, μ. The heat transfer rate is proportional to these parameters according to Equation L12.1.[1]

$$\dot{Q} \propto \frac{\rho \sigma L_V}{\mu} \tag{L12.1}$$

As a result of Equation L12.1, fluids with higher density, higher surface tension, higher latent heat of vaporization and lower dynamic viscosity will have greater heat transfer capability. Equation L12.1 implies heavier, slower-moving fluids will pick up and transport more heat compared to light, fast-moving fluids. These are relative properties such that water should have a better heat transport capacity than alcohol. If the fluid is too thick or dense, then the heat pipe fails due to reaching a limiting property. For operating temperatures ranging from ambient to a couple hundred degrees Celsius, water generally scores high for heat transfer capability, while alcohols tend to score slightly lower.

Vapor pressure is the boiling temperature of a fluid where the pressure of the fluid's vapor is in equilibrium with the liquid phase. The primary tool for selecting a working fluid based on operating temperature is a vapor pressure versus temperature chart, as shown in Figure L12.4.

The vapor pressure of water at 1 atm (101.3 kPa) is 100°C. Reducing the pressure within the heat pipe to approximately 38 kPa will allow water to undergo phase change by boiling at 75°C. Switching to ethanol as a working fluid will reduce the working temperature to approximately 55°C under similar pressure conditions.

Vacuum measurements. Vacuum gauges typically indicate *gauge pressure (vacuum)* rather than *absolute pressure*. By definition, gauge pressure references atmospheric pressure at zero, so a vacuum gauge begins measuring down from the zero atmospheric pressure reference. Conversion is necessary to convert vacuum gauge readings into the absolute pressure typical of most vapor pressure charts.

Sea-level atmospheric pressure is approximately 101.3 kPa. This equates to 29.92″ of mercury (inHg) as a common gauge unit of measure. A vacuum gauge reading ambient conditions indicates *zero* inHg gauge pressure. As vacuum increases, the gauge indicates a *higher* value of *gauge vacuum*. For example, at *8 inHg* displayed on a vacuum gauge, vacuum is *8 inHg* lower than atmospheric pressure, which equates to 21.92 inHg (29.92 – 8.0 = 21.92 inHg). This *8 inHg* gauge vacuum converts to *74.2 kPa (21.92 inHg)* absolute pressure.

Under standard conditions, the lowest value a vacuum gauge can read is *29.92 inHg*. Since standard sea-level atmospheric pressure is *29.92 inHg*, absolute vacuum equates to *0 inHg*. A perfect vacuum displayed in gauge values is therefore *29.92 inHg*, counting down from the zero atmospheric reference to an absolute vacuum.

EXPERIMENTAL METHOD

This experiment takes place over two lab periods. The heat pipes are constructed in the first session and tested in the second session. If the construction process cannot be completed in the first session, extra time is available in the second session to complete all tasks. Wear protective vinyl or latex gloves during the lab to keep hands clean of flux and other trace materials.

PART I: CONSTRUCTION

Prepare two charts for the lab plan similar to Figure L12.12 in the results section. These lab plan charts may be hand sketches to indicate the expected performance of the heat pipes. The charts will be replaced with data acquisition (DAQ) results for the lab report. Follow the instructions associated with Figure L12.12 to include the relevant information.

Construct two pipes. One pipe is the heat pipe (thermosyphon) with a Shrader valve and water as the working fluid. The other pipe is the control pipe without water, which is sealed on both ends. Both pipes will have thermocouples attached near the top. An additional thermocouple is for the heat source (hot water). Refer to Figure L12.11 for the final configuration.

Cut, clean and crimp tubing. Straighten a section of ¼″ copper tubing and complete the activities in Figure L12.5, as described in this section.

Use a tube cutter to cut two equal lengths of copper tubing, approximately 25 cm long. To operate the tube cutter, gently tighten the tube cutter just enough to hold it in position against the copper tubing. Spin the tube cutter around the copper tubing (either direction) while successively tightening the knob to maintain cutting pressure. Do not overtighten the tube cutter against the copper, making it too difficult to spin the cutter around the copper tube. The tube should be cut after multiple turns and successive tightening, not in a single attempt.

After cutting the copper tubing, the ends will be slightly crimped from the cutting process and must be opened back up for cleaning. Insert an awl into a tube end and gently open up the tube end by working the awl around in a circular motion.

Safety note. Never apply force with a tool such as an awl toward your body or hand. If the tool slips or something breaks, an injury is likely. Keep tools pointed away from

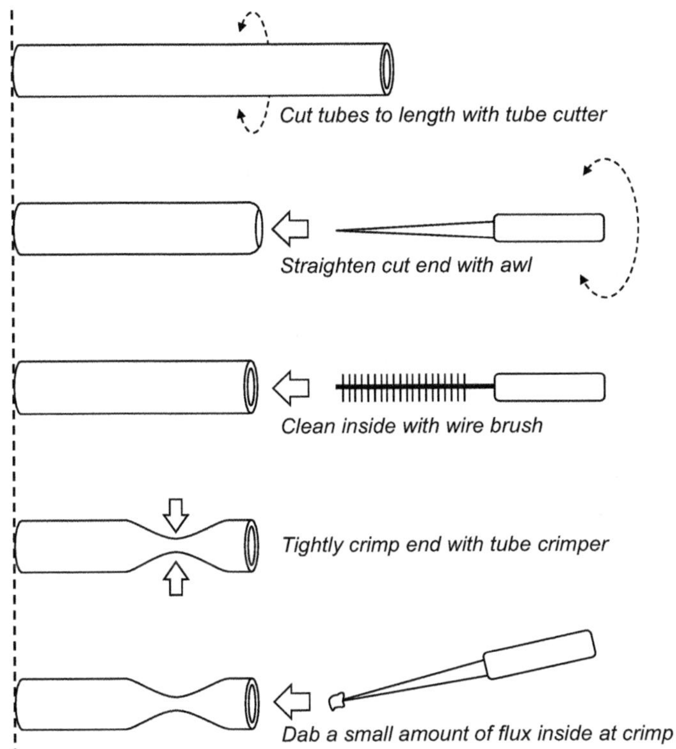

FIGURE L12.5 Visual depiction of preparatory actions.

your body when force is involved. Always let the tools do the hard work and do not force them. If this cannot be done, use a more appropriate tool or setup for the task.

When completed with the previous tasks, the copper tubes should be about 25 cm long and relatively straight, without constricted or flared ends.

Clean the inside of one end of both tubes with the small round wire brush until about a centimeter of the inside copper tubing appears shiny. ***Common failure point #1***: Failure to *properly clean* the inside tube will result in a bad soldering seal.

Use a tubing pinch tool to tightly crimp the cleaned tube end very near the tube end. Crimp the tube in successive stages, releasing the vise-grip pinch tool, tightening the end screw and crimping again. It will take several successive crimp-and-tighten actions to fully crimp the tube end. Do this for both tubes on only the one cleaned-end of each tube.

After crimping, apply a small amount of flux directly on the crimp in the end of the tubes using an awl or other pointed tool. Flux will help to properly solder-seal the crimped ends, but *too much flux* will flow into the tube, contaminating the water within the heat pipe and significantly reducing performance. Both tubes are now ready for sealing their ends with solder.

Solder-sealing crimped tube ends. Working on one pipe at a time, secure the tubing in a vise with the crimped and prepped end slightly elevated above horizontal. Ensure a suitable heat shield is in place below the tubing to prevent inadvertently

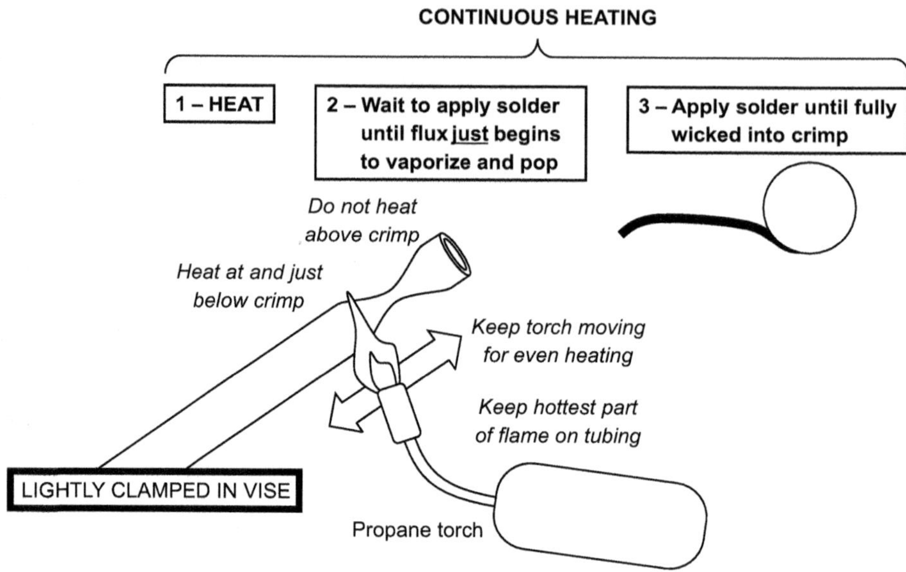

FIGURE L12.6 Solder-sealing crimped tube ends.

burning the table surface. Use solder to seal the crimped ends as shown in Figure L12.6 and described in this section.

Heat the outside of the tube at and just below the crimped portion with the torch. Avoid heating the very end of the crimped tube directly and burning the flux without sufficiently heating the tube itself. Use a medium flame intensity, holding the torch about where the end of the inner blue flame cone touches the copper. Once the copper begins to change color from the heat and the flux just begins to vaporize and boil, apply solder to the crimped joint until solder flows into the crimp and seals the crimp. Keep the torch heating the pipe while applying solder.

When properly heated, the solder should quickly wick into the crimp within a few seconds of touching the solder to the tube. The entire process should take no more than about 20 seconds. Do not overheat the copper prior to applying flux. If the copper tubing is charred and the flux is burned out, the oxidation cannot be cleaned from the crimp, and the tube must be discarded.

Do not melt the solder directly with the torch while attempting to apply it to the copper tube. Do not heat only above the crimp and attempt to melt the solder into the crimp. Do not remove the torch too soon while applying the solder. ***Common failure point #2***: Failure to properly heat the copper tube and wick the solder *into* the crimp will result in a failed cold solder joint.

Turn off the torch and use tongs to grasp the hot tube and carefully cool it under running tap water to expedite cooling as desired. Do not allow water to enter the tube. Both tubes should now have one end sealed.

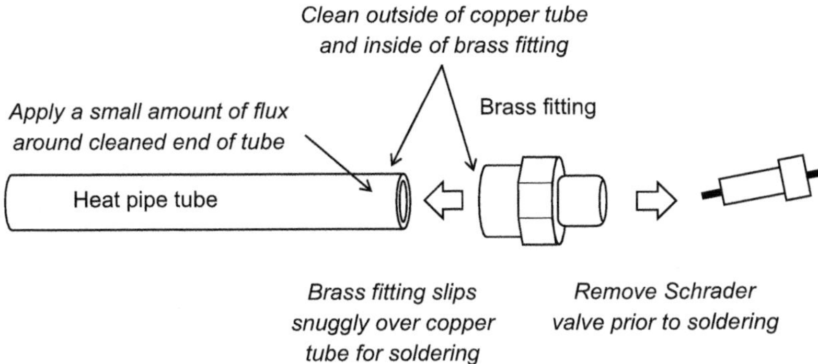

FIGURE L12.7 Brass fitting shown with Schrader valve removed and ready to assemble and solder.

Control pipe. Prepare the open end of one tube for sealing using the same exact procedure as just described. Both ends of one tube should now be crimped and sealed. Once cooled, write "Control" on the outside of this pipe with a marker. This is the control pipe, sealed on both ends with only air on the inside.

Heat pipe. Prepare a brass refrigerant valve by removing the internal Schrader valve with a valve removal tool. Use a wire brush or sandpaper to clean the inside of the brass fitting that will slip over the copper tube. Clean a couple centimeters of the outside open end of the remaining copper tube with sandpaper or a wire brush. The brass fitting should slip reasonably snugly over the end of the copper tubing. An awl can be used to very slightly flare the copper tubing end for a snugger fit, if necessary. Apply a small amount of flux around the outside of the cleaned copper tube end and slip the tube end firmly into the brass fitting. The brass fitting is ready for soldering onto the copper tube, as shown in Figure L12.7.

Place the tube with the brass fitting in a vise, as previously in Figure L12.6. Solder will still wick into the joint in this upward orientation while also preventing excess solder from obstructing the brass threads. Heat the brass fitting directly while also heating the copper tube. The brass fitting will require more heat than the copper tube. Once the flux begins to boil, touch the solder against the copper tube around the joint and keep the torch applied. The solder should melt and wick into the joint. Once solder has wicked completely around the joint, remove the torch heat. Adding too much solder will foul threads for valve re-attachment. Adding too much flux will foul the water working fluid. The heat pipe Is now complete. Allow it to cool from soldering or cool with tap water.

Vacuum check. With the Schrader valve still removed, use a hand vacuum pump to check the vacuum integrity of solder connections, as shown in Figure L12.8. Slip an appropriately sized vinyl tube over the end of the heat pipe brass fitting and the vacuum pump. Squeeze the pump handle several times and ensure the vacuum gauge holds steady. If the heat pipe retains a vacuum, remove the vacuum pump and tubing and proceed to the next step to fill the heat pipe.

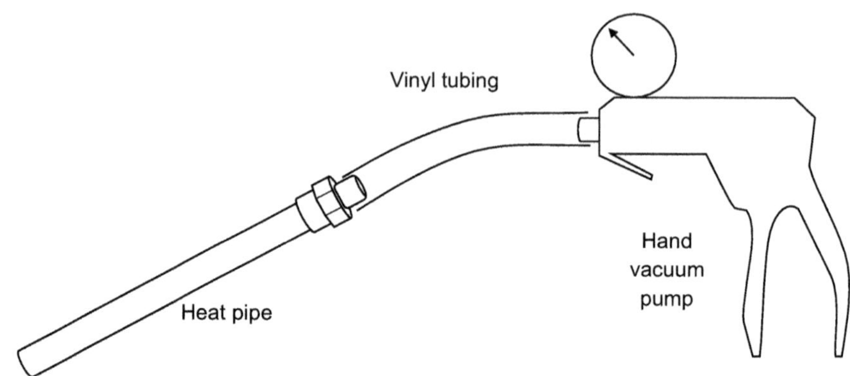

FIGURE L12.8 Vacuum check.

If the heat pipe does not hold a vacuum, remove the heat pipe and ensure the vacuum pump itself holds a vacuum. Place a thumb over the vinyl tube to seal the vacuum pump and squeeze the handle several times. The vacuum force is not harmful when using a thumb or finger with a small handheld pump. If the pump works correctly, the heat pipe must be reworked. Usually, the heat pipe needs to be rebuilt since oxidation prevents successfully repairing a leaking solder joint.

Filling. With the Schrader valve still removed from the heat pipe, use a syringe with a blunt-tipped needle to draw up approximately *0.5 mL* of water into the syringe. Insert the blunt metal needle into the heat pipe as far as possible and quickly inject the water into the heat pipe. The water tends to bead up due to high surface tension and may not properly pool at the bottom of the heat pipe. After injecting the water, *aggressively* tamp the tube a few times against a solid surface (do not damage the table surface) to help pool the water at the bottom of the tube. Do not use more water than recommended. Excess water will require more energy to function, and vigorous boiling action will inhibit the full-length function of the heat pipe. ***Common failure point #3***: Failure to pool water at the bottom of the heat pipe creates an air bubble below the water. If an electric vacuum pump is used to evacuate the tube in the next step, the air bubble will expand and the water will be sucked out of the heat pipe.

If a mistake is made and water must be re-introduced into the heat pipe for any reason, thoroughly drain the heat pipe and aggressively tamp the open end against a solid surface to completely remove any residual water and re-attempt filling with the correct amount of water. Failure to completely drain the heat pipe before refilling results in an uncertain amount of water in the pipe. Much more than about *0.5 mL*, and the pipe will not perform as desired.

Reassemble the valve and evacuate the tube. With the water working fluid in the heat pipe, fit a premium quality Schrader valve core into the brass fitting. The original Schrader valve should be discarded in favor of a high-pressure Schrader valve, usually identified by a red plastic band instead of a black plastic band. Securely snug the Schrader valve into position. ***Common failure point #4***: The Schrader valve needs to be adequately tight to hold a seal. However, do not overtighten to the point of damaging the valve core.

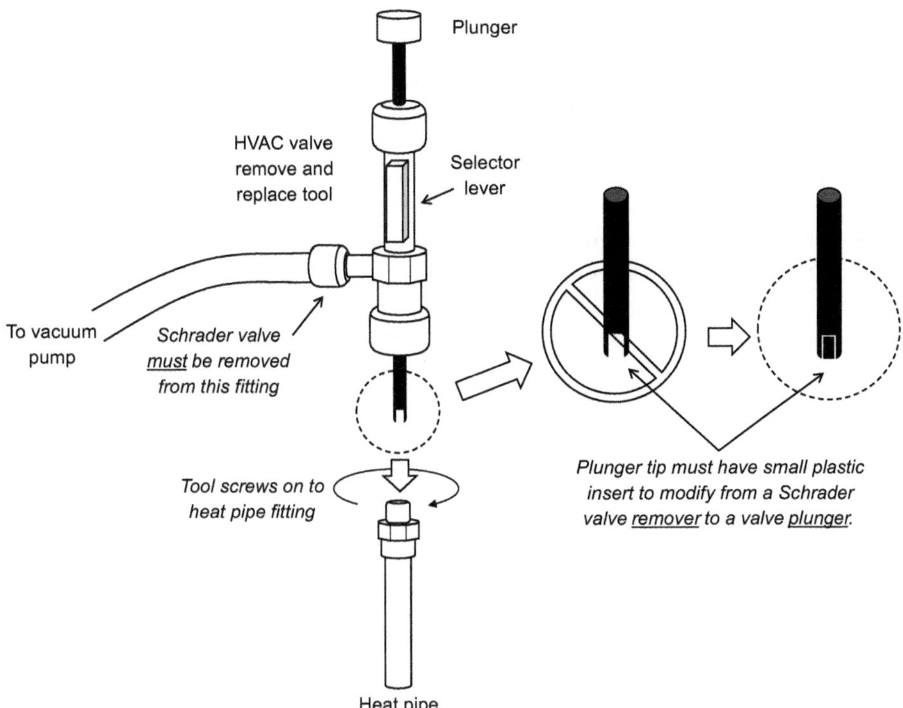

FIGURE L12.9 Use the specialized valve tool as shown in above modified configuration. Connect tool to vacuum pump. Connect heat pipe to tool.

Three options are available for evacuating air from the heat pipe:

1. A hand vacuum pump.
2. An electric vacuum pump.
3. Heating with a torch.

The first two methods require a slightly modified HVAC valve replacement tool for connections. The third heating method requires care not to damage the Schrader valve. Refer to Appendix L12.A for the heating method.

Ensure the water in the heat pipe is completely tamped to the bottom of the heat pipe. Connect the heat pipe to the specialized valve tool and vacuum pump, as shown in Figure L12.9. Ensure the specialized valve tool has the two following modifications as shown in the figure so it can be used as a Schrader valve depressor rather than a removal tool.

1. The Schrader valve should be completely removed from the tool's vacuum attachment fitting.
2. Small insert or other modification at the plunger tip to depress the heat pipe valve.

Follow the steps below to properly evacuate the heat pipe:

1. Connect the valve tool to the vacuum pump.
2. Raise the tool plunger fully up.
3. Screw the heat pipe brass fitting (with Schrader valve in place) onto the tool.
4. Ensure the tool selector valve is upright and inline as shown in Figure L12.9.
5. Hold up the plunger to avoid depressing the heat pump Schrader valve.
6. Turn on the vacuum pump.
7. Slowly and gently depress the tool plunger to depress the heat pipe Schrader valve.
8. Allow the vacuum pump to pull a vacuum on the heat pipe for about 15 seconds.
9. Note the vacuum gauge and record the vacuum *value* in Figure L12.12.
10. Pull the tool plunger all the way up to stop depressing the heat pipe valve.
11. Keep the plunger fully up at all times to avoid accidentally depressing the heat pipe valve.
12. Turn off the vacuum pump.
13. Unscrew the heat pipe from the tool fitting.

The heat pipe should now be sealed under vacuum. Screw a cap fitting over the heat pipe and avoid depressing the heat pipe valve.

Qualitative Testing. With the heat pipe complete, a qualitative test is necessary to verify heat pipe operation. Failure to verify heat pipe function prior to connecting data acquisition equipment will result in wasted effort and time if the heat pipe needs to be reworked.

Safety note. This test requires feeling rapid heat transfer from the heat pipe compared to the control pipe. If working correctly, the heat pipe will get hot quickly. Be prepared to quickly remove the pipes from the boiling water and place them on the table. The pipes can also be quickly transferred to an insulated gloved hand. Use a bit of caution, and this method is completely satisfactory and an excellent demonstration of heat pipe effectiveness.

Heat water in a small beaker to a boil on a hot plate. Hold the heat pipe with the valve end up and control the pipe with a bare hand to feel the heat transfer. Hold the pipes within about the top 2″. Carefully dip the ends of the heat pipe and control pipe into the boiling water at the same time. If the heat pipe functions as expected, the heat transfer will be felt almost immediately compared to the control pipe and the heat pipe cannot be held for more than a short moment before it becomes too hot. Use caution and quickly set aside once the pipe gets hot! If no very obvious difference is noted between the control and working fluid pipes, then the heat pipe must be reworked. Refer to the troubleshooting steps at the end of this section. Sometimes the heat pipe only operates up to about 2″ from the top. If no temperature difference is felt, carefully feel down the heat pipe about an inch or two to confirm the status.

Thermocouple attachment. Once the heat pipe is confirmed to be functional, prepare a total of three thermocouples. One thermocouple is attached to the top of the heat pipe, one to the top of the control pipe and the third thermocouple will measure water temperature. Cut three thermocouple wires at least 15″ long. Strip about one

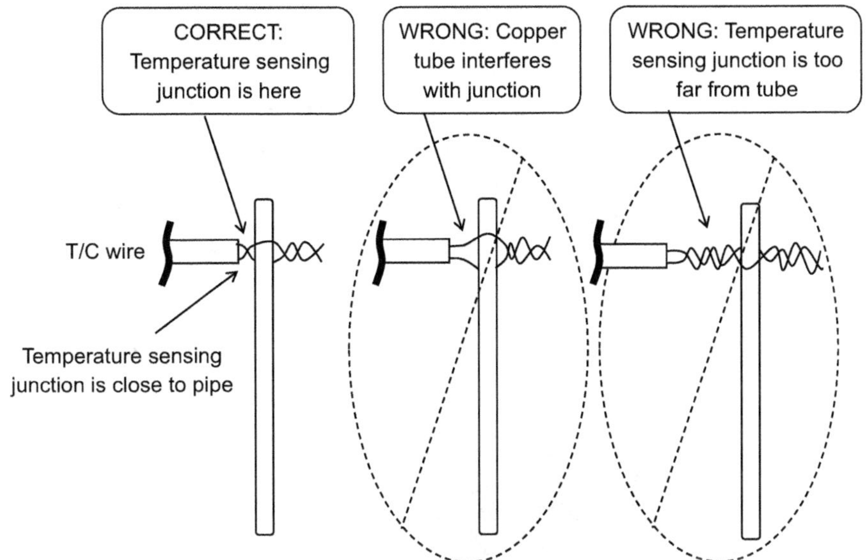

FIGURE L12.10 Twist the bare thermocouple wires together once or twice (for clarity, diagram only shows a half-twist), then wrap around the tube and twist again to hold in place.

inch of insulation off one end of two thermocouples for twisting around the pipes. Strip only about ¼″ insulation from one the end of the water thermocouple. Twist together this end of the water thermocouple to make a sensing junction for the water temperature. From the other end of all three thermocouples, strip about ½″ to 1″ of insulation to allow insertion into a DAQ module.

Attach one thermocouple to the top of the heat pipe and one thermocouple to the top of the control pipe. Locate the thermocouple about 1″ down from the top of both pipes. First, twist the bare thermocouple wires together one or two turns (no more), then wrap the twisted wires around the tubes. Refer to Figure L12.10 to properly twist and attach the thermocouple wires. Use small pliers to snuggly twist the thermocouple wires on the tubes so they remain firmly in place.

Temperature is measured at the point where the thermocouple wires first touch when twisted. If excess twisting occurs away from the heat pipe surface, as shown in Figure L12.10, temperature readings will be inaccurate and likely indicate the steam temperature from the hot plate setup. Also, if the wires are not twisted together first, the copper heat pipe tube interferes with the sensing junction. The heat pipe build is now complete.

PART II: TESTING

The heat pipes should hold a vacuum during the intervening period from build to test. However, the qualitative test should be re-accomplished to confirm status prior to attempting a DAQ test run. If the qualitative test fails, re-accomplish the tube evacuation procedure.

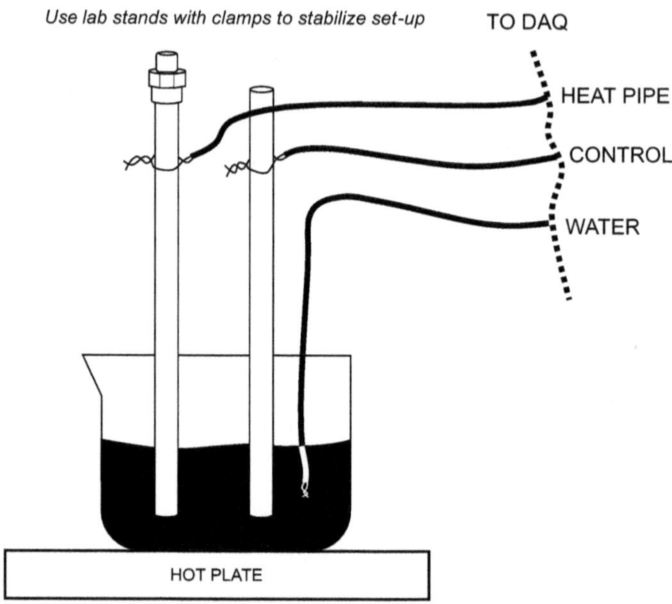

FIGURE L12.11 Data acquisition setup.

Data collection. Prepare two separate data acquisition tests for the two charts in Figure L12.12. The first test run determines the working temperature of the heat pipe, and the second run indicates temperature rise times. During the test runs, use both the heat pipe *and* the control pipe at the same time *together.*

Configure the experiment according to Figure L12.11. Use lab stands with clamps to hold the pipes in position. The water level in the beaker should be *no higher than* about a quarter of the length of the heat pipes. Connect the thermocouple wires to the data acquisition equipment. Note the polarity of the thermocouple wires (red is negative). Use tongs or wear insulated gloves to handle the pipes once the test runs begin.

Working temperature. Configure the data acquisition software to measure at a rate of about one sample per second per channel. Begin with a cooled hot plate in the *off* position. Place a beaker with cool tap water on the hot plate, with just enough water to cover the bottom of the tubes. Water should not be more than one-quarter the length of the heat pipes. Place both pipes (control and heat pipe) into the water, secured with the lab stands. The pipes and thermocouples should not touch the sides of the beaker. The thermocouples at the top of the pipes will record the heat transfer. Place the water sensing thermocouple in water to record water temperatures. Begin DAQ recording and turn on the hot plate. Record temperatures from both pipes as the water heats from room temperature to boiling. Once the water begins to boil, the working temperature test run is complete and data acquisition can cease. Save the data.

Temperature rise time. Prepare the data acquisition system for thermocouple measurements as before. However, increase the data acquisition sample rate to about

2 Hz. Ensure both heat pipes have cooled from the previous experiment. Bring the water in the beaker to a boil on the hot plate, but do not place the heat pipes in the beaker while the water is heating up. When water is boiling, begin data acquisition and recording. Use tongs or wear insulated gloves to hold the heat pipes. Quickly dip both heat pipes simultaneously into the beaker of hot water. Record temperatures with the data acquisition system until temperatures from both pipes stabilize at steady values. Cease data collection and allow the heat pipes to cool. Turn off the hot plate. Do not handle the heat pipes with bare hands until they have cooled. The pipes can be cooled under running tap water. Save the data.

The temperature rise time test should begin and end quickly if working correctly. If the test run takes more than a few moments, cease the experiment and investigate the problem. Do not attempt to hold the heat pipes with insulated gloves over the hot setup for any length of time since it will become too hot.

RESULTS

Figure L12.12 summarizes the heat pipe versus control pipe performance data. For the lab plan, these charts must be included. Estimate a vacuum gauge value of *15 inHg* for the lab plan. Enter this value into the lab plan charts. Refer to the theory section to ensure a correct conversion from gauge vacuum to absolute pressure in kPa. Indicate the absolute pressure in kPa in the *vapor pressure* cell of the chart. Refer to a vapor pressure chart similar to Figure L12.4 to find the temperature associated with the vapor pressure. Indicate this temperature as the *working temperature* in Figure L12.12 chart.

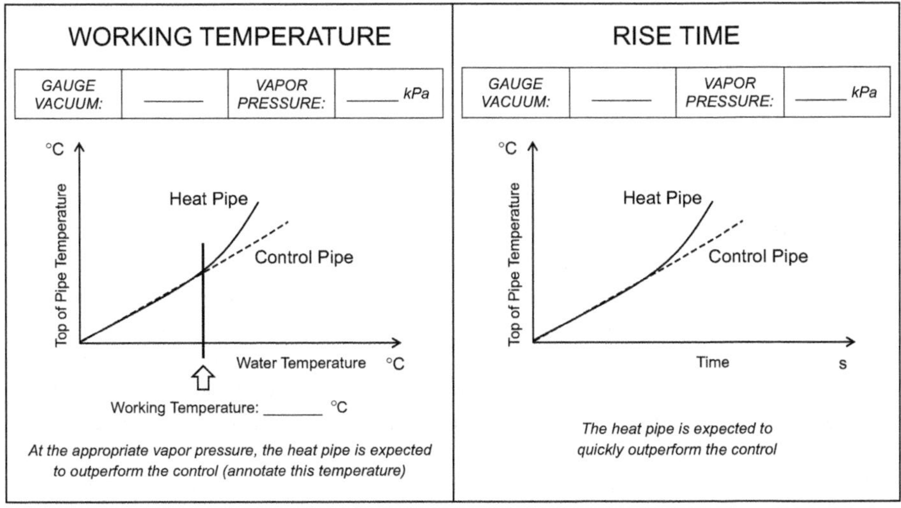

FIGURE L12.12 Heat pipe versus control pipe performance results. (*The line graphs must be extended for additional detail in lab plan.*)

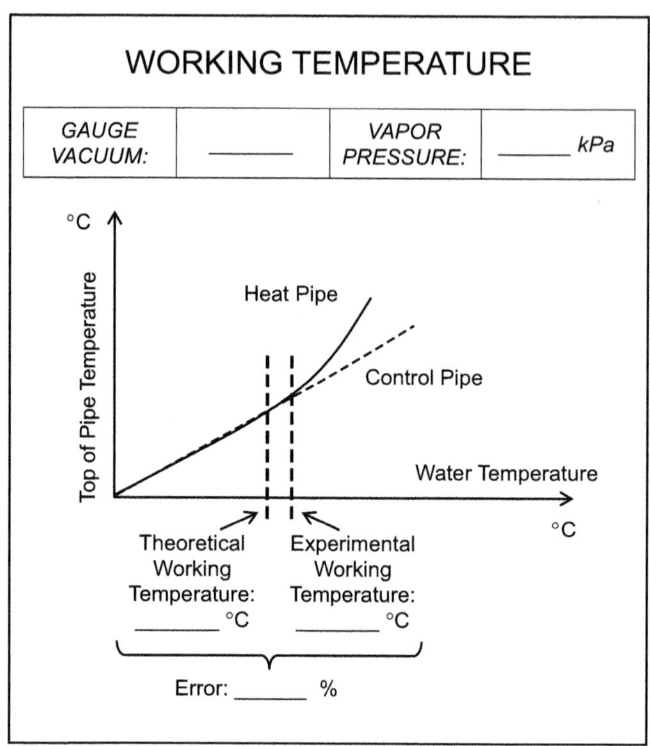

FIGURE L12.13 Working temperature chart with theoretical versus experimental values.

For the lab plan charts, continue the initial sketch lines in the example charts based on expected results. All temperature values will continue to rise to a maximum of boiling water temperature, but the heat pipe and control pipe data should reach this maximum temperature at different rates and times.

For the lab report, use DAQ data to re-create the Figure L12.12 charts, replacing the lab plan sketch charts. The working temperature chart also requires more information for the lab report, as shown in Figure L12.13.

Use the actual vacuum gauge value in the charts obtained from the heat pipe evacuation process. Convert this vacuum gauge value into vapor pressure, and enter this value in the lab report charts. This measured vapor pressure value provides the theoretical working temperature mark on the chart. Compare this theoretical working temperature value to the experimental working temperature value from DAQ data. Mark both temperatures on the chart and compute the error difference as indicated in Figure L12.13.

The working temperature chart is a temperature-versus-temperature chart. Heat pipe data is correlated to control pipe data as a function of water temperature. The chart is not difficult to create in a spreadsheet program. Use water temperature to create the horizontal axis rather than time.

DISCUSSION

Describe what is occurring with the data. Discuss any anomalies in the data along with any limitations or possible sources of error. In particular, discuss the possibility of pressure increasing in the heat pipe with temperature compared to the vacuum measurement taken during the evacuation process. Discuss how this might impact the results.

CONCLUSIONS

Based on the objective, clearly state the theoretical working temperature, the experimental working temperature and the error. Offer a conclusion for any potential differences. Offer a qualitative assessment of a heat pipe's ability to quickly and effectively transfer heat compared to conduction alone in the control pipe.

APPENDIX L12.A: HEAT EVACUATION METHOD

A torch can be used to evacuate air from the heat pipe instead of using a vacuum pump as shown in Figure L12.A1. This method does not require the modified refrigerant valve removal tool. A direct measure of vacuum is also not possible with this method. Experimental results indicate a vacuum of about *15–20 inHg*, and the heat pipe should begin working around 70°C.

The process should be completed with a bit of care to avoid heat-damaging the Schrader valve or desoldering the tube. The Schrader valve must be depressed open during the entire heating process. Heat will expand the air within the tube, and water should *just* begin to boil. The valve must be released at the first indication of water hissing or sputtering from the valve. Steam will displace the air. Upon cooling to

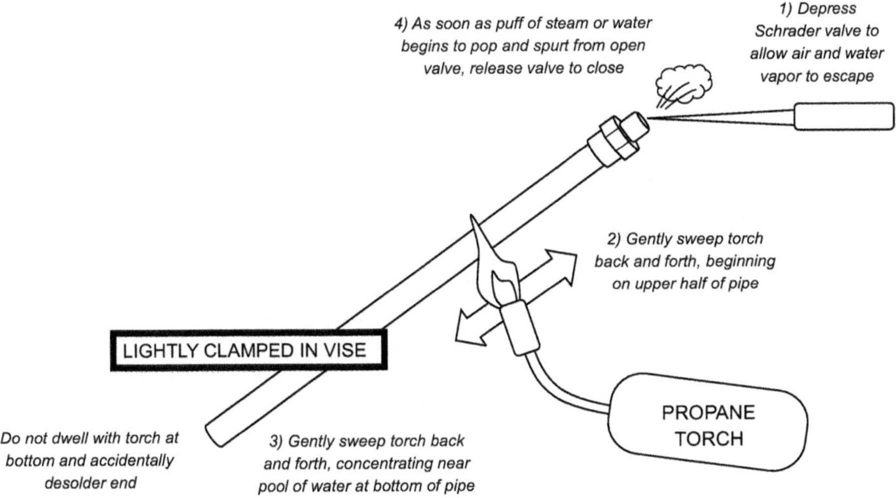

FIGURE L12.A1 Heat method procedure to evacuate air from heat pipe.

room temperature, the sealed tube will contain a vacuum and the heat pipe should function as desired.

If a large glob of water escapes from the heat pipe during the process, discontinue heating and cool the heat pipe. Perform a qualitative check to assess whether the heat pipe is functional. If not, the valve must be removed and the pipe completely drained and refilled.

APPENDIX L12.B: TROUBLESHOOTING

Common failure points are identified in the experimental methods section. Care should be taken to avoid these issues. If the heat pipe fails the qualitative test, begin with the easiest solutions first.

If the heat pipe is not getting hot near the top during the qualitative check, check further down the pipe. Sometimes the pipe will only function normally up to about 2″ from the top. If this is the case, adjust the location of both the heat pipe and the control thermocouples accordingly and run the test normally.

The next most common problem is the Schrader valve failing to hold a vacuum. Before any attempt at removing a Schrader valve from the tube, attempt to listen for a vacuum in the tube. Place the tube valve very close to a *young* ear with good hearing and use an awl to depress the valve. If the tube retained a vacuum, a slight hiss can definitely be heard as pressure equalizes. If there was no vacuum, the valve is suspect. If there was a vacuum, the water is suspect.

Although not designed to hold a vacuum, the high-performance Schrader valves have a very high success rate. Common Schrader valves normally supplied with fittings have a limited 40%–60% success rate and should be replaced with higher quality valves. Ensure that the valve is properly seated and securely tightened. If overtightened, check for damage and replace. The easiest solution is to simply replace any suspect valve. A very small dab of silicone grease can also be used at the valve seat to help ensure a vacuum seal.

Whenever a valve is removed for troubleshooting, it is a good opportunity to check for solder obstructions in the tube. Insert a stiff section of wire, such as Nichrome wire, into the tube and ensure the wire is not hindered. Poor soldering efforts sometimes result in a lump of solder partway down the tube. If this is the situation, the tube must be reconstructed from the beginning.

The next check is to ensure both sufficient and quality water. With the valve removed, use a piece of wood or PVC plastic to aggressively tamp the open end of the tube against a surface and knock out all of the residual water. Several drops should appear, amounting close to the original $0.5\,mL$. If little to no water can be dislodged from the tube, water was likely pulled out during the air evacuation step. Ensure the tube is empty of residual water and re-accomplish the filling procedure. If the water is *muddy*, too much flux was used during soldering and fouled the water. Add some water to clean out the tube. Once the water and flux have been removed, repeat the filling procedure.

The final check is another vacuum integrity check. The Schrader valve should be removed, and a hand vacuum pump with vinyl tube should be slipped securely over the end of the heat pipe. Pull a vacuum and check whether the vacuum is retained. If

TABLE L12.C1
Vacuum Holding Limits for Randomly Selected Schrader Valves

TRIAL:	1	2	3	4	5	6	7	8	9	10
COMMON VALVE FORCE (g):	237	59	167	105	252	0	73	181	265	104
PREMIUM VALVE FORCE (g):	376	364	412	409	306	382	391	352	358	418

not, check the vacuum pump separately since seals can leak in these devices. If the pipe fails the vacuum check, it must be rebuilt.

APPENDIX L12.C: ADDITIONAL NOTES

Heat pipe valve fitting. The brass heat pipe valve fitting is a common *refrigerant access valve*. The specification to fit the ¼″ copper tubing is a ¼″ ODS (Outside Diameter Solder) fitting. A common part number is A31724.

Valve removal tool. The refrigerant valve removal tool is meant to replace Schrader valves in a refrigerant system without losing refrigerant. The tool must be modified according to Figure L12.9. A small piece of stiff plastic or solder can be used to convert the tool into a Schrader valve depressor rather than a removal tool.

Schrader valves. Schrader valves are not designed to hold a vacuum, per this experimental configuration. However, the vacuum force generated on the small surface area is insignificant, and the valves are capable of holding a vacuum. Standard Schrader valves have significantly varied sealing ability, all the way to no ability to hold a vacuum whatsoever. Premium Schrader valves are easily sourced and demonstrate much better capability. Table L12.C1 provides an example of valve holding performance in a vacuum.

NOTE

1 Reay, D., Kew, P. & McGlen, R. (2014). *Heat pipes, theory, design and applications.* Elsevier. p.45.

LAB 13

Pitot Tubes

Safety Requirements

Protective gloves and safety are glasses required at *all times* while using a rotary tool
Protective gloves and safety glasses are required at *all times* while working with the test tube
Never handhold small parts while cutting – use a construction fixture only
Hold the rotary tool securely and do not change hands or set it down while bit is turning
Use SLOW speed on a rotary tool to cut plastic (high speed melts and fractures plastic)
Do NOT assemble parts with bare hands – use protective gloves since plastic could shatter
Do not force parts together – the plastic could shatter

Equipment Requirements

Eye protection, lab coat and protective gloves
Syringe, alcohol with food coloring and ¼″ polyvinyl tubing
Rotary tool, rotary #125 cutting bit 6.4 mm (¼″) diameter
12 mm plastic test tubes with caps, construction fixture and hot glue gun
Duct fan with nozzle, anemometer, needle nose pliers, scissors, calipers and dividers
Tall graduated cylinder, vegetable glycerin, plastic pellet, pellet scoop and stopwatch

Procedural Requirements

Use a construction fixture to work on small parts

Clean-up Requirements

Return all equipment to their proper locations
Clean up spills, plastic fragments and return manometer alcohol to the original container

Emergency Actions

MINOR CUTS: Clean affected area, use bandage kit on tool chest as required.
MAJOR CUTS/INJURY: Stabilize the affected area and apply direct pressure. Call 911.

INTRODUCTION

Vehicles in motion are subject to the effects of the fluid they move within. These effects include lift and drag forces. Lift and drag forces are functions of fluid properties such as velocity, density, viscosity and pressure. An ability to measure these fluid properties facilitates determining the forces acting on the vehicle and therefore the structural design requirements of the vehicle. Fluids include both liquids and gases, such as air.

Viscosity is a measure of a fluid's internal friction. Lower viscosity fluids have less internal friction and therefore flow more freely. Thicker fluids have greater viscosity and flow more slowly than low-viscosity fluids. A vehicle traveling in a higher viscosity fluid will experience greater skin friction drag for a given velocity. Ocean-going vessels are subjected to greater skin friction drag in the more viscous water than aircraft in the less viscous air at similar speeds. As velocity increases, forces also increase for both high- and low-viscosity fluids (non-Newtonian fluids are not considered).

A useful engineering parameter for vehicles moving within a fluid is the Reynolds number. The Reynolds number is a dimensionless parameter relating fluid viscosity, velocity, density and length of an object within the flow. Understanding the Reynolds number associated with fluid motion gives insight into whether the flow is laminar (smooth) or turbulent and where the transition point between laminar and turbulent flow tends to be located. The Reynolds number is also used to *scale* the performance of vehicles modeled at smaller sizes to full-sized production versions. The fluid dynamics of a small test model do not translate well to the dynamics of a larger version of the model. However, by creating flow environments for the smaller model with a Reynolds number similar to the Reynolds number of a larger model, comparable performance is more likely between the larger and smaller versions of the vehicle.

This lab explores both the viscosity and velocity of a fluid as key elements in a fluid dynamics problem. Viscosity is measured with a falling sphere, and velocity is measured with a Pitot tube and u-tube manometer.

OBJECTIVE

Determine the viscosity of a fluid relative to theory. Determine the velocity of a fluid relative to theory.

THEORY

TASK 1: Viscosity. Several methods are available to measure fluid viscosity. A *falling sphere viscometer* is used for this experiment. This method is only appropriate for extremely low Reynolds numbers ($Re < 1$) where a nearly buoyant sphere falls at a constant, terminal rate within a viscous fluid.

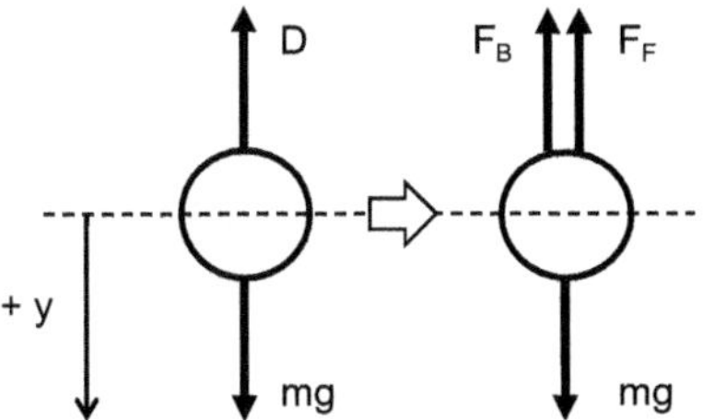

FIGURE L13.1 FBD of falling sphere.

The forces acting on an object falling with constant velocity in a fluid are depicted in the free-body diagram of Figure L13.1. The force due to the weight acting down is balanced by the drag force acting up. The drag force can be decomposed into both a buoyant force, F_B, and a skin friction force, F_F.

Since the sphere is at terminal velocity and not accelerating, the forces sum to zero, as indicated in Equations L13.1a and L13.1b. Each of the three terms in Equation L13.1b can be analyzed separately.

$$\sum F_y = 0 \tag{L13.1a}$$

$$mg - F_B - F_F = 0 \tag{L13.1b}$$

Weight. Mass can be expressed as density times volume so that weight is written as ρVg. Since this is the weight of a *spherical* object, an "*S*" subscript is used, $\rho_S V_S g$. The volume of a sphere is $V_S = \frac{4}{3}\pi r^3$. The first term for weight in Equation L13.1b is now replaced by $\frac{4}{3}\pi r^3 \rho_S g$.

Buoyant force. An object can float when the weight of the fluid displaced is equal to the weight of the object. For a relatively low-density object, the denser fluid is displaced by a fractional volume of the object, causing it to float atop the liquid. For a denser object, a buoyant force is still created, but it may be insufficient to keep the object afloat.

Since buoyant force is a function of displaced fluid volume, it is only a matter of calculating the weight of displaced fluid. If the object is completely submerged within the fluid, the weight of the displaced fluid is equal to the volume of the object, V_S, times the density of the fluid, ρ_F, times the acceleration due to gravity, $\rho_F V_S g$. Substituting the volume of a sphere gives an expression similar to the sphere weight term, differing only by the density of the fluid, $\frac{4}{3}\pi r^3 \rho_F g$.

Skin friction force. Force is pressure times area. The force of friction, F_F, for objects moving through a fluid is the dynamic pressure, $\left(\frac{1}{2}\rho_F v_\infty^2\right)$, times the object's frontal area to the fluid, A. Since an object's shape has an effect, an experimentally determined coefficient of drag, C_D, is also necessary to complete the relationship, $F_F = \frac{1}{2} C_D \rho_F A v_\infty^2$.

All three terms in Equation L13.1b now have greater detail for the result in Equation L13.2.

$$\left[\frac{4}{3}\pi r^3 \rho_S g\right]_{WEIGHT} - \left[\frac{4}{3}\pi r^3 \rho_F g\right]_{BUOYANCY} - \left[\frac{1}{2} C_D \rho_F A v_\infty^2\right]_{FRICTION} = 0 \quad \text{(L13.2)}$$

Reynolds number. Since the coefficient of drag, C_D, in Equation L13.2 is shape-dependent, a sphere in laminar flow demonstrates[1] a drag coefficient of $^{24}/_{Re}$, where "*Re*" is Reynolds number. A Reynolds number is a dimensionless ratio of inertial forces to viscous forces, defined by Equation L13.3, where L is the length of an object, at a velocity, v_∞, and ρ_f and μ are the density and viscosity of the fluid, respectively.

$$Re = \frac{\rho_F v_\infty L \;\; \leftarrow Inertia}{\mu \;\; \leftarrow Viscosity} \quad \text{(L13.3)}$$

A low Reynolds number means there is not enough inertia in the fluid movement to overcome viscosity, and the fluid remains laminar and well behaved. A high Reynolds number means inertia will take the fluid where it desires and there is not enough viscosity to prevent turbulent behavior.

Equation L13.4 is the drag coefficient for a sphere in very laminar, attached flow.

$$C_D = \frac{\rho_F v_\infty L}{24\mu} \quad \text{(L13.4)}$$

Considering only the skin friction force term in Equation L13.2, Equation L13.4 can be substituted for the drag coefficient for a sphere. The frontal area of a sphere is the area of a circle, πr^2, which can replace area, A. After some simplifications, the result for skin friction due to a sphere in laminar flow is Stokes' law in Equation L13.5.

$$F_F = 6\mu\pi r v_\infty \quad \text{(L13.5)}$$

Making the substitution of Stokes' law in Equation L13.5 for the friction force term in Equation L13.2, fluid viscosity can be solved in Equation L13.6.

$$\mu = \frac{2r^2 g(\rho_S - \rho_F)}{9v_\infty} \quad \text{(L13.6)}$$

where: μ is the fluid's dynamic viscosity
r is the radius of the sphere
g is the acceleration due to gravity
ρ_S is the density of the sphere
ρ_F is the density of the fluid
v_∞ is the velocity of sphere falling in fluid

Equation L13.6 is valid only for a sphere falling very slowly at a constant rate in a fluid with laminar, attached flow. To determine fluid viscosity using Equation L13.6, both the fluid and sphere densities are required. These are easily measured terms using the relationship between mass, density and volume, $m = \rho V$. Velocity is also easily measured as distance traveled per time in Equation L13.7.

$$V_\infty = \frac{d}{t} \tag{L13.7}$$

<u>TASK 2:</u> Velocity. Pitot tubes are standard equipment on aircraft to measure the velocity of the aircraft through air. The Pitot tube is a single probe placed within a flowing fluid that differentiates total pressure into both dynamic and static pressure.

A stationary fluid such as water within a lake or the Earth's atmosphere on a calm day has pressure at any given depth or altitude. This pressure is the total pressure of the fluid. For air at sea level on a standard day, this total pressure is 101.3 kPa.

If air is stationary, there is no dynamic component and static pressure is total pressure. If air is flowing, static pressure is reduced by the amount of dynamic pressure such that total pressure remains constant. For incompressible flow with no work or heat transfer, energy is conserved and total pressure must always remain constant. The difference between dynamic and static pressure is a measure of velocity within the fluid. A *u-tube* manometer connected to a Pitot tube probe can measure this pressure difference for conversion to velocity.

Bernoulli's equation. Bernoulli's equation for incompressible flow distinguishes between static and dynamic pressure. One of the simplest derivations begins with the conservation of total energy. By limiting energy possibilities to only kinetic and potential energy with no work or heat transfer, all the energy in a system appears as either kinetic or potential energy in Equation L13.8.

$$KE + PE = TE \tag{L13.8}$$

Kinetic energy is $\frac{1}{2}mv_\infty^2$, where mass is density times volume, ρV. Potential energy is mgh, where "h" is the height of the fluid column. Equation L13.8 can be re-written with these substitutions for Equation L13.9.

$$\left[\frac{1}{2}(\rho V)v_\infty^2\right]_{KE} + \left[(\rho V)gh\right]_{PE} = TE \tag{L13.9}$$

Since volume is constant for incompressible flow, it can be divided out from *every* term in Equation L13.9. The resulting units are pressure, and Equation L13.9 is transformed into Equation L13.10.

$$\left[\frac{1}{2}\rho v^2\right]_{DYNAMIC\ PRESSURE} + \left[\rho gh\right]_{STATIC\ PRESSURE} = Total\ Pressure \tag{L13.10}$$

Dynamic pressure is the kinetic term, and static pressure is the potential term. These two forms of pressure make up the total possible pressure for incompressible flow. Dynamic pressure is therefore the difference between total pressure and static pressure per Equation L13.11.

$$P_{DYNAMIC} = P_{TOTAL} - P_{STATIC} \tag{L13.11}$$

Solving Equation L13.10 for velocity, V, results in Equation L13.12.

$$V = \sqrt{\frac{2(P_{TOTAL} - P_{STATIC})}{\rho_{AIR}}} \tag{L13.12}$$

Based on Equation L13.12, velocity is a measure of the difference between total pressure and static pressure, along with the density of the air. A Pitot tube connected to a u-tube manometer provides this difference by having one port measure total pressure and a second port measure static pressure in the configuration shown in Figure L13.2.

In a stationary body of air with no velocity, a heavier fluid such as water will settle into a u-shaped tube at the same level on both sides of the tube, as shown in Figure L13.2. Static pressure is total pressure, and both the total port on the front of the Pitot tube and the static port on the side of the Pitot tube feel the same total static pressure. This total static pressure is the weight of the atmosphere pressing equally into the Pitot tube ports.

Once air begins to flow with a velocity toward the Pitot tube, the air gains a dynamic pressure component at the loss of total static pressure to conserve total pressure. Air entering the front of the Pitot tube has both the kinetic dynamic pressure components pushing into the tube and the static weight of the atmosphere. Air decelerates to zero velocity when it hits the surface of the water in the tube. This stagnated air is total pressure since it contains both dynamic and static pressure components.

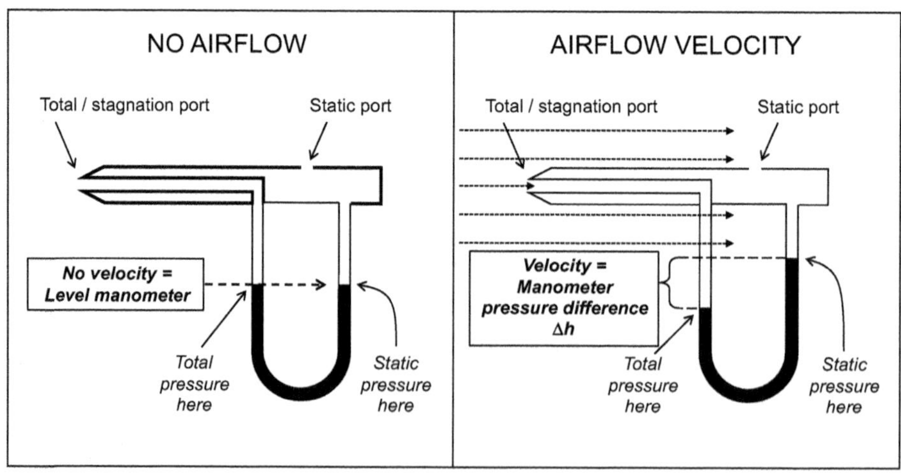

FIGURE L13.2 Pitot tube with u-tube manometer.

Air flowing past the static port on the side of the Pitot tube does not feel the oncoming dynamic pressure component. The static port only feels the static weight of the atmosphere. Since total pressure cannot magically increase and must be conserved, this static *weight* is actually less than total by the amount of the dynamic component.

Since *total pressure* pushes on one column of the u-tube water and *less-than-total-pressure* presses on the other column of u-tube water, there is a height difference in the u-tube fluid as shown in Figure L13.2. This height is the difference between total pressure and static pressure, exactly what is required in the radical of Equation L13.12.

Equation L13.12 must still be modified for use since it was developed considering only the flow of air and not the liquid within the tube. Since air has no velocity at the surface of the u-tube water on both columns, the air is static in both columns, by definition. To clarify terms, only the static port column is referred to as *static* since it contains only the static pressure component. The total pressure port is referred to as the *stagnation* port since, while static, it contains both the dynamic and static pressure components.

Since the u-tube height difference is a static height difference, the static pressure formula in Equation L13.10, $\rho g \Delta h$, applies where Δh dictates the difference in column heights. For equilibrium, where both water and air hold their relative height differences between the two u-tube columns, the air pressure difference is *necessarily* equivalent to the same measure of liquid pressure difference. In other words, the quantity, $p_{TOTAL} - p_{STATIC}$, in Equation L13.12 is exactly the same as the liquid static pressure difference due to height, $\rho_{LIQUID} g \Delta h$. This substitution can be made in Equation L13.12 for the final form of velocity in Equation L13.13.

$$v = \sqrt{\frac{2(\rho_{LIQUID} g \Delta h)}{\rho_{AIR}}} \quad \text{(L13.13)}$$

The velocity of air can be computed based on the change in liquid height per Equation L13.13. Since alcohol is slightly less dense than water, it offers a greater Δh for measurements compared to water.

EXPERIMENTAL METHOD

This experiment is divided into two tasks. The first task is a viscosity experiment, and the second task is a velocity experiment. Prepare a single summarizing results table similar to Table L13.1 in the results section to present results from both experiments.

TASK 1: Viscosity. Prepare a viscosity data collection table similar to Table L13.A1 in Appendix A. The expected results must be calculated for the lab plan. In this case, the expected result is the timing of the sphere falling in the glycerin. Without expected timing, it is uncertain whether a stopwatch with tenth-of-a-second precision is necessary, or whether a smart phone timing in only seconds is adequate.

Look up the viscosity of glycerin and enter this value in Table L13.A1. With the published viscosity value, the table has enough estimated information to complete the calculations for the expected timing in the lab plan. Equation L13.6 must be solved for velocity using the published viscosity of glycerin and estimated sphere parameters. Use Equation L13.7 to calculate the expected time and enter this time value into the first T1 row of Table L13.A1 in parentheses.

Include at least three data run lines to record the fall time of a sphere within a fluid-filled cylinder. Include an entry to convert measured time into velocity based on the distance the sphere drops within the liquid using Equation L13.7. During the lab, update all parameters in Table L13.A1 as necessary.

Fill a tall graduated cylinder with a viscous fluid, such as glycerin. Fill the cylinder slowly and allow the glycerin to settle to release any air bubbles. It is best to prep the glycerin overnight so any air bubbles have a chance to escape.

Enter the density of glycerin in Table L13.1 and Table L13.A1. This truth value density can be from an online source look-up. Alternatively, density can be computed by using a scale to measure mass in a known beaker volume, using the relationship, $m = \rho V$.

Mark and measure an upper- and lower height location on the graduated cylinder of glycerin. The measurement locations should maximize travel distance in the glycerin, but the top mark should be about a couple of centimeters below the surface level to allow for the sphere to stabilize drop velocity. Record this timing distance in Table L13.A1.

Obtain a small plastic sphere that is only slightly negatively buoyant, so it drops quite slowly within the glycerin. Fast-dropping spheres of metal or glass have separated flow and do not adhere to the assumptions of Stokes' law.

Measure the sphere's diameter and calculate the volume using the formula for volume of a sphere. Weigh the sphere and determine the sphere's mass. Calculate the sphere density, ρ_S, using the formula $m = \rho V$. Enter all of these values into Table L13.A1.

Have a timer ready and lower the sphere into the fluid within the graduated cylinder to get past the surface tension. Do not drop the sphere from above the fluid. Release the sphere below the glycerin surface and allow it to drop freely. Begin timing when the sphere drops past the upper timing mark. Stop timing when the sphere passes the lower timing mark. Retrieve the sphere and repeat for a total of three test runs. Record these values in Table L13.A1.

TASK 2: Velocity. Two options are available to construct a Pitot tube. The first option is a traditional construction technique detailed in Appendix L13.B to create a Pitot tube from a plastic test tube. The second option in Appendix L13.C is to 3D print a Pitot tube. Follow the desired construction process, and then complete the u-tube manometer experiment.

Prepare a simple data collection table similar to Table L13.A2 in Appendix A. For expected results, look up the densities of isopropyl alcohol and air and enter these values into the table for the lab plan. Use an estimated *10 m/s* for air velocity. Solve Equation L13.13 for Δh and calculate the expected height difference based on the planned *10 m/s* velocity. Enter this value in parentheses in Table L13.A2 for the lab plan. This expected result provides an idea of the height measurement at *10 m/s*. If the

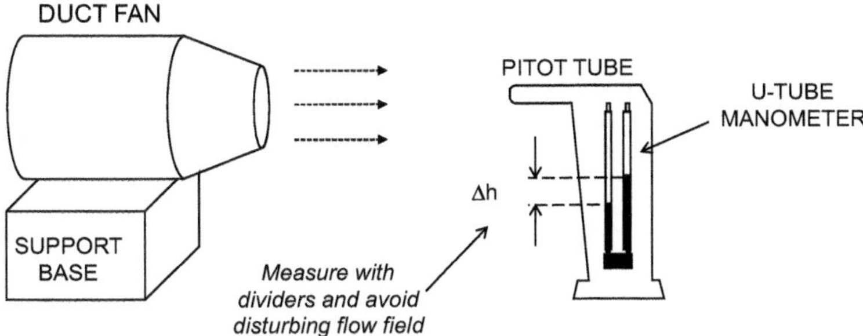

FIGURE L13.3 Experimental setup.

expected height were to be greater than the u-tube manometer, then the experiment must be modified during the lab plan phase.

With the Pitot tube construction complete, partially fill the clear polyvinyl tubing with liquid for the manometer. Standard isopropyl alcohol is readily available and less dense than water. Use a few drops of food coloring in the alcohol to improve visible contrast. It is easiest to insert the plastic tip of a luer slip syringe into one end of the tubing and the other end of the tubing into the alcohol container. Draw the alcohol up into the tubing with suction from the syringe to prevent trapping air bubbles in the tubing. Assemble the Pitot tube with alcohol in the u-tube and place everything into the experimental test fixture according to Figure L13.B5 or L13.C3, depending on the appropriate Pitot tube version.

Place the entire Pitot tube fixture into the airflow path of a duct fan as shown in Figure L13.3. Turn on the fan and adjust the Pitot tube position and fan power to see a height difference in the u-tube manometer.

Measuring the height difference in the u-tube requires care since hands and instruments in the flow field will disturb the flow field and affect the results. Minimize airflow disturbance by using slender dividers to determine the height difference. Then, use calipers or a ruler to measure the span from the dividers and record the height difference. Record this experimental value in Table L13.1 in the results section.

Once a reliable manometer value has been recorded, use the handheld anemometer to record the velocity of the air as truth data at the same location of the Pitot tube.

RESULTS

Table L13.1 summarizes the results of the viscosity and velocity experiments of a fluid along with the percentage error compared to the truth data.

For the experimental viscosity value in Table L13.1, use the velocity calculated from the data collection table, Table L13.A1 in Equation L13.6. Compute the percentage error.

For the experimental velocity, calculate the velocity using Equation L13.13 and the data recorded in Table L13.A2. Enter the experimental velocity from the Pitot tube and the truth value anemometer velocity in Table L13.1. Calculate the error.

TABLE L13.1
Viscosity Parameters and Results

	Viscosity	Velocity
Truth value	*(Look up or measure)*	*(Anemometer)*
Experimental	*(Equation L13.6)*	*(Pitot tube, Equation L13.13)*
Error		

DISCUSSION

Discuss the results and any challenges in the data collection. Note any anomalies during the experiment that might have impacted the data. Discuss how disturbances in the airflow due to measurement attempts affected the airflow and why this happened.

CONCLUSIONS

Based on the objective of the experiment, clearly state the viscosity and velocity values along with the percentage error.

APPENDIX L13.A: DATA COLLECTION TABLES

TABLE L13.A1
Viscosity Parameters and Results

GLYCERIN	PARAMETERS
Density	
SPHERE	
Diameter	*(6 mm)*
Mass	*(0.2 g)*
Density	
Estimated v_∞	
Distance	*(30 cm)*
TEST RUNS	**DATA**
t_1	
t_2	
t_3	
t_{AVG}	
v_∞	*(Equation L13.7)*

TABLE L13.A2
Velocity Parameters and Results

ALCOHOL	PARAMETERS
Density	
Δh	
AIR	
Density	
Velocity	*(10 m/s)*

APPENDIX L13.B: TRADITIONAL CONSTRUCTION TECHNIQUE

This construction technique uses a rotary tool to cut holes into a plastic test tube to serve as the Pitot tube. ***Safety precautions.*** Plastic test tubes can shatter into shards. Never attempt to fabricate small parts using a power tool without the aid of a construction fixture to securely hold the parts. Use the construction fixture recommended in Appendix L13.B7. Eye protection and protective gloves must be worn at all times while constructing the Pitot tube.

Construct a Pitot tube from a 12 mm × 100 mm plastic test tube and cap. Insert the test tube into the construction fixture so the rounded bottom end of the test tube protrudes upward, as shown in Figure L13.B1. The bottom of the test tube will be the front of the Pitot tube with the stagnation port.

Use a #125 cutting bit to carefully cut a ¼″ hole directly into the bottom of the test tube. Use **slow speed** on the tool, and **slowly increase** the hole size by frequency backing out the cutting bit. Take your time, or the plastic tube will fracture. The cutting tool has a max diameter of ¼″, so it needs to cut completely through the bottom of the test tube for the ¼″ vinyl tubing to fit. Clean up the cut hole as necessary with sandpaper. Small fractures around the hole are fine and can be sealed later with hot glue.

Reorient the test tube in the fixture as shown in Figure L13.B2 with the open end of the test tube (top) inserted into the slot. Cut an oblong hole into the side of the test tube between approximately ½″ to 1″ from the open end (top) of the test tube. The ¼″ plastic tubing must be able to fit through this hole at an angle to avoid kinking.

After cutting the oblong hole, reorient the test tube in the slot. Turn the test tube around in the same slot, with the round end inserted first. Use the rotary tool to cut two or three additional small holes on the side of the test tube. These holes are the static ports and should only be large enough to penetrate the side of the test tube and no larger. The holes should also not be obstructed by the Pitot tube fixture or by the vinyl tubing that will be placed within the test tube. Drill these small holes as desired on any side *other than* the oblong hole. If the holes are poorly located and obstructed when the Pitot tube is complete, just drill a couple extra small holes where airflow is not obstructed.

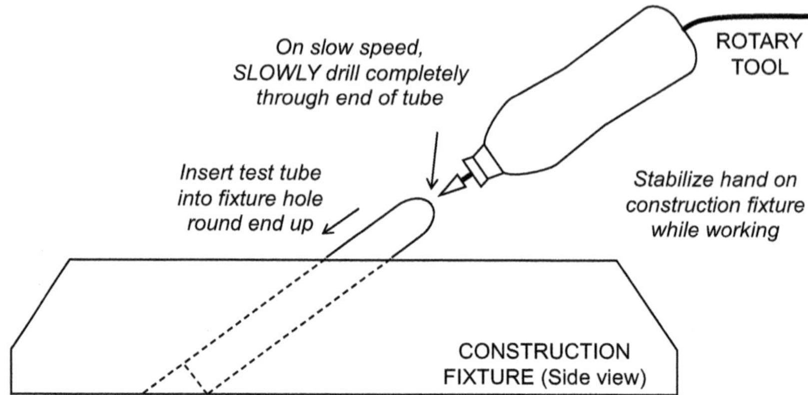

FIGURE L13.B1 Using construction jig to cut hole in the bottom of test tube.

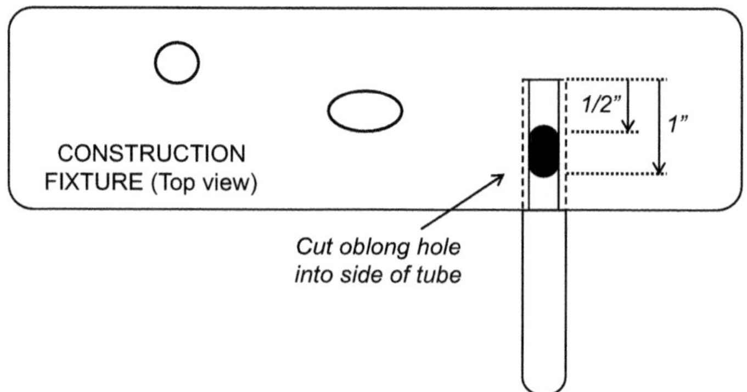

FIGURE L13.B2 Test tube oriented in slot with oblong hole cut into side.

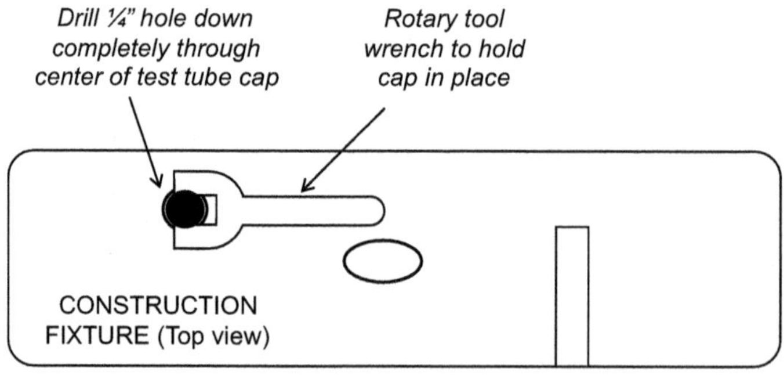

FIGURE L13.B3 Cut one-quarter inch hole through test tube cap.

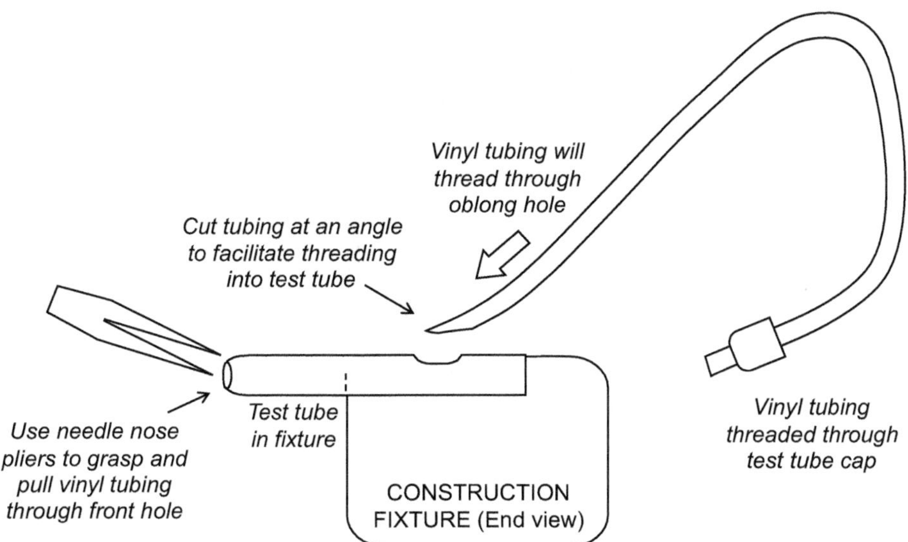

FIGURE L13.B4 Threading vinyl tubing through test tube.

Place the plastic test tube cap into the construction fixture as shown in Figure L13.B3. Use the rotary tool to cut a ¼″ hole through the middle of the cap. Cut completely through the cap so the ¼″ tubing will fit. Do not use hands to directly hold the cap in place. Use a tool such as the rotary tool bit wrench to slip onto the cap and press firmly into the fixture while working.

Cut approximately 26″ length of ¼″ polyvinyl tubing, cutting one end at an angle as shown in Figure L13.B4. Place the test tube into the construction jig as shown in the figure. Feed the cut end of the tubing through the oblong hole in the side of the test tube, working the vinyl tubing through the inside of the test tube toward the hole cut into the bottom of the test tube (front of Pitot tube). Keep the plastic test tube in the fixture to avoid putting too much pressure on the plastic, causing it to shatter. If the vinyl tubing will not go through the oblong hole without kinking, remove the tubing and use the rotary tool with test tube placed back in the construction fixture and adjust the size of the oblong hole.

When the vinyl tubing is inserted all the way through the test tube near the hole at the front of the test tube, use small needle nose pliers to grasp through the hole onto the end of the tubing and pull through the hole. Trim the excess tubing at the front of the Pitot tube. Use hot glue to seal any air leaks at the front of the Pitot tube. When the Pitot tube is in the final test configuration, the assembly should look like Figure L13.B5.

Figure L13.B6 provides an example Pitot tube experimental test fixture. The fixture can be cut from wood or 3D printed. Quarter-inch channels should be routed or milled into the wood to press the ¼″ clear tubing. Dimensions are non-critical. A small clip can be used to hold the Pitot in place in the fixture.

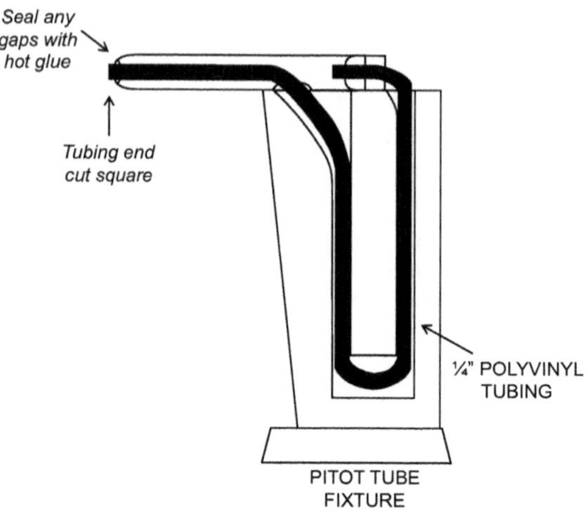

FIGURE L13.B5 Pitot tube fully assembled in experimental test fixture.

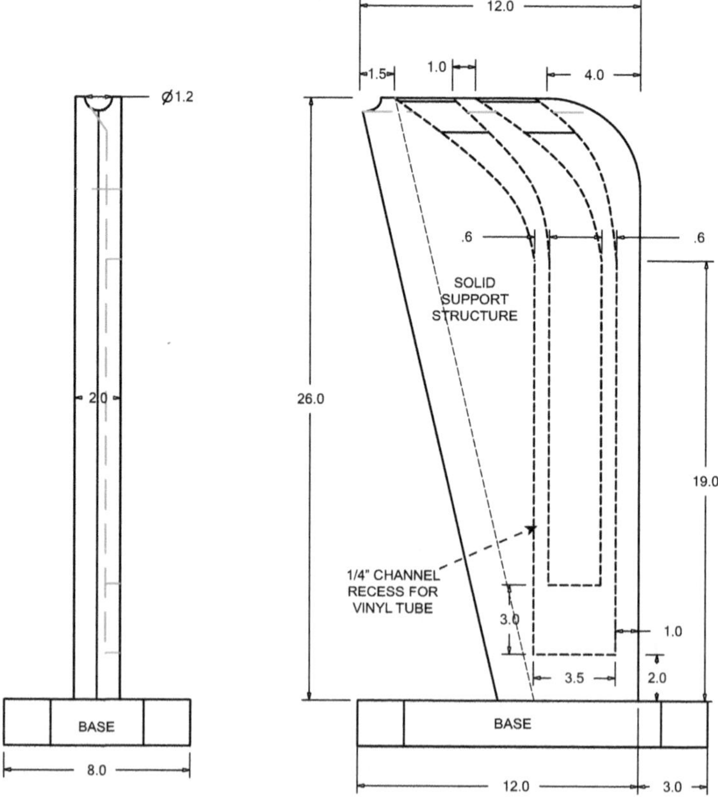

FIGURE L13.B6 Pitot tube experimental test fixture. Dimensions in *cm* are non-critical.

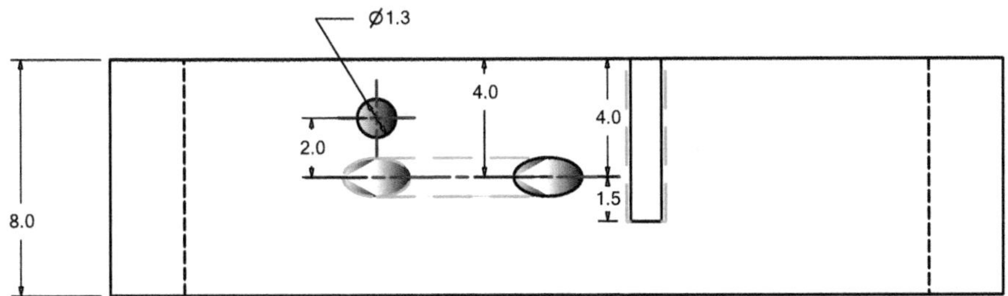

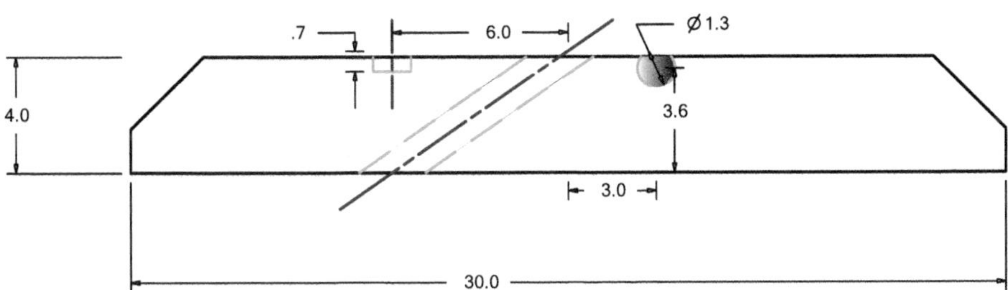

FIGURE L13.B7 Construction fixture block. Dimensions in *cm* are non-critical.

Figure L13.B7 provides an example of the construction fixture. The fixture can be cut from 2×4″ wood or 3D printed. The holes and slots allow for holding the plastic test tube in place for all construction activities.

APPENDIX L13.C: 3D PRINT CONSTRUCTION TECHNIQUE

A Pitot tube can be 3D printed with ports for static and dynamic pressure. Several designs are possible with the following design in Figure L13.C1 suitable for vertical print orientation and ports to fit ¼″ polyvinyl tubing. Using a 3D printed junction at the bottom of the u-tube manometer allows running two closely spaced parallel vinyl tubes to facilitate easier manometer measurements. Dimensions are non-critical. Follow Figures L13.C2 and L13.C3 for complete assembly of the Pitot tube.

Figure L13.C4 provides an example 3D printed Pitot tube experimental test fixture. The fixture can be cut from wood or 3D printed. Quarter-inch channels should be routed or milled into the wood to press the ¼″ clear tubing. Dimensions are non-critical. A small clip can be used to hold the Pitot in place in the fixture.

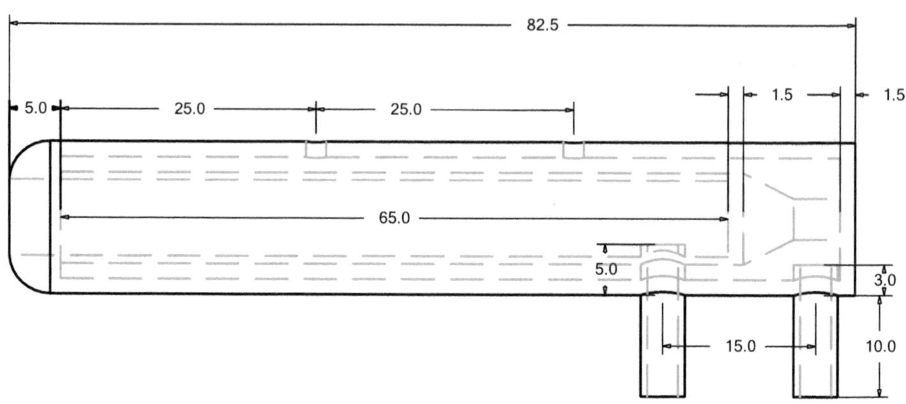

FIGURE L13.C1 Suggested 3D printed Pitot tube design.

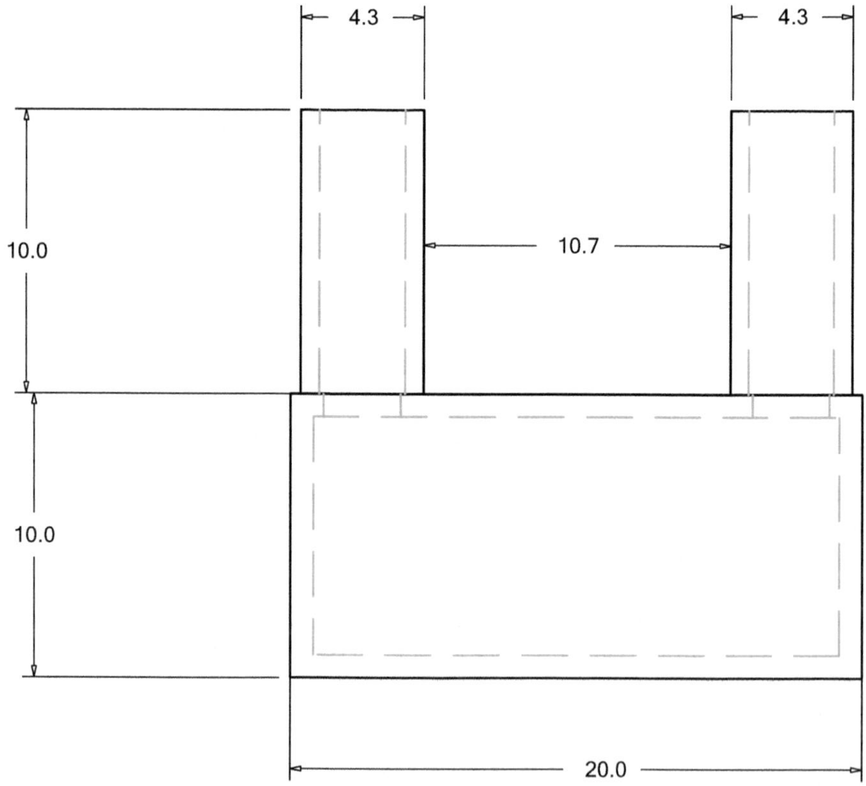

4.3
4.3
10.0
10.7
10.0
20.0

FIGURE L13.C2 Lower fitting to connect polyvinyl tubing.

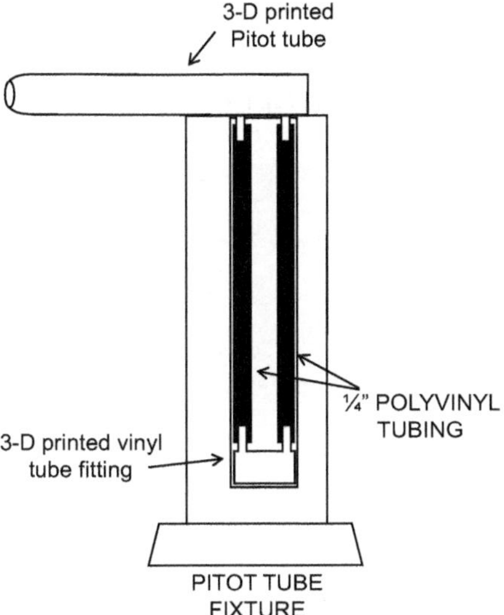

FIGURE L13.C3 3D printed Pitot tube and fitting fully assembled in experimental test fixture.

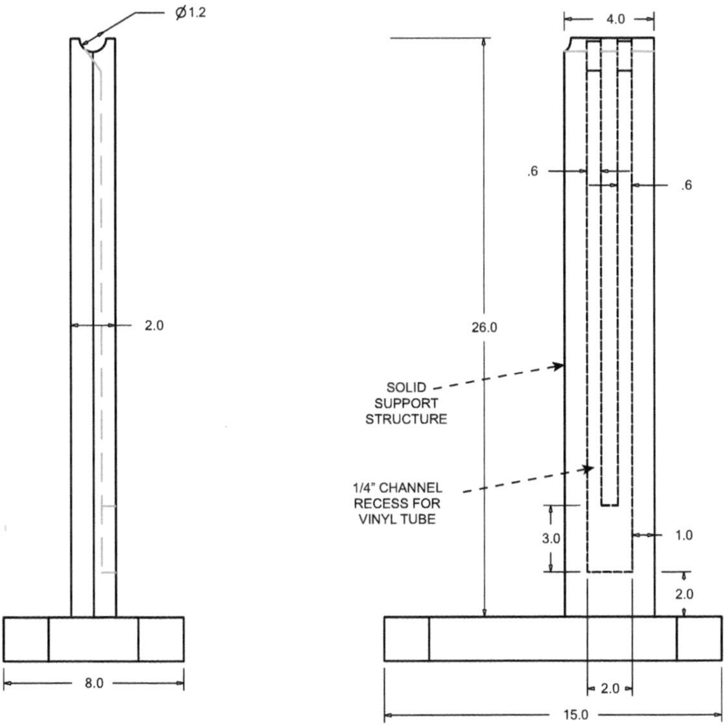

FIGURE L13.C4 3D printed Pitot tube experimental test fixture.

NOTE

1 Eames, I. & Klettner, C. (n.d.). *Stokes' and Lamb's viscous drag laws.* University College London. https://discovery.ucl.ac.uk/id/eprint/10073838/1/Eames_Klettner_.pdf.

LAB 14
Pressure

Safety Requirements

None

Equipment Requirements

Spreadsheet program

Procedural Requirements

No lab plan required
Reports must be *a 100% individual effort* with no collaboration allowed
No length requirements; only length sufficient to cover the material
Report must be uniquely and originally written
Follow the instructions in this lab manual to write the lab report
Do not copy or modify the lab manual introduction; write an original introduction
Report must include one real-life example of a device using pressure measurements
Introduction must introduce and discuss the real-life pressure device
Include all report requirements: cover page, abstract, sections and references

Clean-up Requirements

None

INTRODUCTION

We live near the bottom of a fluid sea called air. When calm, this fluid presses against us with the pressure of one atmosphere on a standard day at sea level. This is total pressure. It is also static pressure when calm. If we move within this fluid or if the fluid is moving locally around us, this total pressure has components of both static and dynamic pressure. Total pressure still remains the same. Only as velocity significantly increases can air compress and pressure begin to increase more than total ambient pressure. This velocity ram-compressed air is called stagnation pressure. Total ambient pressure is still the same, so stagnation pressure is total pressure accounting for increases in pressure due to compressibility effects.

This lab is a computational-only exercise utilizing a spreadsheet to compute the various pressure components of air as a function of velocity. Many devices rely on the ability to measure these various pressure components of air. A pressure measurement device is also discussed as an example.

OBJECTIVE

Define and compute the pressure components of air. Describe a pressure measurement device and how it works.

THEORY

Research ambient, static, dynamic and stagnation pressure concepts. Write an individual and original theory section that includes a minimum of one paragraph describing each of the following concepts:

- Total (far-field) pressure
- Static pressure
- When does static pressure equal total pressure and when it does not
- The relationship between ambient and total pressure
- Dynamic pressure
- Stagnation pressure
- When does stagnation pressure equal total pressure and when it does not
- Copy and paste the following two paragraphs with equations:

Bernoulli's equation provides a relationship between total, static and dynamic pressures according to Equation L14.1. Dynamic pressure is the *difference* between total pressure and static pressure.

$$\left[\frac{1}{2}\rho v^2\right]_{DYNAMIC\ PRESSURE} + \left[\rho g h\right]_{STATIC\ PRESSURE} = Total\ Pressure \qquad (L14.1)$$

where ρ is the density of fluid (air)
v is the velocity through the fluid (air)
g is the acceleration due to gravity
h is the height of the fluid column

Stagnation pressure is isentropically compressible air, given in Equation L14.2.[1]

$$P_{STAGNATION\ PRESSURE} = P\left[1+\left(\frac{v^2}{2C_P T}\right)\right]^{\gamma/\gamma-1} \qquad (L14.2)$$

where P is the total ambient pressure of air
v is the velocity of the air
C_P is the specific heat of air at constant pressure
T is the ambient temperature of air
γ is the ratio of specific heats for air

COMPUTATIONAL METHOD

Use the example in Appendix L13.A as a guide to create a spreadsheet for calculating static, dynamic and stagnation pressures from total pressure at velocities from zero to any desired subsonic airspeed at any desired intervals. Ensure a high enough subsonic airspeed to show a graphical divergence between total and stagnation pressures. Prepare a chart similar to Figure L14.1 in the results section to summarize the tabular pressure data of the spreadsheet.

Consider an aircraft flying at any reasonable altitude and temperature, and look up the total pressure for these ambient conditions. Begin by entering this ambient pressure into the *total pressure* column of the spreadsheet. Use just the dynamic pressure term in Equation L14.1 to calculate *dynamic pressure* in the spreadsheet based on velocity and ambient density. Use the relationship between the three pressure terms in Equation L14.1 to calculate *static pressure* from total and dynamic pressures. Look up the gas constants for air in Equation L14.2. Use Equation L14.2 to calculate *stagnation pressure* from total pressure.

Include the spreadsheet formulas and calculations in Appendix L13.A. Limit the data table to one-page maximum for the appendix. Include all units in the spreadsheet headings and ensure unit consistency (the example in Appendix L13.A does not include units).

RESULTS

Figure L14.1 depicts the relationship between total, static, dynamic and stagnation pressures at the indicated ambient conditions based on the calculations tabulated in Appendix L14.A.

The line chart in Figure L14.1 must *clearly* indicate relationships between pressures. This may require a modification of the chart since a default plot from zero to

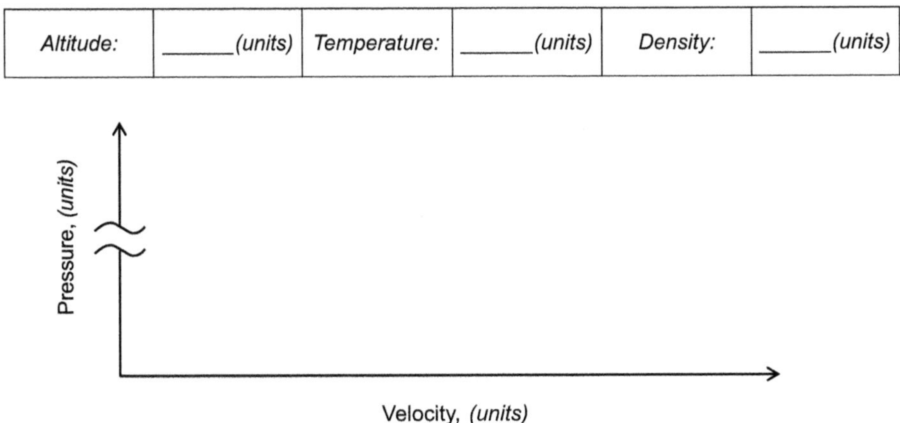

FIGURE L14.1 Graphical relationship of pressure components.

the maximum pressure may visually compress the lines. Do not provide a chart with difficult-to-discern relationships.

DISCUSSION

Describe what is happening with each of the pressures as velocity increases. Refer specifically to the data and indicate how changes in static and dynamic pressures are interrelated. Describe how stagnation pressure can increase beyond total pressure.

CONCLUSIONS

Based on the objective of the experiment, discuss the real-world pressure measurement device and which of the pressure components are measured. Discuss how these pressure components are measured.

APPENDIX L14.A: EXAMPLE SPREADSHEET CONFIGURATION FOR PRESSURE CALCULATIONS

ALTITUDE:	**DENSITY:**
AMBIENT PRESSURE:	**GAMMA:**
AMBIENT TEMPERATURE:	**C_P:**

VELOCITY	**TOTAL PRESSURE**	**DYNAMIC PRESSURE**	**STATIC PRESSURE**	**STAGNATION PRESSURE**
(0)	*(Look up)*	*(Equation L14.1)*	*(Equation L14.1)*	*(Equation L14.2)*
(50)	↓	↓	↓	↓
↓				
(300)				

NOTE

1 Re-derived in alternate form from: Anderson, J. (1978). *Introduction to flight*. McGraw-Hill. p.101.

LAB 15
Aerodynamic Force

Safety Requirements

None

Equipment Requirements

Wind tunnel, computer station with computational fluid dynamics (CFD) software capabilities
Laptop or computer with spreadsheet software

Procedural Requirements

Follow operating instructions for the wind tunnel
Data computation spreadsheet must be complete with all formulas prior to lab
Team members work together to complete the CFD simulation

Clean-up Requirements

None

Emergency Actions

No unusual hazards

INTRODUCTION

An airfoil creates lift from a net pressure difference between the top and bottom of the airfoil as air flows past the surfaces. According to Bernoulli's principle, fluids such as air flowing past a constriction will flow at a greater velocity and a correspondingly lower static pressure. This same principle creates a higher velocity jet of water when a nozzle constriction is used on a garden hose compared to a hose with no nozzle constriction. Rocket engines also develop thrust by increasing the velocity flow through a nozzle constriction. The cambered upper surface of an airfoil creates a constriction in the flow field, forcing air up and over the cambered surface, increasing the flow velocity across the upper surface with a corresponding decrease in static pressure. The lower surface of an airfoil is typically less cambered than the upper surface, ensuring the airflow velocity over the upper surface is greater than the lower surface. The lower surface has an overall higher static pressure under the airfoil compared to the upper surface.

If the pressure distribution around an airfoil can be determined, the lifting force of the airfoil can be calculated and used in the aircraft design process. Since the pressure distribution is dependent on airfoil shape, each airfoil profile has a unique, experimentally determined coefficient of lift. An airfoil's coefficient of lift describes the maximum lift available from the airfoil for a given set of conditions.

This lab uses CFD to validate the published coefficient of lift curve for a given airfoil. Once validated, the CFD input parameters are adjusted to simulate local conditions for a wind tunnel test. A coefficient of lift curve is generated from experimental wind tunnel data and compared to the validated curve from the CFD simulation.

Only the coefficient of the lift curve is explored in this lab, with no consideration given to drag forces. More advanced theories of lift due to the Kutta condition and circulation theory are also not considered here. Finally, no practical distinction is made between an airfoil and a wing in this lab. Wings have 3D effects due to wingtips, while airfoils are assumed to have constant spanwise properties used in the wind tunnel tests of this lab. This lab purposely uses SAE (Standard American English) units common to the aviation industry.

OBJECTIVE

Validate a CFD coefficient of lift curve for a NACA2415 airfoil against published data. Develop a coefficient of lift curve from wind tunnel data and compare it to the CFD curve for similar conditions.

THEORY

Airfoils. An airfoil cross section has several features important to generate lift, as depicted in Figure L15.1. These features are briefly defined as follows:

- Leading edge (LE): front of the airfoil that first contacts and divides the oncoming airflow
- Trailing edge (TE): aft part of airfoil where divided airflow rejoins
- Chord line: a straight line connecting the LE and TE
- Chord (or chord length): length of the chord line
- Thickness: the greatest distance between the upper and lower surface of the airfoil

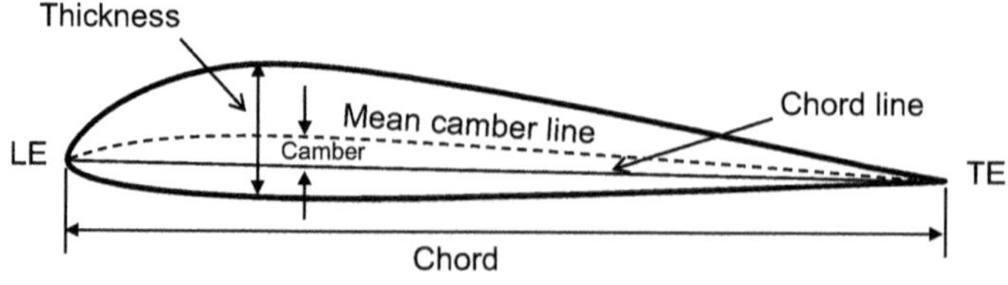

FIGURE L15.1 Airfoil nomenclature.

- Mean camber line: a (typically) curved line equidistant between top and bottom surfaces
- Camber: the distance between the chord line and the mean camber line

The angle between oncoming airflow and the chord line is the *angle of attack* of the airfoil, as shown in Figure L15.2. Lift is a function of both airfoil geometry and angle of attack.

As airflow divides and flows across the upper and lower surfaces of the airfoil, the airflow accelerates across both surfaces. The upper airfoil surface is usually more cambered, so airflow is accelerated to a higher velocity across the upper surface. As a result, low pressure zones are created on both the upper and lower surfaces of the airfoil. The differential velocities ensure a lower pressure zone on the upper surface compared to the lower surface. If the airfoil shape or angle of attack changes, the pressure distribution changes and the overall lift may be positive or negative.

The total pressure distribution across the upper and lower surfaces can be resolved into a single force vector. This total force vector due to the pressure distribution can be broken into y- and x-components, called lift and drag, as depicted in Figure L15.2.

Pressure field distribution. Different (and potentially confusing) sign conventions are used for calculating lift based on pressure distributions. An important consideration is that pressure is not a vector, so it has no direction, despite any arrows often used in descriptions. For incompressible flow, total pressure is conserved, so there is no net negative or positive pressure field around an airfoil. However, for a flowing velocity vector field around an airfoil, conserved total pressure resolves into both a static and dynamic component. While dynamic pressure acts parallel to the airfoil surface, static pressure acts normal to the airfoil surface. It is the static pressure component, normal to the airfoil surfaces, that is responsible for creating lift. Static pressure is reduced locally on both the upper and lower surfaces of the airfoil. Some sign conventions indicate *positive* pressure pushing upward on the lower surface of an airfoil. However, the entire static pressure field around an airfoil is *negative* (reduced pressure), and *tugs outward* on the surface rather than "*pushing up*" on the lower surface.

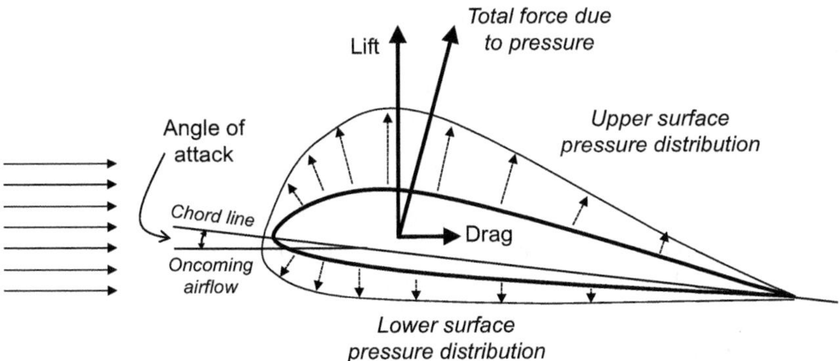

FIGURE L15.2 Aerodynamics of an airfoil.

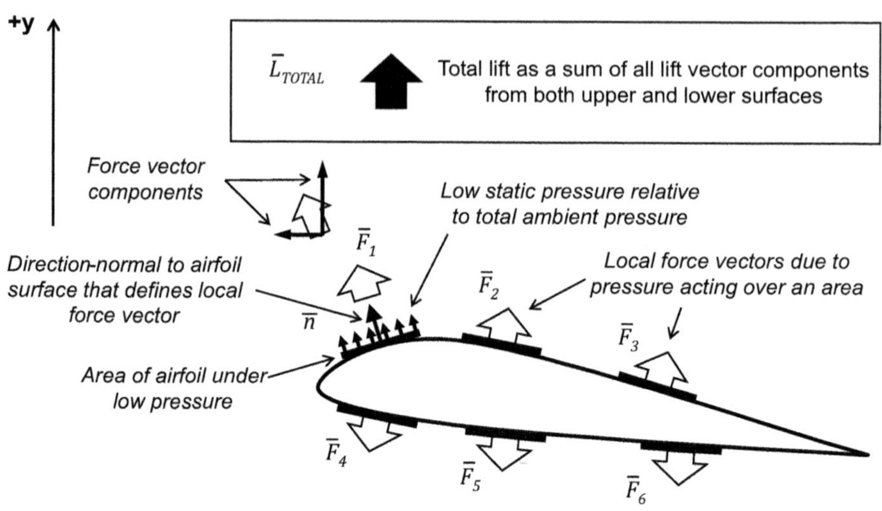

FIGURE L15.3 Force vectors for lift.

Lift vector convention. Lift is a vector that is calculated by multiplying pressure acting on an area. A vector sign convention should always be established, and in this case, positive is up as shown in Figure L15.3. Both the pressure distribution and the airfoil surface orientation vary, so varying local force vectors must be summed (integrated) across the upper and lower surfaces. These localized vectors are defined by the surface-normal unit vectors depicted in Figure L15.3. Based on the sign convention in Figure L15.3, force vectors on the upper surface are positive, and vectors acting along the lower surface of the airfoil are negative.

Total lift. Lift is the integration of all the varying pressure distributions and varying differential areas in Figure L15.3. Given the complexity of integrating over a curved surface, the integration problem is greatly simplified by integrating only along the straight airfoil chord from LE to TE. Also, airfoils do not have a theoretical pressure *function* available for integration. The pressure distributions are unique to each airfoil depending on the shape (along with other parameters) and must be determined experimentally for each airfoil. As a result, airfoils are designed, then configured with pressure-sensing ports and tested in wind tunnels to determine the pressure distributions. A typical configuration of pressure-sensing ports for an airfoil is shown in Figure L15.4.

To determine lift, each of the discrete pressures sensed by ports in Figure L15.4 must be multiplied by the average sensing area and summed for an overall force. Following the sign convention in Figure L15.3, the upper surface force vectors are positive and contribute to lift, while the lower surface force vectors act in the negative y-direction and detract from lift. Total lift is the sum of the upper surface force vectors *plus the negative* lower surface vectors in Equation L15.1. Regardless of the positive or negative sign often used in Equation L15.1, total lift is the *difference* between upper and lower surfaces.

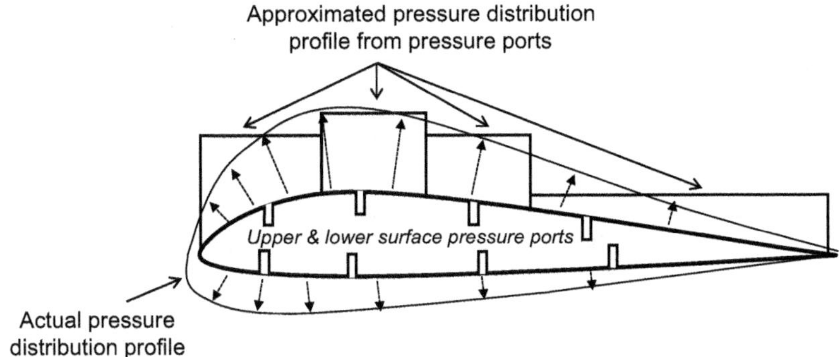

FIGURE L15.4 Discrete pressure profile based on pressure-sensing ports.

$$L = \sum_{n=1}^{\#\ of\ upper\ ports} (P_i * A_i)_{UPPER} + \sum_{n=1}^{\#\ of\ lower\ ports} (P_i * A_i)_{LOWER} \qquad \text{(L15.1)}$$

Lift equation. Since lift is fundamentally a function of airfoil shape, lift must be experimentally determined for every airfoil configuration. Lift is also a function of several other parameters, such as velocity, wing area and angle of attack. It is therefore impractical to experimentally determine lift for every airfoil shape over a range of velocities, wing sizes, and Reynolds numbers. As a result, a dimensionless quantity called the coefficient of lift is developed for each airfoil shape at various angles of attack, independent of velocity and wing size. An engineer can then choose from a database of airfoils, using the coefficient of lift for that airfoil, along with the specific design parameters of the aircraft such as velocity and wing area to calculate expected lift.

To create a coefficient of lift curve for an airfoil, a relationship between coefficient of lift and total lift as a function of several contributing parameters must first be developed. A common approach to developing this classic lift equation is with unit dimensional analysis. Only an abbreviated presentation of this development is presented here, which follows the basic approach of Anderson.[1]

Lift is a function of physical airfoil properties and fluid properties. These properties include the airfoil shape, surface area, S, angle of attack, α, fluid velocity, V, density, ρ, and viscosity, μ. For a given set of parameters, lift is constant for a given airfoil shape and angle of attack combination, so *shape* and *angle of attack* are temporarily combined as the parameter, Z, for convenience. A generalized expression of these parameters for lift is given in Equation L15.2.

$$L = f(V, \rho, \mu, S, Z) \qquad \text{(L15.2)}$$

Lift is a force, so the LHS of Equation L15.2 has units of Newtons (kg-m/s^2). The right-hand side of Equation L15.2 must also result in kg-m/s^2, so units are used to build the actual function on the RHS. The relationships between each parameter on the RHS are not known, so general exponents, "*a*" through "*d*" are used per Equation L15.3. The multiplier, *Z*, is just a constant, so an exponent is not necessary for this value.

$$L = Z\left(V^a \rho^b \mu^c S^d\right) \tag{L15.3}$$

When units are substituted for each of the physical parameters on the RHS of Equation L15.3, the exponents can be resolved to ensure the RHS resolves to Newtons. This effort results in the intermediate Equation L15.4.

$$L = Z\left[\left(\rho V^2 S\right) * \left(\frac{\mu}{\rho V S^{\frac{1}{2}}}\right)\right] \tag{L15.4}$$

The first group of parameters in brackets of Equation L15.4 has units of Newtons, so units are satisfied in this group. By including a "½" with this group, this group matches the definition of dynamic pressure.

The second group of parameters in Equation L15.4 is unitless. Surface area, *S*, has units of m^2, so the square root of "*S*" is chord length, *c*, and "*c*" can be substituted for $\sqrt{S}$. With this simplification, the second group matches the inverse of Reynolds number, *1/Re*.

Since Reynolds number is a constant for a given set of conditions, the second group in the brackets of Equation L15.4 can be combined with the constant, *Z*, and assigned the lift coefficient constant for an airfoil, c_l. Equation L15.4 becomes Equation L15.5 with these substitutions.

$$L = c_l\left[\left(\frac{1}{2}\rho V^2 S\right)\right] \tag{L15.5}$$

Equation L15.5 is the lift equation for an airfoil, where the coefficient of lift, c_l, is experimentally resolved for differing airfoil profiles. The lower-case "*l*" denotes the 2D characteristics of an airfoil compared to an upper-case "*L*" signifying a 3D wing.

Coefficient of lift. All parameters in Equation L15.5 can be directly measured except for the coefficient of lift. When a new airfoil profile is created, the surface area, *S*, is measured, and the airfoil is placed into a wind tunnel. Air density, ρ, is measured, wind tunnel velocity, *V*, is set, and lift force generated by the new airfoil is measured. This process is followed for a range of angles of attack for the airfoil. The result is the unique relationship the airfoil has between lift and dynamic pressure for a given lift area, called the coefficient of lift. Equation L15.5 is re-arranged into Equation L15.6 to create the airfoil-specific coefficient of lift curve based on measurable parameters from the wind tunnel test.

$$c_l = \frac{L}{\left(\frac{1}{2}\rho V^2\right)S} \tag{L15.6}$$

where: c_l is the coefficient of lift for a specific airfoil
L is the lift force generated by airfoil
ρ is the density of air
V is the velocity of air
S is the surface area of airfoil, the "rectangle" of span*chord

Once Equation L15.5 is used to create a coefficient of lift curve for a new airfoil, "c_l" becomes a known parameter for Equation L15.5. An aircraft designer has an idea of aircraft weight and therefore knows how much lift is required. An airfoil can be selected from a database of known profiles for c_l, and lift established with an appropriate wing size, S, and a minimum velocity, V.

With wing size set during design, an aircraft still must be able to change lift during flight. While climbing at heavy fuel loads, the aircraft must generate more lift. While descending at light fuel loads, the aircraft has a much lower lift requirement. Referring to either Equation L15.5 or L15.6, only two options exist for a pilot to adjust lift during flight. The pilot can increase or decrease velocity with engine power to change lift, or the pilot can increase or decrease the coefficient of lift. According to Figure L15.5, the coefficient of lift can be adjusted by increasing or decreasing the

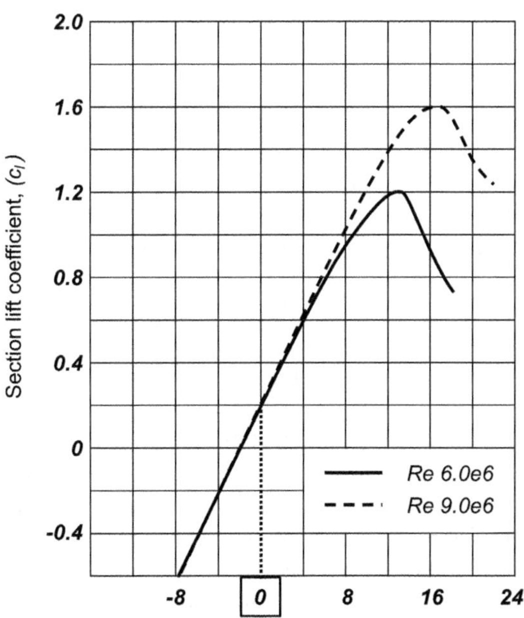

FIGURE L15.5 NACA 2415 coefficient of lift plot. For a given airfoil geometry (and various Reynolds numbers), lift can be increased by higher angles of attack (up to stall). Figure L15.5 recreated from public use NACA Report #824. (Abbott, I., von Doenhoff, A. & Stivers, L. (1945). *Report No.824, Summary of airfoil data.* National Advisory Committee for Aeronautics. Office of Aeronautical Intelligence, Washington, D.C. p.137.)

angle of attack of the wing relative to the oncoming airflow. The pilot changes the angle of attack through pitch control by pulling or pushing the flight controls to move the aircraft elevators. The coefficient of lift can be increased only to the peak limit in Figure L15.5 before the airfoil or wing stalls due to airflow detachment.

Published data. Airfoils have a National Advisory Committee for Aeronautics (NACA) numerical designation and published coefficient of lift plot similar to Figure L15.5. To interpret Figure L15.5, the airfoil angle of attack is selected along the bottom axis, and the corresponding coefficient of lift is read on the left-side vertical axis base on the plot line. Since the NACA2415 airfoil is more cambered across the upper surface, when the airfoil is straight on to the airflow at zero degrees angle of attack, the airfoil still generates positive lift with a coefficient of approximately $c_l = 0.2$. The Reynolds number affects the results to a certain extent, so separate curves are included for different Reynolds numbers. Reynolds number effects are not considered for this exercise.

When analyzing a NACA airfoil coefficient of lift curve or selecting a NACA airfoil for aircraft design, it is quite important to note the profiles were developed with high Reynolds numbers in airflow pressurized to 20 atm.[2] By using a higher Reynolds number, the NACA data obtained from the small airfoil tests scales better to larger aircraft with a similar Reynolds number. The NACA data is too optimistic for small model aircraft at lower Reynolds numbers. The average classroom wind tunnel will demonstrate a much flatter coefficient of lift curve.

COMPUTATIONAL METHOD

Proper unit consistency is essential to success. Begin by choosing an SAE standard set of units, and convert all working units to this chosen standard prior to any computations. Typically, SAE wind tunnel and flow velocities are given in feet per second (fps or ft/s), so pounds, feet and feet-squared are acceptable common units. Airfoil areas should be immediately converted from inches-squared to feet-squared, and manometer inches of water should be converted to feet of water. Working consistently in units of *feet* also eases the challenging unit conversions with *slugs* and acceleration due to gravity.

Confirm the NACA airfoil designation available for wind tunnel use. A common low-speed airfoil is the NACA2415. Ensure all data is based on the correct wind tunnel airfoil.

Prepare a data summary table similar to Table L15.1 along with a corresponding coefficient of lift figure similar to Figure L15.6 in the results section. Enter the published 20-atm coefficient of lift data into Table L15.1 from either Figure L15.5 or an online reference for the lab plan. Figure L15.6 should remain an empty placeholder for the lab plan. Include an NACA airfoil data set for the appropriate airfoil similar to Table L15.B1 in Appendix L15.B. Data sets for other airfoils are readily available online or can be sourced directly from NACA Report #824. Include a table similar to Table L15.C1 and L15.C2 in Appendix L15.C for wind tunnel airfoil parameters. These parameters are specific to the airfoil in use and may be provided by the manufacturer or directly measured. Ensure correct data in these tables prior to any computations and convert to consistent units.

Prepare data collection tables similar to the tables in Appendix L15.D. These tables must be built in a spreadsheet program with all computational formulas in place for the lab plan. Three complete sets of tables are necessary for the CFD simulation at 20 atm, the CFD simulation at ambient atmosphere and experimental wind tunnel data. The tables can either be built and replicated three times for each set of computations or three sets of rows (or columns) created within a single set of tables to contain each of the data collection scenarios.

Choose a wind tunnel velocity between 60 and 85 fps. This is the velocity for both computational and experimental activities. Enter this velocity in Table L15.1 and all tables of Appendix L15.D. Check the local atmospheric density and enter this value into each of the tables except the CFD 20-atm table. Populate all of the Appendix L15.D tables with airfoil area data and formulas. Include all units in all table and spreadsheet row and column headers.

Compute an expected total lift force for each of the angles of attack in the Appendix L15.D tables. Use Equation L15.5 and Figure L15.5. Ensure correct and consistent units. This lift force should be close to the CFD 20-atm results but higher than the ambient density results. These expected lift computation values serve as a reference to help ensure spreadsheet computations are correct.

To begin the CFD simulation, the airfoil must be modeled with pressure port locations clearly identified to record pressure data. The first simulation is run at 20 atm to verify published data, and then the simulation is re-run at ambient atmospheric density to compare with wind tunnel data.

Airfoil data set. The airfoil coordinate set must be loaded into the modeling software to create the solid model. Refer to Table L15.B1 in Appendix L15.B for the coordinate system. The airfoil profile LE begins at $x = 0$, and ends at the TE with $x = 1$. To create a 3D model of the airfoil data, modeling software expects to create a single closed loop. The data in Table L15.B1 is configured in *x-y* coordinate pairs, beginning at the TE, looping continuously around the LE and rejoining back at the TE. If the software expects an *x-y-z* coordinate set, then a third column is required where all z-values are zero.

All values in Table L15.B1 are normalized to a chord length of "*1.0*" and must be scaled to the actual dimensions of the airfoil. For an airfoil with a 3″ chord, use a spreadsheet program to multiply all data from Appendix L15.B by *3″*, then divide by *12* to ensure consistent units of feet. Save the scaled coordinate values as a tab-delimited text file. Import this text file as a *curve from points* to begin a new sketch with this airfoil data. It may be necessary to convert the curve into a spline-based feature for extrusion, depending on the modeling software.

If the modeling software does not have the ability to import the airfoil coordinate points directly, the data set can be entered into a spreadsheet program. Choose a *scatter plot with smooth lines* chart option, and the airfoil profile will be plotted on the spreadsheet chart. Copy the chart and paste into any compatible vector program (such as any presentation software or line art program) and save as a *scaled vector graphic* file (.svg). Most 3D modeling applications can import the *svg* file to create the airfoil model.

Reference geometry is required to identify pressure port locations. The reference geometry must be visible in the global workspace. This may be accomplished with additional sketch planes or points. While still in sketch mode, begin with a centerline reference. Select the centerline tool and click on the LE point of the airfoil and again on the TE point to create a horizontal chord line reference.

Create four vertical centerlines from the chord reference line to the upper surface. Properly space each of these vertical lines using the dimension tool. Click once on the LE and again on the first vertical line. Drag this line above the airfoil, click and enter the appropriate distance given in Table L15.C1 in Appendix L15.C. Repeat this process for each of the upper surface port locations in Table L15.C1 and each of the lower port locations in Table L15.C2. If necessary, create sketch planes or points at these locations for global reference.

The 2D sketch is now complete. The simulation only requires 2D flow characteristics and may be completed in either 2D or 3D CFD software. If running in 3D software, exit the sketch mode and select the extrude feature to create a 3D model of the sketch. Use a midplane extrusion to extrude the airfoil to a 6″ span. Save the model.

CFD simulation. Set up the CFD simulation in the appropriate SAE units with pound-force and feet-per-second units. Use the application wizard, if available. Select degrees for all angle values and not radians, vectors or other options. Use an external analysis that does not include the airfoil internal geometry. Do not include gravity in the simulation. Use air as the fluid with both laminar and turbulent flow options.

Set up the initial thermodynamic properties to 20 atm and 59°F for the first data run. Enter the planned velocity for the flow.

Configure the computational domain as necessary. If running on 2D CFD software, the computational domain may be automatically configured. If so, use the default settings. If running on 3D CFD software, ensure the computational domain encloses the airfoil cross section. However, the airfoil *extrusion* of 6″ should *protrude* from the domain space. The x and y dimensions of the computational domain should be about four or five times the size of the airfoil, with sufficient computation space downstream from the airfoil. For an airfoil with a chord of 3″, about plus-and-minus 15″ square should be sufficient. For the *z-direction* extrusion, only an inch is required. The airfoil span was extruded to 6″, so the airfoil will protrude past both sides of the domain.

Use a default global mesh for the flow field. A local refining mesh can be added for better resolution around the airfoil. Select the surface of the airfoil and double the local cell refinement as desired. Check the mesh prior to running the flow field simulation. Depending on the computer, the mesh cells should be limited to no more than about 100,000 to ensure the simulation completes within a few minutes. Ensure the mesh does not extend past the airfoil, so 3D wingtip effects are not modeled. If the airfoil was properly extruded past the computational domain, this should not be an issue.

If possible, the coefficient of lift should be computed using two methods from the CFD simulation. The first method uses Equation L15.1 to sum the pressure times the area at each port location. This is the same technique for the wind tunnel. The second technique is to allow the CFD simulation to determine an overall lift force exerted on the airfoil from the flow field. This method used Equation L15.6. To construct Equation L15.6 in the CFD software, use global goals (GG) from the flow field data. Select the software global goal option and choose the following parameters of interest: minimum static pressure, maximum dynamic pressure, average fluid density, maximum and bulk average velocity and y-direction force. Build the equation similar to Equation L15.7, and assign it a name of "C_L" for coefficient of lift.

$$\{\mathbf{GgforceY}\}/\left(\mathbf{0.5}*\left(\{\mathbf{GGAvgFluidDen}\}/\mathbf{32.2}\right)*\left(\{\mathbf{GGBulkAvgVel}\}\,^\wedge\,\mathbf{2}\right)*(\mathbf{0.021})\right) \quad \text{(L15.7)}$$

where: {GgforceY} is the CFD computed lift force in *y*-direction
{GGAvgFluidDen} is the CFD ambient fluid density
{GGBulkAvgVel} is the CFD computed average flow velocity
32.2 is the acceleration due to gravity for mass correction
0.021 is the surface area of airfoil with unit conversion factors

The denominator in Equation L15.7 must evaluate to pound-force to cancel with pound-force in the numerator. The software may return pound-force units for fluid density when pound-mass is required in SAE units of *slugs*. Acceleration due to gravity, *32.2 ft/s²*, is divided out of the fluid density term in Equation L15.7 to ensure units of *slugs/ft³*. If CFD density is returned in proper units, the *32.2 division* is not required. This should be verified.

The surface area of the wing, *S*, in Equation L15.7 is based on the 3D computational domain of a 3″ airfoil chord and 1″ domain span, converted to consistent units (*3 × 1 in²/144 in²*). If the airfoil chord is different or the computational domain space in the extrusion dimension is different, this value requires modification.

Run the simulation. The default airflow orientation should be a zero degrees angle of attack relative to the modeled airfoil and 20 atm. This first simulation provides only the zero-degree angle of attack data, so the flow field requires reorientation for each angle of attack in the Appendix L15.D tables. When the first simulation is complete, record the C_L equation goal results into the last row of Table L15.D4 for the 20-atm data in the zero-degree angle of attack column. Display and save a cut plot image of the pressure flow field by selecting "*relative pressure*," if not already visible.

The Appendix L15.D tables must now be populated with pressure port data to complete the summed lift force computations. Begin by entering the fixed-condition parameters into Table L15.D1. Use the global reference geometry created during the sketch to identify the port locations on the airfoil within the resolved flow field. It may be necessary to click on the actual sketch to highlight reference geometry associated with the sketch. Use a probe tool to click very near the pressure port locations within the flow field. Record the relative pressure at each of the eight pressure port locations into Tables L15.D2 and L15.D3. A properly built spreadsheet used for this purpose should automatically update the total lift and coefficient of lift for this angle of attack data point.

Depending on time available to complete the lab activities, the flow field should be re-oriented to each of the angles of attack in the Appendix L15.D tables. This requires a total of six simulation runs at 20 atm. If time is limited, at least one more data point at four or six degrees is required to establish a line to compare with published data. The 12-degree data point is likely in the stall region for slower velocities and should not be used if only two data points are planned. To reorient the flow field, modify the CFD input data initial conditions and change the flow field angle relative to the airfoil.

When the 20-atm data runs are complete, modify the CFD input data initial conditions again using local ambient pressure or atmospheres. For most locations above sea level, a value less than 1 atm should be used. Enter an appropriate angle of attack for the flow field, and run the simulation. Complete this task for each of the angles of attack to create a full coefficient of lift plot for comparing to actual wind tunnel data. View and save at least one cut plot of the pressure flow field to include in the final report.

EXPERIMENTAL METHOD

Measurements and conversions. Wind tunnel pressure measurements may be either water-filled manometers or computer-displayed transducer data. For wind tunnels using water-filled manometers, refer to Appendix L15.A for necessary calculations. For wind tunnels using transducer data, confirm displayed data units. Data is most likely provided as *pressure* in units of lb/in^2. Sometimes displays will mimic water-manometers and indicate units of $in\text{-}H_2O$. Pressure transducer transfer functions are correlated for pressure, so conversion from $in\text{-}H_2O$ to pressure is likely unnecessary. Pressure is the primary wind tunnel data output for all computations. Ensure consistent unit conversions from lb/in^2 to lb/ft^2 as necessary.

If velocity data is provided for each pressure port, do not use this velocity information in the calculations. Use only the single test section velocity data for all computations. If wind tunnel pressures are given in *negative* pressure values, review the discussion associated with Equation L15.1 as necessary. Regardless of any sign conventions, ensure total lift is computed as the difference between lift created on the upper surface and lift created on the lower surface. Both surfaces create lift due to the upper and lower cambers, but the lower surface creates less lift than the upper surface. Ensure signs are properly configured for all spreadsheets, regardless of differing sign conventions for Equation L15.1.

Confirm the wind tunnel airfoil configuration to be consistent with Appendix L15.B, C and D tables. Update any information as required and ensure all CFD simulations correspond to the actual wind tunnel configuration.

Wind tunnel operation. Follow all operational instructions to ensure the wind tunnel is ready for use. Follow instructions to connect the local computer to the wind tunnel client software. Enable the wind tunnel controls and select the velocity chosen for the lab plan and CFD simulations. Adjust the airfoil angle of attack to the first planned angle from the Appendix L15.D tables. Once the wind tunnel has stabilized at the appropriate velocity and angle of attack, click the "*record*" data option to take a snapshot of the pressure port data. Record each of the pressure port pressures into the Appendix L15.D tables. Repeat this process for each of the angles of attack.

RESULTS

The coefficient of lift values from published data, CFD data at 20 atm and ambient, and wind tunnel data are summarized in Table L15.1. Errors are computed between the validating CFD data and published data, and also for CFD ambient results compared to wind tunnel results. The wind tunnel and ambient CFD results from Table L15.1 are also plotted in Figure L15.6. Include at least one cut plot from the CFD pressure simulation in this section of the lab report.

TABLE L15.1
Summary of Coefficients of Lift

	COEFFICIENTS OF LIFT					
	AIRSPEED:					
	PUBLISHED C_L	COMPUTATIONAL C_L			EXPERIMENTAL C_L	
AoA	20 ATM	20 ATM	ERROR	1 ATM	1 ATM	ERROR
−2°						
0°						
2°						
4°						
6°						
12°						

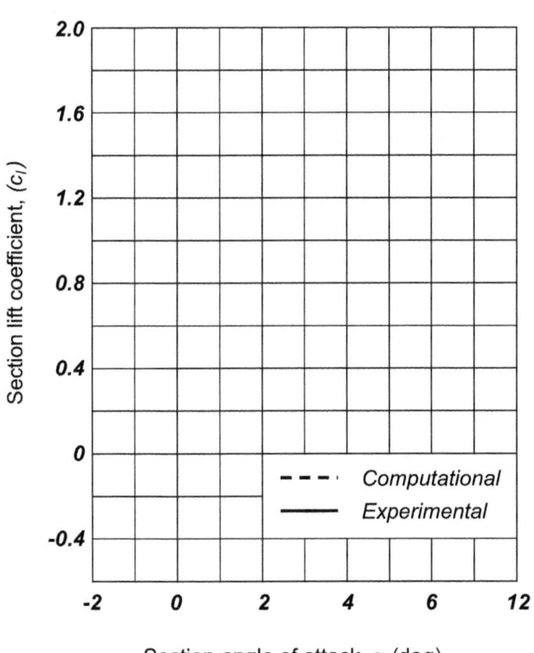

FIGURE L15.6 Summary of coefficients of lift.

DISCUSSION

Comment on the differences between computational, experimental and published results. If the higher (or lower) angle of attack data indicates a loss of lift in the plot, this is likely due to stall behavior. Analyze the experimental data and the computational data at similar data points to verify behavior. Discuss what is occurring with the data.

Discuss possible sources of error for the coefficients of lift. In particular, discuss the simplifications of an average pressure distribution measured over the pressure port area versus an actual pressure distribution. Perform some basic research about skin friction effects and boundary layer effects, and discuss how these may affect results in the data. In particular, both skin friction and boundary layer issues can slow velocity locally over the pressure ports. In the most severe limiting case, the velocity is slowed to zero and the pressure ports would not sense any fluid velocity. Discuss whether pressure data influenced by these effects would result in a higher or lower coefficient of lift.

Use the coefficient of lift curve to describe how an aircraft can maintain level flight at different speeds by changing its angle of attack.

CONCLUSIONS

Based on the objective of the experiment, clearly state a bottom-line error summary between published, computational and experimental results for overall lift. In very general terms, discuss how flow field pressures can generate force and how this must be designed into structures.

APPENDIX L15.A: WATER MANOMETER COMPUTATIONS

For water manometer measurements, record the following:

Inches of water for four pressure port locations on the upper and lower surface
Inches of water for the wind tunnel test section (velocity determination)
Include columns or rows to convert inches of water into pressure (Equation L15.A2)

If the airfoil ports are connected to manometers, the water height is <u>measured in inches</u> and must be <u>converted into pressure</u>.

A measure of static pressure is just the potential energy of a column of water as developed from Bernoulli's equation, given in Equation L15.A1, where "**ρ**" is the density of water, "**g**" is acceleration due to gravity and "**h**" is height of the water column against a zero atmospheric reference.

$$\mathbf{P}_{\mathrm{STATIC}} = \rho_{\mathrm{WATER}}\mathbf{gh} \qquad \text{(L15.A1)}$$

Since pressure is force divided by area, English units can be worked out to convert inches of water into pounds per square foot based on the density of water for Equation L15.A2 as follows:

Density:	$\rho_{WATER} = 1.94$ slugs/ft^3	
Force:	1 lb_f = slug-ft/s^2	
Pressure:	1 lb_f/ft^2 = slug/(ft s^2)	
Convert: ____ inches of water to ___ lb_f/ft^2	(__ in) (1 ft/12 in) (1.94 slugs/ft^3) (32.2 ft/s^2) = ___ lb_f/ft^2	(L15.A2)

The above calculations are in standard English units (lb/ft^2) but the airfoil area is given in units of in^2. Convert airfoil area to ft^2 using the conversion factor (1 ft/12 in)^2 to ensure unit consistency for lift in units of lb_f.

Bernoulli's equation can also be solved for velocity as shown in Equation L15.A3 where "h" is the height of column of water measured from a static pressure port in the wind tunnel.

$$V = \left[2\left(\rho_{WATER} g h\right)/\rho_{AIR}\right]^{1/2} \quad \text{(L15.A3)}$$

APPENDIX L15.B: NACA 2415 DATA SET CONFIGURED FOR CAD CURVES, NACA REPORT NO.824, P.100

TABLE L15.B1
NACA 2415 Airfoil Coordinate Points

	Percentage Chord (x-Position)		y-Position
TE	1.0000		0.0000
	0.9000	UPPER SURFACE	0.0245
	0.8000		0.0441
	0.7000		0.0610
	0.6000		0.0750
	0.5000		0.0857
	0.4000		0.0925
	0.3000		0.0938
	0.2500		0.0917
	0.2000		0.0870
	0.1500		0.0797
	0.1000		0.0683
	0.0750		0.0606
	0.0500		0.0507
	0.0250		0.0371
	0.0125		0.0271
LE	0.0000		0.0000

(*Continued*)

TABLE L15.B1 (*Continued*)
NACA 2415 Airfoil Coordinate Points

	Percentage Chord (*x*-Position)		*y*-Position
	0.0125	LOWER SURFACE	−0.0206
	0.0250		−0.0286
	0.0500		−0.0384
	0.0750		−0.0447
	0.1000		−0.0490
	0.1500		−0.0542
	0.2000		−0.0566
	0.2500		−0.0570
	0.3000		−0.0562
	0.4000		−0.0525
	0.5000		−0.0467
	0.6000		−0.0390
	0.7000		−0.0305
	0.8000		−0.0215
	0.9000		−0.0117
TE	1.0000		0.0000

APPENDIX L15.C: TYPICAL AIRFOIL PARAMETERS

TABLE L15.C1
Wind Tunnel Upper Surface Airfoil-Specific Data

UPPER SURFACE

Port #	Distance from LE	X/C Location	Surface Area Per Port	Center Point between Ports
LE	0			
1	0.227 in	7.4%	3.84 in^2	
				0.665 in
2	1.102 in	36.1%	4.63 in^2	
				1.467 in
3	1.832 in	60.0%	3.89 in^2	
				2.141 in
4	2.450 in	80.3%	5.24 in^2	
TE	3.050 in			
Total upper surface area:			17.60 in^2	

TABLE L15.C2
Wind Tunnel Lower Surface Airfoil-Specific Data

LOWER SURFACE				
Port #	**Distance from LE**	**X/C Location**	**Surface Area Per Port**	**Center Point between Ports**
LE	0			
1	0.200 in	6.6%	3.61 in^2	
				0.792 in
2	1.050 in	34.7%	4.57 in^2	
				1.680 in
3	1.783 in	59.0%	3.92 in^2	
				1.923 in
4	2.410 in	79.8%	5.33 in^2	
TE	3.020 in			
Total lower surface area:			17.43 in^2	

APPENDIX L15.D: DATA COLLECTION TABLES (THREE SETS REQUIRED FOR CFD 20-ATM, 1-ATM AND WIND TUNNEL)

TABLE L15.D1
Global Fixed Parameters

FIXED LIFT PARAMETERS	
DENSITY:, ρ:	**DYNAMIC PRESSURE:**
VELOCITY, V:	**TOTAL AIRFOIL AREA, S:**

TABLE L15.D2
Upper Airfoil Port Data

UPPER SURFACE			ANGLE OF ATTACK					
PORT	**AREA**		**−2**	**0**	**2**	**4**	**6**	**12**
1		**PRESSURE**						
		FORCE						
2		**PRESSURE**						
		FORCE						
3		**PRESSURE**						
		FORCE						
4		**PRESSURE**						
		FORCE						
UPPER SURFACE TOTAL FORCE								

TABLE L15.D3
Lower Airfoil Port Data

LOWER SURFACE			ANGLE OF ATTACK					
PORT	AREA		−2	0	2	4	6	12
1		PRESSURE						
		FORCE						
2		PRESSURE						
		FORCE						
3		PRESSURE						
		FORCE						
4		PRESSURE						
		FORCE						
LOWER SURFACE TOTAL FORCE								

TABLE L15.D4
Computed Results from Airfoil Port Data

	ANGLE OF ATTACK					
NET FROM UPPER AND LOWER	−2	0	2	4	6	12
LIFT FORCE						
COEFFICIENT OF LIFT						
Expected lift force at 20 atm						
CFD computed coefficient of lift						

NOTES

1 Anderson, J. (1978). *Introduction to flight.* McGraw-Hill. p.149.
2 Jacobs, E., Ward, K. & Pinkerton, R. (1935). *Report No.460, The Characteristics of 78 Related Airfoil Sections from Tests in the Variable-Density Wind Tunnel.* National Advisory Committee for Aeronautics, Navy Building, Washington, D.C. p.6.

Appendix A
Recommended Lab Plan Template

The following template is recommended for academic lab plans and reports. In lieu of alternate requirements, lab plans and reports should be double-spaced with 12-point Times New Roman font and 1-inch margins. Left-justify headings and paragraphs with no indentation. Use an extra line return to separate paragraphs. Center the abstract and appendix headings. Center all tables and figures. Tables and figures have their own continuous numbering schemes. Table captions are centered above the table, and figure captions are centered below the figure. Do not orphan headings or captions separately between page breaks. The writing style may be passive or active, but it should be succinct with short sentences.

University Capstone Course

LAB PLAN

TENSILE TEST EXPERIMENT

(Date of Document)

Submitted to Dr. Rusher, Soulk

Team Members:

* Lead, Team
Fish, Go
Swimmer, Notta
Name, Anudder

Point of Contact:
(Name, email, phone)

Planned Date of Experiment:
February 2, 20XX

INTRODUCTION

Our team proposes to determine the tensile stress of 3D printed plastic test specimens. These tests are in direct support of a national lab-sponsored capstone course project. The results of the test will be published and presented as part of the course requirements and retained by our national lab sponsors.

We intend to test three specimens each of polylactic acid (PLA), polyethylene terephthalate glycol (PETG) and acrylonitrile butadiene styrene (ABS) for a total of nine experiments. The test specimens will be sized for an expected failure load of less than 2000 N. Test specimen configurations are included in Appendix A and subject to re-configuration based on test equipment requirements. We plan four team members in the lab to complete the experiments during a single 2-hour session.

The objective of our experiment is to confirm existing published data on the failure stress for each material type. 3D printed materials are increasingly used in design projects, and verification of failure stress is necessary prior to our design efforts.

THEORY

Tensile stress is determined by load per cross-sectional area, given in Equation A.1.

$$\sigma = \frac{P}{A} \tag{A.1}$$

Failure stress occurs at the load when the test specimen breaks. For 3D printed material, we expect this stress to be different based on the orientation of print layer lines relative to load direction. For this experiment, intra-layer print lines will be perpendicular to load direction, as shown in Figure A.3 in Appendix A1. No consideration is given to different load orientations.

EXPERIMENTAL METHOD

This experiment utilizes the following equipment and software:

- Desktop tensile test system
    - Load capacity requirement of at least 3000 N
    - Stress and strain data logging capability
- Computer interface with USB port for recording and saving data
- Spreadsheet for tabulating key data points during experiment

The test specimens and equipment will be configured as shown in Figure A.1.

Planned test specimen dimensions are included in Appendix A1.

A lab team of four personnel will conduct the test. Two personnel will secure the test specimen in the test machine while two personnel will start the capture software and ensure computer connectivity. Results will be downloaded to a USB drive. All personnel will ensure equipment is returned to its proper location and facilities are cleaned. Material disposal is non-hazardous and will be disposed of in the trash.

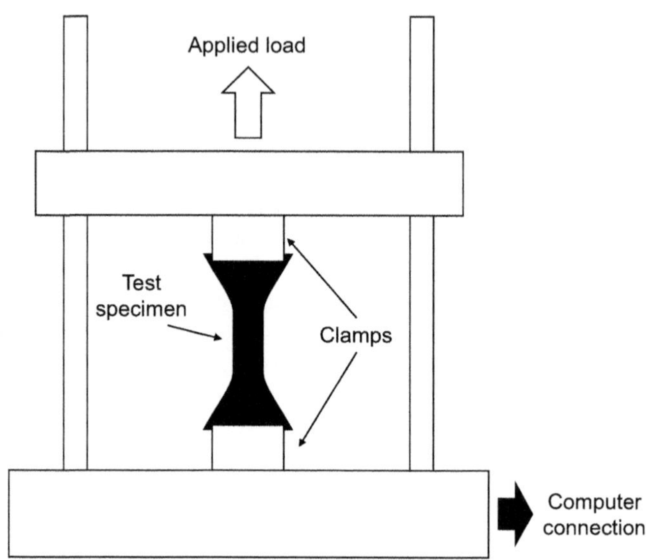

FIGURE A.1 Experimental setup.

TABLE A.1
Tensile Test Data

Material	Published Ultimate Stress (σ_{ULT})	Area (A)	Expected Failure Load (P_{ULT})	Experimental Failure Load (P)	Percentage Difference
PLA	26 MPa	2e−5 m^2	520 N		
PETG	20–70 MPa	2e−5 m^2	1400 N		
ABS	22–74 MPa	2e−5 m^2	1480 N		

EXPECTED RESULTS

Published values for the test plastic materials are included in Table A.1. These values are used in conjunction with the cross-sectional area in Appendix A1 and Equation A.1 to calculate the expected failure load. As shown in Table A.1, these values are within the capabilities of the planned test equipment.

Stress-strain data will be recorded and plotted as shown in Figure A.2. The materials are expected to exhibit somewhat ductile behavior. The notional plot in Figure A.2 will be replaced with actual data.

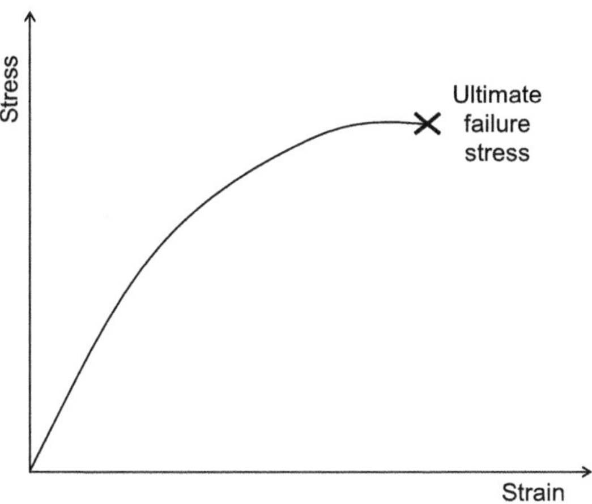

FIGURE A.2 Stress-strain plot.

LAB PLAN APPENDIX A1: TEST SPECIMEN CONFIGURATION

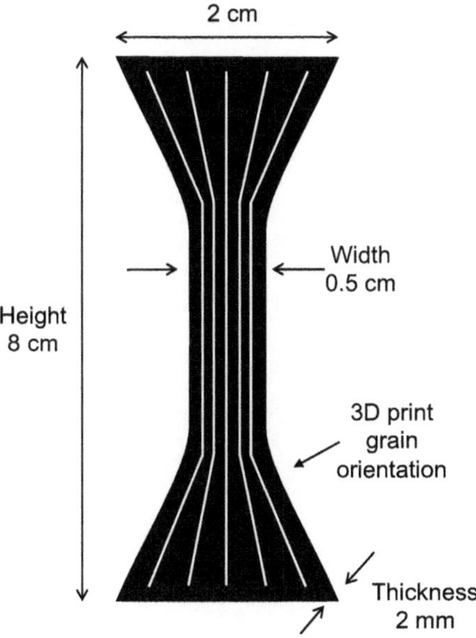

FIGURE A.3 Test specimen configuration and dimensions.

Appendix B

Recommended Academic Team Performance Documentation Tables

Consider Table B.1 to identify the written contributions of individual student team members. Include a copy of Table B.1 in both lab plans and reports. All teams must have an *internal* deadline *prior* to the actual assignment deadline. Include the agreed upon internal deadline as indicated. Team leads are responsible for filling out the form completely and accurately. The falsification penalty for all team members is −50% of the submission grade.

Consider Table B.2 to identify the in-lab performance for each team member. The lab instructional assistant should fill out Table B.2 and include it in an appendix of the lab report.

TABLE B.1
Team Participation Summary

ASSIGNMENT DUE DATE AND TIME:
TEAM INTERNAL DUE DATE AND TIME:
NAME:

	YES:	NO:	YES:	NO:	YES:	NO:	YES:	NO:	YES:	NO:
Met group internal due date and time:										
Equitable contribution:										
Quality contribution:										
RUNNING TALLY OF ALL REPORT SECTIONS WRITTEN, INCLUDING THIS REPORT:										
Abstract										
Introduction										
Theory										
Experimental Method										
Expected Results										
Results										
Conclusions										

TABLE B.2
Lab Team Member Performance

LAB #: TEAM #:

LAB NAME: TEAM LEAD:

NAMES (LAST, FIRST):	TARDY:	ABSENT:	EXCUSED:	SAFETY INFRACTION:	CLEAN UP ISSUE:	PHONE USE:	OTHER ISSUE:	EXCELLED: (1 ONLY)
*								

TEAM LEAD SIGNATURE: TA SIGNATURE:

COMMENTS FROM ABOVE CHECKMARKS:

LAB NOTES/SAFETY OR PROCEDURAL ISSUES / RECOMMENDATIONS:

Appendix C
Course Organizational Tools

Experimental lab courses require managing multiple tasks to maintain a professional program. Tasks include managing inventory, arranging instructional content, choosing grade schemes and managing lab activities. A few organizational course tools are offered here as a reference.

ACADEMIC LAB STANDARD OPERATING PROCEDURES

Every lab program has a unique set of policies and guidelines. Regardless of the program, policies must be clearly established at the beginning of the endeavor to ensure success. The following are a set of lab-based standard operating procedures and policies to consider.

- Professionalism is always required in the classroom and lab
    - Personal timeliness is always a sign of respect for others' time and efforts
    - Personal electronic device use must never interfere with activities
- Documents must be submitted as scheduled or directed
- Lab plans must *always* be submitted prior to lab facility attendance
    - After experimental activities, lab plans develop directly into the lab report
- Arrive to lab on time
- All team members prepare by reading lab manuals and completing the expected results
    - As a minimum, understand the *experimental method* section
    - Know lab tasks and the flow of activities prior to beginning the lab
- Have all spreadsheets and recording documents ready prior to lab attendance
    - Spreadsheets must have all computational formulas complete
- Lab workspace must remain organized during activities
- Remain actively engaged during all lab activities
    - Phone use in lab must be limited
    - Quick texts or brief calls are acceptable
    - Absolutely no watching or listening to extraneous content during the lab session
- No food in the lab
- Personal drink bottles are okay
    - Drinks must be in a sealed container and kept with personal items
    - No open cups with plastic lids, nor fountain drink cups
- No extended breaks, no smoking breaks and no snack breaks during lab
    - Once the session begins, do not depart the lab except for restroom breaks
- No sitting on lab tables

- Mandatory clean-up prior to departing lab session
  - No one is excused prior to clean-up without special permission
  - Lab teams should depart as a team when complete

Laboratory experimentation involves hands-on activities. No matter how low-risk an activity might be, experiments do not always go as planned and hazards exist. Understanding and following proper procedures helps minimize risk while maximizing the chances of success.

ACADEMIC LAB SAFETY POLICIES

Since most organizations have a dedicated safety professional, these policies and unique hazards should be *presented by the appropriate safety personnel.* The following summary provides a basis for safety presentations.

- Follow all lab-specific safety procedures outlined in the lab manual
  - Each lab requires different procedures and equipment
- Immediately stop any unsafe activities, procedures or actions in real-time
  - Safety issues require follow-up documentation and resolution
- Know the locations of the fire alarm, fire extinguisher, wash station and exit
- Personal attire requirements during lab sessions
  - Long pants only; no shorts, skirts or dresses
  - No loose-sleeved tops or scarfs
  - Long or flowing hair must be gathered
  - No open-toed footwear, such as sandals
- Personal protective equipment (PPE) requirements
  - Lab coats are required during all labs unless specifically stated otherwise
  - Gloves and other PPE required as specified
  - Safety glasses must be worn as specified in lab manuals
- Eye Protection
  - Safety glasses protect from flying particles, mild chemical splashes, dust and debris
  - Face shields are necessary when working with harsh chemicals or when there is a high risk of flying debris
  - Ensure ANSI Z87.1 designation
  - Corrective vision eyewear is not suitable as a substitute for eye protection
- Soldering
  - Remove all power to the circuit and components
  - Use a well-ventilated and lighted work space
  - Assume the solder tip is hot and never touch
  - Watch for flying wire when clipping excess leads
  - Always wash hands after soldering
  - Avoid inhaling fumes
- Torches, burners and open flames
  - Assume a flame is present since it can be difficult to see
  - Never point a torch toward yourself or another, lit or unlit

- Gas cylinders
  - Must be secured from tipping
  - Cylinder must have torch removed and be capped when not in use
- Gloves
  - Always use gloves specified by the lab manual or activity requirements
  - Nitrile gloves for irritants and general handling
  - Protective and heavy-duty gloves for use with breakable parts or rotary tools
- Electrical equipment
  - Cords must not pose a trip hazard
  - Do not overload outlets
  - Do not pull plug by the cord; grasp plug to unplug

SYLLABUS PLANS

Several approaches are possible for studying an experimental methods course. The labs covered in this text can generally be divided between mechanics and dynamics for junior-level topics and heat transfer and fluid dynamics for senior-level topics. While the text chapters are arranged in a sequential flow, topics should be arranged to match the flow of lab activities for each course. The emphasis for junior-level courses should be on basic approaches, hands-on lab activities and data analysis. The emphasis for senior-level courses should be on speaking, writing and designing experiments. Tables C1 and C2 provide useful programs for considering course organizations.

PRESENTATION PROGRAM

Team presentations provide an opportunity for students to present a particular lab in a professional manner to the class. Team presentations require practice together to meet timing requirements and handoffs between speakers. Individual presentations provide an opportunity to introduce the class to a large variety of experimental lab equipment, instruments and sensors. Individual and team presentations provide important public speaking practice. Public speaking is a learned skill, and every opportunity should be taken to practice it in the academic environment. Every profession has an opportunity for public speaking, but not everyone is required to speak. Those who rise to the challenge of public speaking will be called upon to speak more and will receive recognition and career advancement beyond those who choose not to speak.

Table C3 offers suggested grading programs, individual presentation topics and critiques. Everyone should actively participate in the presentation program. Honest and anonymous peer critiques are an incredibly valuable resource for identifying areas of improvement. Critiques should be anonymous when presented to the speaker but submitted for critiquer points to ensure the motivation of a well-considered critique. The following guidelines should be considered for presentation programs.

- Grading plan is *independent* of natural ability
  - Grade based only on preparation and practice
  - Presentation begins with maximum points

TABLE C1
Mechanics and Dynamics Suggested Syllabus Topic Content and Flow

WK	LSN	Topics	Assignment
1	1	Intro, Policies and Safety	
2	*Lab 0*	*Choose teams of 4, select team leads*	No lab sessions
	2	Approach to Experimentation	Lab 0 report due
	3	Approach (cont.), The Team	
3	*Lab*	*Inventory Check*	No documents due
	4	The Team (cont.), Ethics, Deliverables	
	5	Deliverables (cont.)	
4	*Lab 1*	*Vibrations – Discrete*	Lab 1 plan due
	6	Lab 1 Prep, System Modeling	
	7	Second order systems	
5	*Lab 2*	*Vibrations - Continuous*	Lab 2 plan due
	8	Lab 2 Prep, Continuous Systems	Lab 1 report due
	9	Continuous Systems & V-M Diagrams	
6	*Lab 3*	*Waves*	Lab 3 plan due
	10	Lab 3 Prep, Waves	Lab 2 report due
	11	Waves (cont.), Simulation Software	
7	*Lab 4*	*Stress & Strain (prep)*	Lab 4 plan due
	12	Lab 4 Prep, Stress-Strain	Lab 3 report due
	13	Wheatstone bridge	
8	*Lab 4*	*Stress & Strain (test)*	No documents due
	14	Stress-Strain (cont.), Photoelasticity	
	15	Photoelasticity (cont.)	
9	No Lab	Study	No lab room sessions
	16	EXAM	Lab 4 report due
	17	Analyzing Data Sets	
10	*Lab 5*	*Fractures*	Lab 5 plan due
	18	Lab 6 Prep, Fractures	
	19	Chauvenet's Criterion	
11	*Lab 6*	*Non-Destructive Inspection*	Lab 6 plan due
	20	Lab 6 Prep, NDI	Lab 5 report due
	21	Evaluating the Hypothesis	
12	*Lab 7*	*Individual Monte-Carlo Simulation*	No lab room sessions
	22	Lab 7 Prep, Virtual Experiments	Lab 6 report due
	23	Presentation Techniques	
13	*Lab*	*Clean up & Inventory*	Mandatory lab sessions Lab 7 report due
	24	Team Presentations	
	25	Team Presentations	
14	No Lab	Study	No lab room sessions
	26	Team Presentations	
	27	Team Presentations	
15	No Lab	Study	No lab room sessions
	28	Team Presentations	
	29	EXAM	

TABLE C2
Heat Transfer and Fluid Dynamics Suggested Syllabus Topic Content and Flow

WK	LSN	Topics	Assignments
1	1	Intro, Policies and Safety	
2	*Lab 0*	*Choose teams of 4, select team leads*	No lab sessions
	2	Ethics, Deliverables, Presentations	Presentations sign-up
	3	Design of Experiments	Lab report 0 due
3	*Lab*	*Inventory check*	No lab room sessions
	4	Design of Experiments (cont.)	No documents due
	5	Individual Presentations 1–6	
4	*Lab 8*	*Thermocouples*	Lab 8 plan due
	6	Lab 8 Prep, Thermocouples	
	7	Individual Presentations 7–12	
5	*Lab 9*	*Thermal Loading Simulation*	Lab 9 plan due
	8	Lab 9 Prep, Active Writing	Lab 8 report due
	9	Individual Presentations 13–18	
6	*Lab 10*	*Thermal loading*	Lab 10 plan due
	10	Lab 10 Prep, Sample Sizes	Lab 9 report due
	11	Individual Presentations 19–24	
7	*Lab 11*	*Fins*	Lab 11 plan due
	12	Lab 11 Prep, First Order Systems	Lab 10 report due
	13	Individual Presentations 25–30	
8	*Lab 12*	*Heat pipe (build)*	Lab 12 plan due
	14	Lab 12 Prep, Managing Errors	Lab 11 report due
	15	Individual Presentations 31–36	
9	*Lab 12*	*Heat pipe (test)*	No documents due
	16	Managing Errors (cont.)	
	17	Individual Presentations 37–42	
10	No Lab	Study	No lab room sessions
	18	EXAM	Lab 12 report due
	19	Individual Presentations 43–48	
11	*Lab 13*	*Pitot Tubes*	Lab 13 plan due
	20	Lab 13 Prep, Unexpected Results	
	21	Individual Presentations 49–54	
12	*Lab 14*	*Individual Pressure Simulation*	No lab sessions
	22	Lab 14 Prep, Reporting Numbers	Lab 13 report due
	23	Individual Presentations 55–64	
13	*Lab 15*	*Aero Forces*	Lab 15 plan due
	24	Lab 15 Prep, Lift equation	Lab 14 report due
	25	Team Presentations	
14	Lab	Clean up and Inventory	Lab 15 report due
	26	Team Presentations	Mandatory lab sessions
	27	Team Presentations	
15	No Lab	Study	No lab room sessions
	28	Team Presentations	
	29	EXAM	

TABLE C3
Presentation Grading Program and Critique Guideline

Category	Max Deduction
Total possible points	+25
PREPARATION	
Technical or personal issues causing delay of 15 sec or more	−1
Small font sizes less than 24 point	Up to −3
Lengthy bullets exceeding two lines or excess verbiage	Up to −2
Unreadable equations	−1
Inappropriate attire (jeans, athletic clothes, no collar and caps)	−1
Unorganized content	−1
Insufficient material	−1
Failure to follow important guidelines	−1
Lack of clarifying diagrams for technical information	−1
No back-up or recovery plan if lost	−1
PRACTICE	
Exceeded time limit	−3
Under time limit	−2
Failure to introduce self	−1
Occasional *um, uh, you know, like* or other filler words	−1
Several *um, uh, you know, like* or other filler words	−2
Difficult to follow	−1
Talks to slides and not audience	−1
More than 50% of time reading from note cards	−1
Too quiet or inaudible	−1
Bogged down while explaining diagrams or technical info	−1

  - Points are subtracted based on the stated criteria
  - Criteria are a guideline and subject to adjustments as necessary
- Individual presentations on instruments, equipment and sensors as suggested in Table C4
  - If any of the prescribed points do not fit, create a suitable alternative point
- Three main points for Student #1 topic
  - What it is
    - Introduce the device
    - Explain what it is
    - Explain what it looks like
  - How it works
    - Describe the buttons, displays and interface
    - Describe how it is connected for use
    - Describe the internal components

TABLE C4
Individual Presentation Topics

Topic	Date	Student 1	Student 2
Galileo thermometer			
Ideal gas thermometer			
Liquid-in-glass thermometer			
Bi-metallic thermometer			
RTDs			
Thermocouples			
Thermistors			
Pyrometers			
Quartz-crystal temperature			
Liquid crystal thermography			
Digital vs Analog devices			
Infrared thermometers			
Thermal imagers			
Ammeters			
Voltmeters			
Ohmmeters			
Multimeters			
Frequency generators			
Oscilloscopes			
Lock-in amplifiers			
Vector network analyzers			
U-tube manometers			
Bourdon tube gauges			
Bellows pressure gauges			
Diaphragm pressure gauges			
Piezoelectric pressure gauges			
Electrical pressure gauges			
Cup/vane anemometers			
Thermal anemometers			
Doppler anemometers			
Flow meters			
Obstruction flow meters			

- Application
  - Explain what it is used for
  - Explain how to use it in an actual application
  - Explain the limitations of its use

- Three main points for Student #2 topic
  - Historical development
    - Explain how it was invented or who invented it
    - Describe earlier versions of the device
  - Theory of operation
    - Explain conceptual theory
    - Explain any mathematical equations
  - Future development
    - Describe any possible future refinements
    - Describe alternative technology for the same device
    - Offer any interesting facts about the device

GRADING PLANS

Course grading should reflect the appropriate mix of report writing and testable material. Point allocation should ensure appropriate motivation for all activities. Report and presentation scores should generally start at a maximum value, with deductions up to a specified limit for each category. Tables C5 and C6 offer suggested grading plans.

TABLE C5
Suggested Point Allocations

Activity	Percentage	Points
Lab plans	8%	8 points per plan
Lab reports	45%	26 points per report
Exams	35%	100 points per exam
Quizzes, exercises, HW	6%	
Team presentations (10 min ± 1 min)	6%	25 points per presentation
Individual presentations (6 min ± 1 min)	(7% – when individual presentations are implemented, reduce report and team presentation percentages above to retain a 100% grading scheme.)	25 points per presentation

TABLE C6
Written Report Grading Program

Category	Max Deduction
Total possible points	+26
Late turn in	−3
Late turn in and tardy to class	−7
Format issues	Up to −3
Inadequate sources of error	Up to −2
Inadequate or inappropriate conclusion	Up to −3
Inadequate tables or charts	Up to −3
Poor spelling, grammar and punctuation	Up to −3
Poorly worded or difficult to follow	Up to −2
Inadequate content coverage	Up to −1
Frivolous content (padding)	Up to −1
Lack of original content and mostly paraphrased content	Up to −3
Missing units in data presentations	Up to −1
Missing legends or titles on charts	Up to −1
Crowded or illegible content	Up to −1

Index

For Product Safety Concerns and Information please contact our EU representative GPSR@taylorandfrancis.com Taylor & Francis Verlag GmbH, Kaufingerstraße 24, 80331 München, Germany

Batch number: 10397790

Printed by Printforce, the Netherlands